Lecture Notes in Computer Science 8710

Commenced Publication in 1973
Founding and Former Series Editors:
Gerhard Goos, Juris Hartmanis, and Jan van Leeuwen

Weihong Han   Zi Huang   Changjun Hu
Hongli Zhang   Li Guo (Eds.)

Web Technologies and Applications

APWeb 2014 Workshops
SNA, NIS, and IoTS
Changsha, China, September 5, 2014
Proceedings

 Springer

Volume Editors

Weihong Han
National University
of Defense Technology
Changsha, China
E-mail: hanweihong@gmail.com

Zi Huang
University of Queensland
Brisbane, QLD, Australia
E-mail: huang@itee.uq.edu.au

Changjun Hu
University of Science and Technology
Beijing, China
E-mail: huchangjun@ies.ustb.edu.cn

Hongli Zhang
Harbin Institute of Technology
Harbin, China
E-mail: zhanghongli@hit.edu.cn

Li Guo
Chinese Academy of Sciences
Beijing, China
E-mail: guoli@iie.ac.cn

ISSN 0302-9743   e-ISSN 1611-3349
ISBN 978-3-319-11118-6   e-ISBN 978-3-319-11119-3
DOI 10.1007/978-3-319-11119-3
Springer Cham Heidelberg New York Dordrecht London

Library of Congress Control Number: 2014947545

LNCS Sublibrary: SL 3 – Information Systems and Application, incl. Internet/Web
and HCI

Typesetting: Camera-ready by author, data conversion by Scientific Publishing Services, Chennai, India

Printed on acid-free paper

Springer is part of Springer Science+Business Media (www.springer.com)

Message from the APWeb 2014 Workshop Chairs

It is our great pleasure to welcome you to the proceedings of the 16th APWeb workshops. APWeb is a leading international conference on research, development, and applications of Web technologies, database systems, information management, and software engineering. This year, we have three workshops held in conjunction with the main conference, which include

- First International Workshop on Social Network Analysis (SNA 2014)
- First International Workshop on Network and Information Security (NIS 2014)
- First International Workshop on Internet of Things Search (IoTS 2014)

The goal of these workshops is to promote the new research directions and applications, especially on social network analysis, security, and information retrieval against the heterogeneous big data. After a series of evaluation process conducted by the workshop Program Committee members, the workshop program features 34 papers with two invited papers among 59 submissions. All papers were presented in the workshop session at the main conference, held in Changsha, China on September 5, 2014.

We would like to thank the authors for choosing APWeb workshops as a venue for publishing their high quality papers and the Program Committee members for their timely reviews of the papers. We are also very grateful to the workshop organizers from the University of Science & Technology Beijing, Harbin Institute of Technology and Institute of Information Engineering CAS, China for their essential efforts on paper selecting and program organizing. Furthermore, we also gratefully acknowledge the support of the main conference organizers for their great effort in supporting the workshop program.

Finally, we hope that you enjoy reading the proceedings of the APWeb 2014 workshops.

July 2014   Weihong Han
                                                                              Zi Huang

Organization

Executive Committee

Workshops Co-chairs

Weihong Han	National University of Defense Technology, China
Zi Huang	University of Queensland, Australia

First International Workshop on Social Network Analysis (SNA 2014)

Steering Scientists

Fang Binxing	Beijing University of Posts and Telecommunications, China
Jia Yan	National University of Defense Technology, China

First International Workshop on Network and Information Security (NIS 2014)

Steering Scientists

Fang Binxing	Beijing University of Posts and Telecommunications, China

First International Workshop on Internet of Things Search (IOTS 2014)

Program Committee Chairs

Hu Changjun	University of Science & Technology Beijing, China
Zhang Hongli	Harbin Institute of Technology, China
Guo Li	Institute of Information Engineering, Chinese Academy of Sciences, China

Program Committee

SNA 2014

Li Yuxiao	Beijing University of Posts and Telecommunications, China
Qi Jiayin	Beijing University of Posts and Telecommunications, China
Wu Bin	Beijing University of Posts and Telecommunications, China
Cheng Xueqi	Institute of Computing Technology, Chinese Academy of Sciences, China
Yu Zhihua	Institute of Computing Technology, Chinese Academy of Sciences, China
Li Jianhua	Shanghai Jiaotong University, China
Xie Gengtan	Shanghai Jiaotong University, China
Pan Li	Shanghai Jiaotong University, China
Huang Heyan	Beijing Institute of Technology, China
Liao Lejian	Beijing Institute of Technology, China
Zhang Huaping	Beijing Institute of Technology, China
Wu Xindong	University of Vermont, USA and Hefei University of Technology, China
Liu Yezheng	Hefei University of Technology, China
Xu Jin	Peking University, China
Shi Peng	University of Science & Technology Beijing, China
Zhu Yan	University of Science & Technology Beijing, China
Sun Chang-ai	University of Science & Technology Beijing, China
He Xiao	University of Science & Technology Beijing, China
Hu Yue	Institute of Information Engineering, Chinese Academy of Sciences, China

NIS 2014

Wang Jilong	Tsinghua University, China
Hu Liang	Jilin University, China
Du Yuejin	CNCERT/CC, China
Chen Xingshu	Sichuang University, China
Zhou Bin	National University of Defense Technology, China
Du Xiangjiang	Temple University, USA
Chen Tingting	Oklahoma State University, USA

Wu Zhigang	Beijing University of Posts and Telecommunications, China
Shi Jinqiao	Institute of Information Engineering, Chinese Academy of Sciences, China
Jin Shuyuan	Institute of Computing Technology, Chinese Academy of Sciences, China
Xue Jingfeng	Beijing Institute of Technology, China
Zhang Zhaoxin	Harbin Institute of Technology at Weihai, China

IoTS 2014

Aiping Li	National University of Defense Technology, China
Bin Wang	Institute of Computing Technology, Chinese Academy of Sciences, China
Fenghua Li	Institute of Information Engineering, Chinese Academy of Sciences, China
Hanhua Chen	Huazhong University of Science & Technology, China
Jiaheng Lu	Renmin University of China
Jiangang Ma	Victoria University, Australia
Lan Chen	Institute of Microelectronics of China Academy of Sciences
Lihua Yin	Institute of Information Engineering, Chinese Academy of Sciences, China
Limin Sun	Institute of Information Engineering, Chinese Academy of Sciences, China
Nenghai Yu	University of Science and Technology of China
Peng Zhang	Institute of Information Engineering, Chinese Academy of Sciences, China
Xiangzhan Yu	Harbin Institute of Technology, China
Xiaotong Zhang	University of Science & Technology Beijing, China
Xiaoyong Li	Beijing University of Posts and Telecommunications, China

Table of Contents

First International Workshop on Network and Information Security (NIS 2014)

First International Workshop on Internet of Things Search (IOTS 2014)

Sentiment Analysis Based Online Restaurants Fake Reviews Hype Detection

Xiaolong Deng[1] and Runyu Chen[2]

[1] Key Laboratory of Trustworthy Distributed Computing and Service,
Beijing University of Posts and Telecommunications
10 Xitucheng Road, Beijing, 100876, China
shannondeng@bupt.edu.cn
[2] International School, Beijing University of Posts and Telecommunications
10 Xitucheng Road, Beijing, 100876, China
4007821@qq.com

Abstract. In our daily life, fake reviews to restaurants on e-commerce website have some great affects to the choice of consumers. By categorizing the set of fake reviews, we have found that fake reviews from hype make up the largest part, and this type of review always mislead consumers. This article analyzed all the characteristics of fake reviews of hype and find that the text of the review always tells us the truth. For the reason that hype review is always absolute positive or negative, we proposed an algorithm to detect online fake reviews of hype about restaurants based on sentiment analysis. In our experiment, reviews are considered in four dimensions: taste, environment, service and overall attitude. If the analysis result of the four dimensions is consistent, the review will be categorized as a hype review. Our experiment results have shown that the accuracy of our algorithm is about 74% and the method proposed in this article can also be applied to other areas, such as sentiment analysis of online opinion in emergency management of emergency cases.

Keywords: Sentiment analysis, Hype review, Multi-dimension analysis, Bayes judgment.

1 Introduction

In nowdays, most consumers have the habit of scanning the online reviews before purchasing the commodity. Products with many positive reviews often make good impact to customers. Therefore, for merchants in e-commerce time, it is important to maintain the online reputation of their products. However, some of the merchants, in order to enhance the popularity of their products, they hired some groups of people who is called internet "Water Army" to post positive reviews of their products online, and which can greatly affect the choice of consumers.

In this article, we have established collaborative relationship with dianping.com and have obtained fake review dataset collected by the company. Through serious observation and analysis, we concluded that most of the fake reviews are hypes. And they are very similar to authentic reviews in many aspects including length, tone and

W. Han et al. (Eds.): APWeb 2014 Workshops, LNCS 8710, pp. 1–10, 2014.

wording and so on which is the reason that is why they are most misleading to consumers.

We have done some further investigation on fake reviews of hype we sort out. This kind of reviews is most distinctive in its text, i.e. totally positive or negative. Some reviews of this kind, although not suspicious according to analysis on other attributes, possess little reference value because of extreme sentiment.

Aiming at the problem described above, we have applied sentiment analysis methods to detection of such kind of fake reviews. To obtain a desirable accuracy, we first find out all the subject words by matching the sentiment words we extracted from all the fake reviews data set with the corresponding features. We manually judged the subject words and divided them into four dimensions: taste, environment, service, overall attitude. To evaluate the authenticity of a review about restaurant, our algorithm will conduct sentiment analysis from the four dimensions of it based on Bayes classifier. If the analysis results of the four dimensions are the same (all positive or negative), the review is hype. And to our surprise, the algorithm has an good accuracy of about 74%.

2   Related Work

Relevant research is always focus on spam detection [1] and rubbish website diction [2]. And in recent years, researchers started to identify spam reviews.

Researchers have concentrated on some different characteristics of spam reviews. Jindal proposed the algorithm concerns unusual score [3][4].Wang and Xie paid more attention on store reviews to detect spam.[5][6].Arjun considered a number of indicators to find fake reviewer groups[7]. Some mainly considered relevant, Song defined the relevance of their basic characteristics and features among other comments [8]. Myle proposed the combination of language features SVM modeling [9].

In this paper, a innovative method is proposed to classify the content itself from the marked spam reviews, for which account the largest number and most influential speculation reviews. Then we draw the appropriate model program to solve this problem. By using sentiment analysis in more special areas of multi-dimensional analysis, our method gained greater accuracy of identification of such spam reviews. In comparison, our algorithm requires much more manual handling, but the recall rate reflects better performance.

3   Sentiment Analysis

3.1   Bayes Classifier

Through investigation on the possibility of a specific event happening in the past, Bayes Theorem can calculate the approximate the probability of the event happening in the future. And Bayes theorem only requires a few parameters for estimation and is not sensitive to the lost data. Algorithms that implemented from this theorem can run more faster due to the simplicity of Bayes theorem [10].

Bayes classifier is the application of Bayes Theorem to classification of text. By calculating the probability of each category, the algorithm categorizes the text to the one with largest probability.

Bayes equation:

$$P(c_j \mid x_1, x_2, ..., x_n) = \frac{P(x_1, x_2, ..., x_n \mid c_j)}{P(c_1, c_2, ..., c_n)} \tag{1}$$

Since each condition is independent and identically distributed, we have:

$$c_{NB} = P(c_j) \prod_i P(x_i \mid c_j) \tag{2}$$

Because there is no apparent feature in the labeled text, we apply multi-event model based on word frequency here:[11]

$$P(t_i \mid c = spam) = \frac{1 + Count(t_i)}{n + Counts(\sum_1^n t_i)} \tag{3}$$

3.2   Training Set Obtainment

Training data is always needed in calculating Bayes prior probability, and we manually selected 500 totally positive and negative reviews respectively. We have proved that the training set is enough since by enlarging the size from 300 to 500 we did not find noticeable change in the result. When selecting the sample of training set, we followed the principle that the review is as comprehensive as possible.

Table 1. Training set data

Positive	Negative
三人行骨头锅太给力了，多天啃绝对过瘾，料很足，好多肉，人气爆满，装修美观，去晚了就要排队，钵钵鸡，个人觉得不错，喜欢吃辣的朋友可以尝试下，味道十足。	服务态度极差提前一周打去说只能提前两天预订等到提前两天打去又没位子了还说没人接过电话说要提前两天预订领班态度也差还挂客人电话。
昨天请朋友来吃饭，听说这里的蟹宴菜挺有名的，于是就点了几只大闸蟹，点了个蟹粉豆腐。还有几道特色菜，片皮鸭，5A牛柳粒，古法局鱼。朋友都说不错我很高兴。这里环境也挺舒服的，服务态度很好，注重细节，我们想要的在未开口之前服务员都想到了。	第一次去就被咖喱牛腩的牛腩震惊了，那么老的牛腩，都快赶上牛肉干了，牛腩也完全不入味…询问服务员还说他们的牛腩一直是这样的，那请问前面点评里面说牛腩嫩、烂的吃的是什么…强烈不推荐！服务员态度很不好，下次绝对不会去了。蟹粉小笼十分一般，根本不值这个价格。

Table 1. (*Continued.*)

同事们商量着一起去吃点东西，大家一致赞同，冬天里吃火锅应该是再好不过的选择了吧，就这样我们一伙人就过来了，店里的生意很好的，但还是井然有序的，我们找了个地坐下，点了个火锅，一会的功夫就上来了，菜很丰富，话不多说，大家就开动了，有菜有肉，真是一顿营养大餐呢，汤鲜味美啊，大家都吃的很高兴，一边聊着天，一边吃着饭，真是享受呢，这才是生活啊，以后有时间还要常来啊，朋友们有空也可以过来尝一下，相信不会让你失望的。	锅贴的品种不多，也没什么可挑选的，贡丸汤一份，偌大一只碗，里面晃悠着两个贡丸；中午时分也没有坐满，三两个服务员在休班闲聊，毫不避讳的高谈阔论着；我们点的锅贴好了，得自己去台面上找，就几个单就有点混乱了，服务水平可见一斑。
...	...

3.3   Sentiment Evaluation

To find out the sentiment of a review, we use the method of multiplying the conditional probability of each word. Whether a review is positive or negative depends on the value of the result of multiplying process. We multiply the probability of each word with 10 first in case the result is too small.

4   Multidimensional Discrimination

4.1   Establishment of Sentiment Word Library

First we sorted out 56483 reviews about restaurants from our original data set. With the help of ICTCLA50 algorithm [12], we divided all the Chinese contents into words. We extracted all the sentiment words and set up our sentiment word library. In the same time, we calculated the frequency of each adjective, which enables us to get rid of some undesirable words with low frequencies. The final version of our sentiment library contains 1590 adjectives.

Table 2. Emotional word library

Words/Frequencies		Words/Frequencies	
很好/a	12.389356882973637	实在/a	7.453503647044743
好吃/a	11.14669772771281	舒适/a	7.406438314516372
不错/a	11.052315097990974	丰富/a	7.3550685280099195
实惠/a	9.742061863915502	亲切/a	7.350785294189377
干净/a	9.651343822312056	安静/a	7.289251208561696
值得/a	9.466860974306062	确实/a	7.262434541770648
周到/a	9.096459212424548	可口/a	7.18616605729332
特别/a	9.000203638575638	贴心/a	7.141585285545526
舒服/a	8.995655552806497	独特/a	7.044336876237392
便宜/a	8.929070356945344	合理/a	7.03207042683553
最好/a	8.40933391245834	一样/a	6.988856149668119
优惠/a	8.213934707582403	美味/a	6.957181112804157
优雅/a	8.193907482484725	重要/a	6.9435164471041055
温馨/a	8.078017307503831	合适/a	6.795440558011954
地道/a	8.023967758424863	有味/a	6.752448672847059
划算/a	8.008506960447109	适中/a	6.7334198176058955
一般/a	7.982753540338287	整洁/a	6.654735396670123
方便/a	7.9191048016013506	一流/a	6.63665330359303
热情/a	7.905370323918099	主动/a	6.636631314797421
鲜美/a	7.887444208296432	过瘾/a	6.61751253180918
地方/a	7.806704408558014	失望/a	6.612146670170864
入味/a	7.788232816199374	随便/a	6.533682077002941
精致/a	7.6268743931156555	耐心/a	6.46625147932886
其实/a	7.513344086828133	开心/a	6.452121735931351
...		...	

4.2   Feature Finding

Feature refers to the corresponding none phrase of existing sentiment words. We managed to find them by searching words before the adjectives that match those in our library; the nearest noun phrase is the matching result. Finally, we obtained 2303 subject words of reviews about restaurants.

4.3   Classification of Subject Words

After finding all the subject words, we analyzed the dimension each of them belongs to and finally limited the dimension into four types: taste, environment, service and overall. Then we classified these subject words. There are 1314 words belong to dimension "taste", including those that describe the look, flavor and taste of foods. 222 words describing the environment, geographical conditions and traffic conditions of the restaurants are categorized into "environment". 286 words are categorized into "service", most of them describe the quality of service, prices. And there are 151 words classified as "overall", including name of restaurants, names of places and holistic description etc.

Table 3. Main body of word library

Taste	Service	Environment	Overall
口味	阿姨	摆设	安溪
阿拉斯加蟹	按摩	包房	安阳
爱尔兰咖啡	按摩师	包房环境	澳门豆捞
安格斯牛肉	包装	包房装饰	澳门街
安徽菜	包装盒	包间	澳洲
安排的菜品	保安	包间风格	八佰伴
八宝辣酱	杯子	包间环境	巴贝拉
八宝年糕	菜的价格	包厢	巴蜀风
八宝养生茶	菜价	包厢环境	斑鱼府
八宝鱼	菜价格	背景音乐	北京烤鸭
霸王蛙	菜肴的服务	布局	餐厅
霸王鱼头	菜肴的价位	布置	大酒店

Table 3. (*Continued.*)

Taste	Service	Environment	Overall
白菜	菜肴的质量	餐馆生意	店
白菜粉丝	菜肴服务	餐厅	店气氛
白肉丝	餐厅的厨师	餐厅的布置	店人气
白水鱼	餐厅的服务人员	餐厅的环境	店生意
白汤	餐厅的老板	餐厅的设计	店味道
白汤牛奶	操作过程	餐厅印象	东西总体
白糖	茶水	厕所	饭店
…	…	…	…

5　Hype Identification

5.1　Determination of Dimension

By observing the sample data we find that spans of reviews are usually very large, thus using period to divide the sentences brings many errors. To avoid this, we use comma as monitoring sign to divide the sentences first.

When categorizing each sentences, we first set the default dimension as "overall". Then we matching each subject word in the sentence with our library established before, the sentence will be categorized into the dimension with most matching words.

5.2　Detection of Hype

In our process of detecting reviews of hype, our model analyzed the sentiment of all sub sentences of each dimension. If the results of four dimensions are same (all positive or negative), we will label the review as hype. In the following statistics, we find that positive reviews of hype are far more than negative reviews. Thus we assume that the default sentiment of a review is positive when there is information loss in 1 or 2 dimensions. Result shows that this preprocessing brings error less than 0.1%.

6   Experiment Result

6.1   Algorithm Difficulty Analysis

Through the analysis and category on the original data provided by dianping.com, we obtain 17681 fake reviews of hype about restaurants, and we applied our model to test these data. With the increase of reviews being tested, the result accuracy remained stable around 68%.

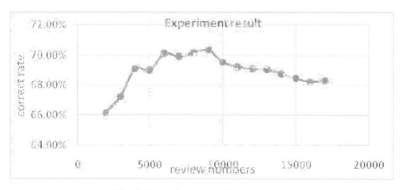

Fig. 1. Experiment result on raw data

Afterwards, we randomly sampled 2000 reviews to analyze. We found some error in labeling among the original data by checking the content of each review. Possible reasons are summarized below: (1) reviews are not about restaurants; (2) reviews are apparently not hype. After correcting the label of the 2000 reviews one by one, we witnessed the accuracy increased from 67% to 73%. Therefore we can estimate that same labeling errors exist in other parts of original data set, and the accuracy of our algorithm could reach around 74%.

Table 4. A random sample of 2,000 reviews

Review numbers	Original correct rate	Non-restaurant reviews	Available reviews	Error mark	Revised correct rate
1-500	68.4%	34	466	7	73.2%
501-1000	62.8%	29	471	11	70.3%
1001-1500	65.6%	38	462	9	71.4%
1501-2000	71.8%	21	479	5	76.2%
All	67.15%	122	1878	32	72.79%

In Figure 2, sample 1 to 6 is respectively stands for review numbers 1-500,501-1000,1001-1500,1501-2000,1-2000 and 1-17681. It is hard for us to clean all the error marked data, but the ultimate accuracy can be predicted about 74%.

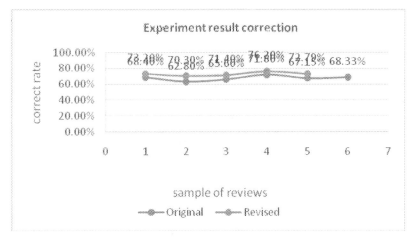

Fig. 2. Experiment result on revised data

6.2   Algorithm Difficulty Analysis

During the error analyzing process, we found that nearly 60% of errors occur for mainly two reasons. (1) Using comma as delimiter can cause incomplete sentence-breaking. (2) Some neutral reviews can be sorted wrongly by different training set, and it is difficult to judge whether they are hype or not.

7   Application of Emergency Management

The method proposed in this paper can also be used in emergency management for sentiment analysis. We can divide comments into different dimensions to get more specific sentiment analysis. We can also base on individuals, considering some of its comments together, to discover some associations with highly consistent in the emotion.

8   Conclusion

In this paper, we proposed a method to detect fake reviews of hype based on sentiment analysis. Reviews of hype take large part of fake reviews and are influential. The model we established can detect this type of review with an accuracy of around 74%. We set up our own sentiment word and multi-dimensional subject word library focus on reviews about restaurants as well. This method can also be applied to field other than reviews about restaurants.

Acknowledgement. Thanks to the support of National Natural Science Foundation of China (NNSF) (Grants No.90924029), National Culture Support Foundation Project of China (2013BAH43F01), and National 973 Program Foundation Project of China (2013CB329600, 2013CB329606).

References

1. Drucker, H., Wu, D., Vapnik, V.N.: Support vector machines for spam categorization. IEEE Transactions on Neural Networks 10(5), 1048–1054 (2002)
2. Ntoulas, A., Najork, M., Manasse, M., et al.: Detecting spam web pages through content analysis. In: Proceedings of the 15th International World Wide Web Conference (WWW 2006), Edinburgh, Scotland, May 23-26, pp. 83–92. ACM, New York (2006)
3. Jindal, N., Liu, B., Lim, E.P.: Finding Unusual Review Patterns Using Unexpected Rules. In: The 19th ACM International Conference on Information and Knowledge Management (CIKM 2010), Toronto, Canada, October 26-30, pp. 625–628 (2010)
4. Jindal, N., Bing, L.: Opinion spam and analysis. In: Proceedings of the 1st ACM International Conference on Web Search and Data Mining (WSDM 2008), February 11-12, pp. 137–142. ACM, New York (2008)
5. Wang, G., Xie, S., Liu, B., Yu, P.S.: Review Graph based Online Store Review Spammer Detection. In: 2011 IEEE 11th International Conference on Data Mining (ICDM), Vancouver, BC, December 11-14, pp. 1242–1247 (2011)
6. Xie, S., Wang, G., Lin, S., Yu, P.S.: Review Spam Detection via Temporal Pattern Discovery. In: Proceedings of the ACM SIGKDD International Conference on Knowledge Discovery and Data Mining, Beijing, China, August 12-16, pp. 823–831 (2012)
7. Mukherjee, A., Liu, B., Wang, J., Glance, N., Jindal, N.: Detecting Group Review Spam. In: Proceedings of the 20th International Conference Companion on World Wide Web (WWW 2011), New York, NY, USA, pp. 93–94 (2011)
8. Haixia, S., Xin, Y., Zhengtao, Y., et al.: Detection of fake reviews based on adaptive clustering. Journal of Nanjing University: Natural Sciences 49(4), 38–43 (2013)
9. Ott, M., Choi, Y., Cardie, C., Hancock, J.T.: Finding Deceptive Opinion Spam By Any Stretch of the Imagination. In: Proceedings of the 49th Annual Meeting of the Association for Computational Linguistics: Human Language Technologies (ACL HLT), Portland, OR, United States, June 19-24, pp. 309–319 (2011)
10. Wang, J., Wang, L., Gao, W., Yu, J.: A research on the keywords extraction of naive Bayes. Computer Applications and Software 31(2), 174–181 (2014)
11. Zhang, F., Wu, Z., Yao, F.: Research of spam filter based on Bayes. Journal of Yanshan University 33(1), 47–52 (2009)
12. Xia, T., Fan, X., Liu, L.: Implementation of ICTCLAS system based on JNI. Computer Application 24(z2), 945–950 (2004); Deng, J.: Control problems of grey system. Systems & Control Letters 1, 288–294 (1982)

A Hybrid Method of Sentiment Key Sentence Identification Using Lexical Semantics and Syntactic Dependencies

Chong Feng, Chun Liao, Zhirun Liu, and Heyan Huang

Department of Computer Science and Technology
Beijing Institute of Technology, Beijing 100081, China
{fengchong,cliao,zrliu,hhy63}@bit.edu.cn

Abstract. Many news articles in the Internet portal, blog and forums always have their own emotional orientations. Considering sentiment key sentence plays an important role in supervising social trends and public sentiment state, there has been a significant progress in this area recently, especially the lexicon-based method. However, the lexicon-based method totally dependents on lexical semantics and does not excavate the implied syntactic structure. We propose a new method which integrates lexical semantics and syntactic dependencies. And the method performs dependency parsing on the basis of a novel lexicon-based algorithm. Experimental results on COAE 2014 dataset show that this approach notablely outperforms other baselines of sentiment key sentence identification.

Keywords: sentiment key sentence, lexical semantics, syntactic dependencies, SVM, TextRank, LDA, PMI.

1 Introduction

With the rapid development of Internet, more and more news and blog articles contained emotional orientations are emerging online. Therefore, it is important for network monitoring and analysis to identify sentiment key sentences.

Sentiment key sentence, also called topical sentiment sentence, is characterized by two parts. One is topic related keyword, and the other one is sentiment related keyword. Existing sentiment key sentence identification methods maintain a single score for each sentence and most of them are lexicon-based approaches. To address the shortcomings of lexicon-based method which loses the implied syntactic structure, it is intuitive to consider the combination of lexical semantics and syntactic dependencies. In this paper we propose to acquire lexical semantics by emotion lexicon expansion and keywords lexicon construction, and get syntactic dependencies through dependency analysis. Then, we regard sentiment key sentence identification as a classification task, in which a model is trained to determine whether a candidate sentence is a sentiment key sentence. Finally, by using SVM classifier and choosing different groups of features to finish the accurate identification. In experiments on the COAE 2014 dataset we find that our method can substantially extract sentiment key sentence more effectively under different evaluation metrics.

W. Han et al. (Eds.): APWeb 2014 Workshops, LNCS 8710, pp. 11–22, 2014.

2   Related Work

Sentiment key sentence identification is a new subject which is proposed recently. Starting with topic sentiment sentences, researches on sentiment key sentence identification are drawing more and more people's attention. Sun[1] designed a novel Bi-segment method to extract topic words and converted the problem of topic sentiment sentences identification into Chinese chunking by using CRFs. Yang[2] computed the semantic similarity of a candidate sentence with the ascertained topics which were identified using an n-gram matching approach and meanwhile determine whether the sentence was topic sentiment sentence. Lin[3]presented a new algorithm which takes three attributes into account: sentiment, position and special words attributes. Until 2014, the 6th Chinese Opinion Analysis Evaluation proposed the task named the extraction and determination of sentiment key sentences. It required to extract the sentiment key sentence in a collection of news articles which had been cut into sentences.

However, existing sentiment key sentence identification methods only considered to analyze lexical semantic information using lexicon-based methods, and the methods of topic words extraction are not effective enough. Consequently, we propose a new method which incorporates syntactic dependencies within lexical semantics in this paper through emotion lexicon expansion, keywords lexicon construction and syntactic dependencies analysis.

3   Lexical Semantics Analysis for Sentiment Key Sentences

Since emotional words and topic-related keywords are two primary ingredients of sentiment key sentence, we perform expansion of emotion lexicon and construction of keywords lexicon for lexical semantics information.

3.1   Expansion of Emotion Lexicon

Beginning with semantic lexicon construction[4,5], we start trying to build an emotion lexicon. Currently, there is not a complete and universal emotion lexicon in sentiment analysis. In this paper, we select the positive, negative emotional words and evaluation words in Hownet[1], with simplified Chinese NTUSD[2] words together for addition to the basic emotion lexicon.

Since the emotional words in the basic emotion lexicon are limited and the means of Chinese expression are varied, it is necessary for us to expand the basic emotion lexicon. There are many approaches to expand emotion lexicon such as PMI[6], semantic similarity[7], TongyiciCilin[8] and Synonyms-based method[9]. Through investigation, we adopt the method based on mutual information to expand emotion lexicon, and thus build a field-related emotion lexicon.

---

[1] http://www.keenage.com/html/c_index.html
[2] http://www.datatang.com/data/11837

In expansion of emotion lexicon, we calculate the probability of two words co-occur in the same sentence using the point mutual information (PMI)which is mainly used to calculate semantic similarity between words. For every two words w_1, w_2, the PMI value between them is computed as follows:

$$\text{PMI}(w_1, w_2) = \log\left(\frac{P(w_1 \& w_2)}{P(w_1)P(w_2)}\right) \tag{1}$$

Where $P(w_1 \& w_2)$ represents the co-occurrence probability in one sentence of word w_1 and word w_2. $P(w_1)$ and $P(w_2)$ are respectively defined as the separately appearance probability of each word.

The algorithm based on PMI is composed of following 3 steps:

1. Select noun, verb, adjective from the corpus as candidate emotional words.
2. Calculate the PMI value between basic emotion lexicon and candidate emotional words using Eq. (1) and choose the highest 600 as expansion of basic emotion lexicon.
3. Add expanded words with their appearances together into basic emotion lexicon and generate the field-related emotion lexicon.

This PMI algorithm not only makes up for the lack of small-coverage basic emotion lexicon, but also enriches the field-related emotional words. Finally, we can obtain an integral emotion lexicon which has stronger field-applicability.

3.2   Construction of Keywords Lexicon

The task of keywords lexicon construction is automatically extracting certain meaningful words from a given article. It can be achieved by building a model through training corpus[10,11], as well as analyzing the internal relationship between words. The latter is widely used because of its unsupervised characteristic. In general, there are three kinds of unsupervised methods: TF-IDF, topic model[12][13] and graph model[14,15,16]. In this paper, we employ the graph model which combines LDA with TextRank to acquire the ranking scores of words. Considering the fact that the quality of keywords lexicon construction varies substantially with different kinds of weighting methods, we propose a novel weighting approach namely PCFO.

In fact, the weight is mostly related with following four factors:

- Position Influence
  Generally, words in title are more important than others.
- Coverage Influence
  A vertex is important if the number of vertexes pointing to it is large.
- Frequency Influence
  A vertex is important if the vertexes pointing to it have high appearance frequency.
- Co-occurrence Influence
  The co-occurrence number represents for the cohesion relation between words.

In graph-based model, the score of a vertex is obtained from its neighboring vertexes mutually, which is named diffusion of influence. Consequently, we design a new weighting method which takes the above four factors into consideration. Let

$\alpha,\beta,\gamma,\delta$ be the proportions of the four influence factors, where $\alpha + \beta + \gamma + \delta = 1$. The weight value between two vertexes is defined as

$$w_{ij} = \alpha w_p(v_i,\ v_j) + \beta w_c(v_i,\ v_j) + \gamma w_f(v_i,\ v_j) + \delta w_o(v_i,\ v_j) \qquad (2)$$

- $w_p(v_i,\ v_j)$ represents for the position influence of v_i , it can be computed using

$$w_p(v_i,\ v_j) = \frac{P(v_j)}{\sum_{v_t \in Out(v_i)} P(v_t)}$$

$P(v_j)$ is the importance score of vertex v_j. In this paper, we only consider the importance of title information. It can be calculated by

$$P(v) = \begin{cases} \lambda, v \text{ in title} \\ 1, v \text{ is not in title} \end{cases}$$

where $\lambda > 1$. In experiment, through investigation and evaluation, we set $\lambda = 1.5$.

- $w_c(v_i,\ v_j)$ is regarded as the coverage influence of v_i , it can be calculated by

$$w_c(v_i,\ v_j) = \frac{1}{|Out(v_i)|}$$

where $|Out(v_i)|$ is the out-degree of vertex v_i.

- $w_f(v_i,\ v_j)$ shows the frequency influence of v_i , its function is

$$w_f(v_i,\ v_j) = \frac{f(v_j)}{\sum_{v_t \in Out(v_i)} f(v_t)}$$

where $f(v_j)$ is the appearance frequency of v_j in a document.

- $w_o(v_i,\ v_j)$ indicates the co-occurrence influence of v_i , it can be expressed as

$$w_o(v_i,\ v_j) = \frac{Co(v_i,\ v_j)}{\sum_{v_t \in Out(v_i)} Co(v_i, v_t)}$$

where $Co(v_i,\ v_j)$ is the co-occurrence number of v_i and v_j within a sliding window w. Here we assign $w = 15$ based on the result demonstrated in [14].

Finally, we summarize and conclude the keywords extraction algorithm as follows:
1. Construct graph model [14] and weight them using PCFO weighting method.
2. Compute Eq. (3) and Eq. (4) in [14] to obtain the final ranking scores of words using LDA model[3] and TextRank. Noticing that the final values obtained are not affected by the choice of the initial value[16],we set initial value = 1.
3. Select the top ranking ones with final ranking scores together for addition to the keywords lexicon.

---

[3] https://www.cs.princeton.edu/~blei/topicmodeling.html

4   Mining Sentiment Key Sentence Structures on Syntactic Dependencies

Dependency analysis is primarily mining the implicit syntactic structure through analyzing the various components of language units in a sentence. Dependency parsing[17][18] regards the verb as the center of a sentence that it can dominate or be restricted by other ingredients. The dependencies reflect the semantic dependency relations between core word and its subsidiaries words[19].

An example of dependency analysis result is illustrated in Fig.1.

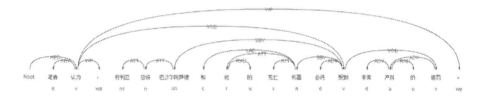

Fig. 1. Example of dependency analysis result

For sentiment key sentence identification, we expect to exploit the implicit syntactic structure relations in one sentence. Consequently, we perform HIT-LTP[4] to extract dependency templates as follows:

Algorithm 1. Dependency templates extraction

```
Input:   Preprocessed corpus T, Dependency analysis result D.
Output:  Dependency knowledge base DB.

for word in sentences of T:
    if word in expanded emotion lexicon, Hownet advocating words:
        CoreWord = word
    if word.relate=="HED" in D:
        CoreWord = word
    for word in sentence:
        if word.parent == CoreWord:
            add word and word.relation into dpWords.
    for word in dpWords:
        if relation == WP:
            delete from dpWords.
    //Such as "笔者(SBV)认为(HED)受到(VOB)"
    // ForeRelations = SBV and BackRelations = VOB.
    for word in dpWords:
        if word.ip < CoreWord.ip:
            ForeRelations += relation
        else:
            BackRelations += relation
    //Using following approach extract templates, such as template.
    // SBV+认为+VOB
    for forerelation in ForeRelations:
        for backrelation in BackRelations:
            templates += forerelation + CoreWord + backrelation
```

---

[4] http://www.ltp-cloud.com/

```
//Select the final templates
for template in sentiment key sentences and other sentences:
    calculate the appearance frequency of each template
    if frequency in sentiment key sentences > other sentences:
        Final_templates += template
//Combine the final templates with their frequencies in sentiment key
//sentences for addition to DB ultimately.
DB += Final_templates + frequencies.

return DB
```

5   Sentiment Key Sentence Identification Using SVM

Through Section 3 and 4, we have acquired expanded field-related emotion lexicon, keywords lexicon and dependency knowledge base. Relying on all these above work, we select four kinds of features as the candidate feature vectors for SVM: sentiment feature, keyword feature, dependency feature and position feature. For each sentence, we select the top n appearance frequencies or keywords ranking scores with the sum of this feature together for addition to SVM through expanded emotion lexicon, keywords lexicons and dependency knowledge base. If the sum is smaller than n, we set the excessed dimension as 0. Finally we assign n as 8 through experiments. Moreover, considering the fact that sentences in the beginning or ending of a document are more important than other sentences, we design two scoring functions as follows:

1. Improved Gaussian distribution

$$score_{sen}\big(pos(sen)\big) = \frac{1}{\sqrt{2\pi}\sigma}\left(1 - e^{-\frac{(pos(sen)-\mu)^2}{2\sigma^2}}\right) \tag{3}$$

where $\mu = \frac{n}{2}$, pos(sen) is the position of sentence in a document.

2. Parabola

$$score_{sen}\big(pos(sen)\big) = a \times pos(sen)^2 + b \times pos(sen) + c \tag{4}$$

where $-\frac{b}{2a} = \frac{n}{2}$, a > 0, b < 0, pos(sen) is the position of sentence in a document.

6   Experiments and Analysis

6.1   Experiments System

The process of sentiment key sentence identification consists of following five steps which is also illustrated in Fig.2:

1. Preprocess the articles in dataset: tokenization, part of speech, remove stop words.
2. Expand emotion lexicon, construct keywords lexicon and extract dependency templates.

3. Filter out some sentences based on the expanded emotion lexicon and keywords lexicon to obtain candidate sentiment key sentences.
4. Select four features for candidate sentiment key sentences: emotional feature, keyword feature, dependency feature and position feature.
5. Use support vector machine classifier to determine whether a sentence is a sentiment key sentence.

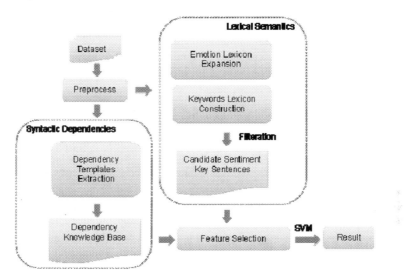

Fig. 2. The process of sentiment key sentence identification

6.2 Preparation of Dataset and Emotion Lexicon

To evaluate the performance of our method, we conduct experiments on dataset provided by COAE2014. The dataset is roughly composed of news articles which have been divided into sentences from some news and blogs. In this paper, we adopted NLPIR2014[5] to make tokenization and part of speech.

This dataset contains 1,994 news articles. After filtration by expanded emotion lexicon and keywords lexicon, there are almost 38,797 sentences with 5,019 manually annotated sentiment key sentences. Finally, we select 4,047 sentiment key sentences and 5,000 normal sentences for training, 972 sentiment key sentences and 7,325 normal sentences for testing to perform classification. Standard precision(P), recall(R) and F-measure(F) were adopted in this paper to evaluate the performance in each kind of experiment.

As the significant factor of the sentiment key sentences, we test expanded emotion lexicon on two aspects of completeness and adaptability. We separately investigate the coverage rate of the basic emotion lexicon of 2,201 words and the expanded field-related one of 2,801 words. The dataset in this paper includes 5,119 sentiment key sentences and 43,699 normal sentences before filtration by emotion and keywords

---

[5] http://ictclas.nlpir.org/

lexicons. In this experiment, we analyzed the dataset and respectively computed the coverage rate of the basic emotion lexicon and expanded field-related one.

The result shows that there are 3,721 sentiment key sentences among 5,119 tested sentences when using the basic emotion lexicon with 72% coverage rate. On the other hand, using the expanded field-related emotion lexicon, we find there are 5,019 sentiment key sentences among 5,119 tested sentences with a larger coverage rate of 98%. Consequently, the expanded emotion lexicon improves the coverage rate to a high level and compensates for the shortage that the basic emotion lexicon may be not field-related. However, it is not enough for sentiment key sentence identification only depending on emotion lexicon. So we should combine emotion lexicon with other methods to make the quality of sentiment key sentence identification better.

6.3 Comparison of Different Approaches of Keywords Lexicon Construction

As keywords are an important ingredient of sentiment key sentence, the method of keywords lexicon construction will greatly influence the accuracy of sentiment key sentence identification. In this part, we employed four methods to extract keywords for keywords lexicon construction such as TF-IDF and three other ones based on graph model. In methods of graph-based model, we mainly adopted three weighting ways: the reciprocal of distance, the times of co-occurrence and PCFO proposed by us. We added the four methods of keywords extraction with sentiment, dependency and position of Eq.(4) to the SVM classifier. The comparing results of the four methods are represented in Tables 1.

Table 1. Comparing results in sentiment key sentence identification of different keywords extraction methods

Methods	P/%	R/%	F/%
Tf-idf	31.49	48.63	38.22
1/distance	32.51	49.57	39.26
Co-occurrence	33.84	50.10	40.39
PCFO	**36.29**	**52.35**	**42.87**

The results show that the method PCFO proposed by this paper performs notablely than the other three ones. The effectiveness of PCFO primarily is because its combination of four kinds of important information: position, coverage, frequency and co-occurrence. To optimize the performance of PCFO, we investigated the influence of the parameters $\alpha,\beta,\gamma,\delta$ to PCFO for sentiment key sentence identification. In this experiment, we selected five parameters combinations of $(\alpha,\beta,\gamma,\delta)$ in which the first three ones only considered a part of weighting influential factors and the other two took all this four weighting influential factors into consideration with different proportion distribution. They are represented as 1,2,3,4,5 of $(0,1,0,0)$, $(0.5,0.5,0,0)$, $(0.3,0.4,0.3,0)$, $(0.2,0.3,0.2,0.3)$, $(0.25,0.25,0.25,0.25)$ in Fig. 3.

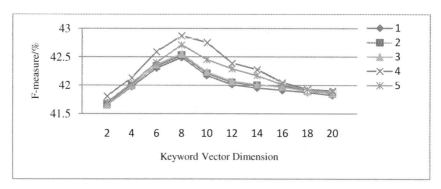

Fig. 3. F-measure of sentiment key sentence identification when $\alpha,\beta,\gamma,\delta$ are assigned with different values and different keyword vector dimensions are selected for SVM

From this figure we find that, it performs best when we select 8 dimensions as keyword vector for SVM and use the fourth approach of (0.2,0.3,0.2,0.3). In the research of keyword vector dimensions, we found too large dimensions decreased classify ability inversely. And in the investigation of the five combinations of $\alpha,\beta,\gamma,\delta$, we found the integration of the four weighting influential factors performs obviously better than just a part of the factors in the four. Moreover, the coverage and co-occurrence influential factors β,δ occupy more important places than the other two, which is probably because they mainly reflect the relations among words and not the importance of the word itself just as position and frequency do.

Consequently, this experiment not only demonstrated the effectiveness of PCFO, but also revealed the importance of keywords lexicon construction to sentiment key sentence identification.

6.4 Analysis of Groups of Features

In this section, we adopted four kinds of different groups of features for SVM classifier. As emotional words and keywords are the two basic ingredients of sentiment key sentences, we chose four groups of features: emotional words and keywords; emotional words, keywords and dependency; emotional words, keywords, dependency and position of Eq. (3); emotional words, keywords, dependency and position of Eq. (4). The below Table 2 shows the results of them:

Table 2. Results of SVM with different groups of features

Methods	P/%	R/%	F/%
Sentiment+Keyword	23.04	50.02	31.54
Sentiment+Keyword+dp	33.24	50.79	40.49
Sentiment+Keyword+dp+Pos_(3)	35.13	51.76	41.85
Sentiment+Keyword+dp+Pos_(4)	**36.29**	**52.35**	**42.87**

It can be seen that the effect of sentiment key sentence identification is highly improved after adding dependency analysis and position information, which is probably because the dependency analysis can explore the implied syntactic structure information and the position is a representation for the structure of an article. Meanwhile, we also investigated the effect of two scoring functions and found that the parabola form performed better than improved Gaussian distribution form with the reason that Gaussian distribution is too smooth at both beginning and ending of the article, which could not reflect the importance of beginning and ending obviously.

6.5   Combination of Lexical Semantics and Syntactic Dependencies

We compared combination of lexical semantics and syntactic dependencies method with other two basic ones as follows: best result of COAE2014 task 1 and lexicon-based method[3]. Besides, we also investigated combination of lexical semantics and syntactic dependencies method using COAE-500labelled, statistics-based and combination of rules and statistics approach for contrast. The COAE-500labelled approach employed 500 human-assigned sentiment key sentences as training corpus using combination of lexical semantics and syntactic dependencies method proposed in this paper as to compare with the COAE baseline which did not offer training set in the contest. Moreover, differing from our combination of rules and statistics approach, statistics-based method performed classification without filtering by emotion and keywords lexicon before. The results are presented in Table 3 as follows:

Table 3. Results of different baselines

Methods	P/%	R/%	F/%
COAE	10.41	38.88	16.42
Lexicon[3]	12.18	29.13	17.19
COAE-500labelled	16.74	39.09	23.44
Lexicon+Syntax(Statistics)	30.70	50.79	38.27
Lexicon + Syntax(Rules+Statistics)	**36.29**	**52.35**	**42.87**

As we can see, the lexicon-based method such as Zheng Lin, 2012 did not improve the result substantially as the method based on lexical semantics and syntactic dependencies especially using combination of rules and statistics approach. The main reason for this is that this method integrated lexical semantics with syntactic dependencies and when we used the emotion and keywords lexicon filtering out some sentences in rules and statics approach, it was equal to reduce the noise of corpus and so as to reach a higher precision, recall and F-measure. Furthermore, even though we only selected 500 human-assigned sentiment key sentences as training corpus, it outperformed the best result of COAE2014 task 1 and lexicon-based method considerably. So this experiment strongly demonstrated the effectiveness and applicability of combination of lexical semantics and syntactic dependencies method.

7   Conclusions and Future Work

In this paper we proposed a novel method for sentiment key sentence identification. It was regarded as a classification task, in which we incorporated sentiment, keyword, dependency and position features into SVM classifier to judge whether a sentence was a sentiment key sentence. The experiment results showed that it performed better than other baseline approaches with different keywords lexicon construction methods and groups of features.

In the future work, we will take the following points into consideration:

1. Perform phrase structure analysis on sentences and combine with dependency analysis for sentiment key sentences.
2. For dependency templates extraction algorithm, we plan to conduct synonym expansion on the core word of the template so as to improve the quality of algorithm.
3. In this paper, we use SVM for classification. We will investigate the impact on other classifiers.

Acknowledgements. The work described in this paper was supported by the National Basic Research Program of China (973 Program, Grant No. 2013CB329605, 2013CB329303) and National Natural Science Foundation of China (Grant No. 61201351). We would like to thank COAE[20] for offering this dataset. We would also acknowledge the help of HIT-IR-Lab for providing the Chinese dependency parser[21].

References

1. Sun, H., Lu, Y.: Study of topic sentiment sentences auto- extraction in Chinese blogs. Computer Engineering and Applications 44(20), 165–168 (2008)
2. Yang, J., Peng, S., Hou, M.: Recognizing sentiment polarity in Chinese reviews based on topic sentiment sentences. Application Research of Computers 28(2), 569–572 (2011)
3. Lin, Z., Tan, S., Cheng, X.: Sentiment Classification Analysis Based on Extraction of Sentiment Key Sentence. Journal of Computer Research and Development 49(11), 2376–2382 (2012)
4. Riloff, E., Shepherd, J.: A corpus-based approach for building semantic lexicons. In: Proceedings of the Second Conference on Empirical Methods in Natural Language Processing, pp. 117–124 (August 1997)
5. Hatzivassiloglou, V., McKeown, K.R.: Predicting the semantic orientation of adjectives. In: Proceedings of the 35th Annual Meeting of the Association for Computational Linguistics and Eighth Conference of the European Chapter of the Association for Computational Linguistics, pp. 174–181 (July 1997)
6. Turney, P.D., Littman, M.L.: Measuring praise and criticism: Inference of semantic orientation from association. ACM Transactions on Information Systems (TOIS) 21(4), 315–346 (2003)
7. Zhu, Y., Min, J., Zhou, Y., Wu, X.H.L.: Semantic Orientation Computing Based on How-Net. Journal of Chinese Information Processing 20(1), 14–20 (2006)

8. Lu, B., Wan, X., Yang, J., Chen, X.: Using TongyiciCilin to Compute Word Semantic Polarity. In: Processings of International Conference on Chinese Computing, pp. 17–23 (2007)
9. Wang, S., Li, D., Wei, Y., Song, X.: A Synonyms Based Word Sentiment Orientation Discriminating. Journal of Chinese Information Processing 23(5), 68–74 (2009)
10. Frank, E., Paynter, G.W., Witten, I.H., Gutwin, C., Nevill-Manning, C.G.: Domain-specific keyphrase extraction. In: Proceedings of 16th International Joint Conference on Artificial Intelligence, pp. 668–673 (1999)
11. Turney, P.D.: Learning algorithms for keyphrase extraction. Information Retrieval 2(4), 303–336 (2000)
12. Blei, D.M., Ng, A.Y., Jordan, M.I.: Latent dirichlet allocation. The Journal of Machine Learning Research 3, 993–1022 (2003)
13. Pasquier, C.: Task 5: Single document keyphrase extraction using sentence clustering and Latent Dirichlet Allocation. In: Proceedings of the 5th International Workshop on Semantic Evaluation, pp. 154–157 (July 2010)
14. Liu, Z., Huang, W., Zheng, Y., Sun, M.: Automatic keyphrase extraction via topic decomposition. In: Proceedings of the 2010 Conference on Empirical Methods in Natural Language Processing, pp. 366–376 (October 2010)
15. Page, L., Brin, S., Motwani, R., Winograd, T.: The PageRank citation ranking: Bringing order to the web, pp. 1–17 (1999)
16. Mihalcea, R., Tarau, P.: TextRank: Bringing order into texts. In: Proceedings of Empirical Methods in Natural Language Processing, pp. 404–411 (2004)
17. Hermjakob, U.: Parsing and question classification for question answering. In: Proceedings of the Workshop on Open-Domain Question Answering, vol. 12, pp. 1–6 (July 2001)
18. Hu, B., Wang, D., Yu, G., Ma, T.: An Answer Extraction Algorithm Based on Syntax Structure Feature Parsing and Classification. Chinese Journal of Computers 31(4), 662–676 (2008)
19. Li, X., Roth, D.: Learning question classifiers. In: Proceedings of the 19th International Conference on Computational Linguistics, vol. 1, pp. 1–7. Association for Computational Linguistics (August 2002)
20. http://www.liip.cn/CCIR2014/pc.html
21. Che, W., Li, Z., Liu, T.: Ltp: A chinese language technology platform. In: Proceedings of the 23rd International Conference on Computational Linguistics:

Influence Maximization in Independent Cascade Model with Limited Propagation Distance

Shunming Lv and Li Pan

National Engineering Laboratory for Information Content Analysis Technology,
Department of Electronic Engineering, Shanghai Jiao Tong University, China
{shunming_lv,panli}@sjtu.edu.cn

Abstract. Influence Maximization (IM) is the problem of finding k most influential users in a social network. In this paper, a novel propagation model named Independent Cascade Model with Limited Propagation Distance (ICLPD) is established. In the ICLPD, the influence of seed nodes can only propagate limited hops and the transmission capacities of the seed nodes are different. It is proved that IM problem in the ICLPD is NP-hard and the influence spread function has submodularity. Thus a greedy algorithm can be used to get a result which guarantees a ratio of $(1 - 1/e)$ approximation. In addition, an efficient heuristic algorithm named Local Influence Discount Heuristic (LIDH) is proposed to speed up the greedy algorithm. Extensive experiments on two real-world datasets show LIDH works well in the ICLPD. LIDH is several orders of magnitude faster than the greedy algorithm while its influence spread is close to that of the greedy algorithm.

Keywords: influence maximization, propagation models, limited propagation distance, local influence, heuristic algorithm.

1 Introduction

With more and more popular utilization, the online social network currently serves as a fundamental communication platform and ties individuals closely in daily life. Due to its powerful information propagating capability, the social network becomes a good carrier of viral marketing for the real world transactions. In viral marketing, the seller usually wants to choose a small group of influential people such that they can influence the maximum people through the word-of-mouth effect in a social network. The Influence Maximization (IM) is just the problem of how to find such an influential group. Kempe et al. [1] first formulated IM problem as a discrete optimization problem. In the formulation, a social network is modeled as a directed graph $G = (V, E)$ where each vertex in V represents an individual and each edge in E represents the relationship between individuals. A propagation model M also known as the diffusion model describes the entire propagation process in a social network. Given M, it aims to find a seed set S including k nodes such that the influence spread of S denoted as $\sigma_M(S)$ is maximized.

W. Han et al. (Eds.): APWeb 2014 Workshops, LNCS 8710, pp. 23–34, 2014.

Most of the propagation models studied before assume that influence of seed nodes spreads endlessly and the transmission capacity of each node has no difference which is too idealized. In viral marketing, information can't diffuse endlessly and different information sources have different transmission capacities. Firstly, information/product may be time-sensitive, so after propagating certain steps, no one will be interested in the information/product. For example, a company intends to promote a product designed for the coming holiday via the method of viral marketing. It is obvious that no one will accept the product when the holiday comes to the end. Furthermore, the source of the information also affects individuals' choice. One will not accept the information if he doesn't trust the source of the information even though his friend has already accepted it. So after propagating several steps, the rest nodes usually terminate to forward the information because they tend to distrust the sources. As the sources have different reputation, then they have different transmission capacities.

Taking into account the above situations, we establish a new propagation model named Independent Cascade Model with Limited Propagation Distance (ICLPD) in this paper. In the ICLPD, each node has an important parameter called limited propagation distance. Such a distance means the maximum hops that the influence of the node can propagate when it is chosen as a seed and of course it represents the transmission capacity of the node. It is proved that IM problem in the ICLPD is NP-hard and the objective function has monotonicity and submodularity. Thus a greedy algorithm can be applied to solve IM problem in the ICLPD, which can guarantee a ratio of $(1 - 1/e)$ approximation. In order to speed up the greedy algorithm, an efficient heuristic algorithm called Local Influence Discount Heuristic (LIDH) is proposed. The LIDH can select influential nodes with high efficiency by utilizing local structure of a social network. Experiments show that LIDH runs several orders faster than the greedy algorithm while its influence spread is close to that of the greedy algorithm and it significantly outperforms other compared heuristic algorithms in terms of the effectiveness.

2   Related Work

The problem of information diffusion in a social network is well studied in researches. The most popular and classic propagation models studied in the literatures are the Independent Cascade (IC) model and the Liner Threshold (LT) model [1]. The IC model is the simplest cascade model [2] and both the models are successful at explaining propagation phenomena in social networks. Based on these work, many sophisticated propagation models are proposed [3–6], which consider more factors affecting information diffusion in social networks. For example, Li et al. [5] observed that not only positive relationships but also negative relationships exist in social networks. Considering this situation, they studied influence diffusion and influence maximization in signed networks where two opposite opinions propagate. In this paper, a novel propagation model based on the IC model is proposed with the restriction that the information can only propagate limited hops which is analyzed before. It is called Independent Cascade

Model with Limited Propagation Distance (ICLPD). Chen and Liu studied the time-sensitive IM problem [7, 8] where influence also diffuses limited hops. They solve the IM problem in the IC model under the condition of limited time and find out the set of nodes which have the maximum influence for a certain period of time. However, different from their work, our work aims to solve the IM problem in the ICLPD in which different seed nodes have different transmission capacities.

Domingos et al. [9] are the first to study IM problem as an algorithm problem. Following their work, Kempe et al. [1] formalized the problem as a discrete optimization problem. They proved that IM problem in both the IC model and the LT model is NP-hard and the influence spread function is submodular and monotonic. Therefore, a climbing-hill greedy algorithm can be applied to solve the IM problem in both models and it guarantees an approximation within $(1 - 1/e)$. Due to the inefficiency of the greedy algorithm, a lot of work has been done either on improving the original greedy algorithm [3], [10, 11] or proposing efficient heuristics [12–15]. For instance, Chen et al. [13] proposed the fast heuristic algorithm PMIA which is based on the most influential path between two nodes. The PMIA needs enormous amount of memory to store arborescence for each node, which makes PMIA not applicable to large-scale networks. In this paper, we prove the NP-hardness of the IM problem in the ICLPD and the monotonicity and submodularity of the objective function. Then a greedy algorithm is applied to solve the IM problem in the ICLPD. In addition to the greedy algorithm, an efficient heuristic algorithm called LIDH is designed to speed up the solution. The LIDH selects influential nodes with the maximum local influence which represents the node's ability to influence its 2-hops neighborhood.

3   Indecent Cascade Model with Limited Propagation Distance

We first review the classic Independent Cascade (IC) model, and then introduce the new propagation model named ICLPD which is an extension to the IC model. It is proved that the IM problem in the ICLPD is NP-hard and the influence spread function has monotonicity and submodularity.

The IC model is one of the most popular propagation models. In the IC model, nodes have two states, i.e. active and inactive, which represents whether an individual accepts the information. The state of nodes can transform from inactive to active but not in the opposite direction. The influence of seed nodes spreads in discrete steps. At step 0, the seed set S are set active. Nodes in S try to activate each of their inactive neighbors. If the activation succeeds, the activated node becomes active at step 1. Then the newly active nodes try to activate their own inactive neighbors and the process goes on. In the activation process, if a node u first gets activated at step t, u will have a single chance to activate each of its inactive neighbors, for example, the node v. The activation action succeeds with a probability $P(u, v)$ which is the propagation probability of the edge (u, v), and then the node v becomes active at step $t + 1$. This process ends when no more activation happens.

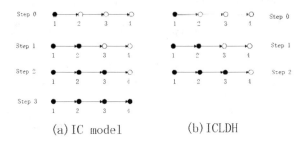

(a) IC model   (b) ICLDH

Fig. 1. Difference of propagation process between the IC model and the ICLPD

Before introducing the ICLPD, the maximum propagation distance of a seed node in the cascade model [2] is defined below.

Definition 1. *Given a social network $G = (V, E)$ and seed set S, let the influence spread in the cascade model, then the maximum propagation distance of the seed node k ($k \in S$) denoted as $d_m(k)$ is defined as the maximum step t at which the last inactive node gets activated by node k.*

According to Definition 1, the maximum propagation distance of any seed node k ($k \in S$) in the IC model is not fixed. If the network G is large enough, the distance $d_m(k)$ can be arbitrarily large. As we analyzed in the Introduction, this assumption is too idealized to model the influence propagation in online social networks. So we extend the IC model with the restriction condition that the seed nodes can only propagate limited hops. It is named Independent Cascade Model with Limited Propagation Distance (ICLPD). In the ICLPD, each node u has a new parameter called limited propagation distance denoted as θ_u which means influence of u can only propagate up to θ_u steps when it is selected as a seed. The parameter θ_u represents the transmission capacity of the node u and its value is in a limited range, i.e. $\theta_u \leq 10$.

In the ICLPD, influence of seed nodes spreads in discrete steps. At step 0, seed nodes S are set active. Each node u which first gets activated at step $t - 1$ has a single chance to activate their inactive neighbors, for example, the node v. The activation action succeeds with the propagation probability $P(u, v)$. If succeed, the node v becomes active at the step t. The activation process starting from each seed node k ($k \in S$) terminates by up to θ_k steps. Then it is clear that the maximum propagation distance of node k ($d_m(k)$) is not larger than θ_k in the ICLPD whatever the activation process is.

A simple example to demonstrate the difference of propagation process between the IC model and the ICLPD is shown in Fig. 1: (a) activation process in the IC model; and (b) activation process in the ICLPD and $\theta_1 = 2$. In this example, node 1 is the only seed node and propagation probability of each edge is 1 in both models. To the same simple network, the results of influence spread in the IC model and the ICLPD are different. After propagating two steps, node

3 becomes active at step 2. Then activation process will end in the ICLPD because influence of node 1 can only propagate up to 2 steps ($\theta_1 = 2$), while that goes on in the IC model until the last node is active at step 3.

IM problem in the ICLPD is stated as follows: Given a social network $G = (V, E)$ and the parameter θ_u for all the node u, it aims to find a set nodes S ($|S| = k$), such that influence of S propagating in the ICLPD influences the maximum number of nodes.

Consider the difference between the IC model and the ICLPD. When all the limited propagation distance θ_u is large enough, saying $\theta_u \to \infty$, the ICLPD is just the classic IC model. Thus the IC model is one special case of the ICLPD. As is known, IM problem in the IC model is NP-hard which is proved by Kempe et al. [1]. It is clear that IM problem in the ICLPD is also NP-hard.

To prove the monotonicity and submodularity of the influence spread function in the ICLPD, the t-step-paths of node u is defined first. For a directed path $P =< v_0 \to v_1 \to \cdots \to v_n >$ in a network, where each node on the path is unique, the distance of path P is the number of edges in the path. Thus the distance of such a path P is n.

Definition 2. *Suppose t is an integer ($t > 0$), the t-step-paths of node u denoted as $\Gamma_t(u)$ is defined as the set of paths starting from the node u with a distance not exceeding t.*

Theorem 1. *Given an instance of the ICLPD, the influence spread function $\sigma()$ is monotone, submodular and non-negative.*

Proof. We prove the theorem above using the similar way as Kempe et al. [1] done in the proof of Th. 2.2. The spread function $\sigma()$ is monotone and submodular iff $\sigma(S \bigcup \{v\}) - \sigma(S) \geq \sigma(T \bigcup \{v\}) - \sigma(T) \geq 0$ whenever $v \notin T$ and $S \subseteq T$. View each edge in the graph as live and blocked links by flipping a coin with related probability. Let X be one of the possible outcome after flipping all coins, and the possibility of sample X is $P(X)$. Let θ_v be the limited propagation distance of node v in the ICLPD and $\Gamma_{\theta_v}(v)$ denote θ_v-step-paths of node v according to Definition 2. For a fixed sample X, let $L(v; X)$ be the set of all live link paths starting from v and $R(v; X)$ denote the nodes that belong to the paths in $\Gamma_{\theta_v}(v) \cap L(v; X)$. According to the activation process in the ICLPD, it is clear that $\sigma_X(\{v\}) = |R(v; X)|$ and $\sigma_X(S) = |\bigcup_{u \in S} R(u; X)|$. Then the quantity of $(\sigma_X(S \bigcup \{v\}) - \sigma_X(S))$ is the number of nodes in $R(v; X)$ that are not in the union $\bigcup_{u \in S} R(u; X)$. Mathematically, the inequality is shown below.

$$\sigma_X(S \bigcup \{v\}) - \sigma_X(S) = |R(v; X)| - |R(v; X) \bigcap (\bigcup_{u \in S} R(u; X))|$$

$$\geq |R(v; X)| - |R(v; X) \bigcap (\bigcup_{u \in T} R(u; X))|$$

$$= \sigma_X(T \bigcup \{v\}) - \sigma_X(T) \geq 0. \tag{1}$$

By now, it is proved $\sigma_X()$ is monotone and submodular. With $\sigma(S) = \Sigma_{outcome X} P(X) \sigma_X(S)$, it obvious that $\sigma(S \bigcup \{v\}) - \sigma(S) \geq 0$, so $\sigma()$ is monotone. Furthermore, $\sigma()$ is also submodular because the non-negative linear combination function of submodular functions is also a submodular function. Besides, with $\sigma(\emptyset) = 0$ and the monotonicity of $\sigma()$, it is proved that $\sigma()$ is non-negative.   □

With Theorem 1, IM problem in the ICLPD can be solved by a greedy algorithm which can guarantee a $(1-1/e)$ approximation according to the conclusion by Nemhauser [16]. The specified greedy algorithm is given in Algorithm 1.

Algorithm 1. The Greedy Algoriithm

Input: $G = (V, E)$, k, M
Output: seed set S
1: $S \leftarrow \emptyset$
2: **while** $|S| < k$ **do**
3:     Select $v = arg \ max_{u \in V \backslash S}(\sigma_M(S \bigcup \{u\}) - \sigma_M(S))$
4:     $S \leftarrow S \bigcup \{v\}$
5: **end while**
6: return S

---

4 Local Influence Discount Heuristic Algorithm

As is known, the low efficiency is one of the most serious drawbacks of the greedy algorithm for the IM problem. Algorithm 1 is a typical greedy algorithm. Chen et al. [13] proved that the computation of influence spread $\sigma_M(S)$ is #P-hard, so computing the exact influence spread of seeds is nearly impossible in a network. Greedy algorithms select the most competitive node by evaluating $\sigma_M(S \bigcup \{u\}) - \sigma_M(S)$ with Monte Carlo Simulations (MCS) in each round. However, it takes too much time to do MCS in a social network, which makes greedy algorithms not applicable to large-scale networks. Finding competitive nodes (perhaps not the most competitive) with high efficiency is a good way to solve IM problem in large-scale networks.

Mathematically, let $\gamma_u(t)$ denote the number of nodes get activated by node u at step t, given node u is a seed node. Suppose $d_m(u)$ is the maximum propagation distance of node u in a cascade model, then $\sigma(\{u\}) = \sum_{t=0}^{d_m(u)} \gamma_u(t)$. Algorithm 1 selects the most influential node with maximum $\sigma(\{u\})$ with MCS which is too time-consuming. As analyzed before, it is a good way to select nodes with some parameter which approximates to $\sigma(\{u\})$ with high efficiency. So we give the definition of local influence of u as follows.

Definition 3. *Given a social network $G = (V, E)$, Let $N_2(u)$ denote the 2-hops-neighborhood of node u. $N_2(u)$ is a set of nodes which contains node u and neighbors of u and neighbors of $u's$ neighbors. Local influence of u, denoted as $\phi(u)$, is the expected number of nodes in $N_2(u)$ that get activated by node u.*

With $\sigma(\{u\}) = \sum_{t=0}^{d_m(u)} \gamma_u(t) = \phi(u) + \sum_{t=3}^{d_m(u)} \gamma_u(t)$, we declare the quantity of $\sum_{t=3}^{d_m(u)} \gamma_u(t)$ is limited. Firstly, it is shown that $\gamma_u(t)(t \geq 3)$ is not large in a social network. Suppose P is a directed path starting from u and ending at v with a distance of t, and path P is $u = q_0 \to q_1 \to \cdots \to q_t = v$. The possibility that v gets activated by u through path P is $\prod_{i=0}^{t-1} P(q_i, q_{i+1})$. This probability is at most 10^{-t} as the propagation possibility of an edge in the cascade model is among $0.01 \sim 0.1$. When $t = 3$, the value is at most 0.001. In a sparse network, the number of paths from u to v with a distance of 3 is limited. So $\gamma_u(t)(t \geq 3)$ is not large. Furthermore, notice the variable $d_m(u)$ in the ICLPD is generally a small value because $d_m(u)$ is up to θ_u which is the limited propagation distance of node u. Thus, in the ICLPD, the local influence of u defined as $\phi(u)$ which equals to $\sum_{t=0}^{2} \gamma_u(t)$ is a good approximation to $\sigma(\{u\})$.

Local influence represents the ability one can influence his two hops neighborhood. Intuitively, if one can influence a great number of people in the whole social network, he is sure to have a great influence over his surroundings. On the contrary, if the nodes are chosen with maximum local influence, they are sure to have great influence over the whole network. Local influence can be calculated with high efficiency, so it is a good measurement to select competitive nodes with maximum local influence in a heuristic algorithm.

Given a social network $G = (V, E)$ and $P(u, v)$ is the propagation probability of each directed edge (u, v). Let PP be the 2-hops-propagation-matrix of the network that $PP(u, v)$ is the probability that node v gets activated by node u at step 2. Node u can activate v through various directed paths starting from u and ending at v with a distance of 2. So $PP(u, v) = 1 - \prod_{k=1}^{n}(1 - P(u, k)P(k, v)) \approx \sum_{k=1}^{n} P(u, k)P(k, v)$. It is clear that $PP(u, v) = P^2(u, v)$. Notice each node on a path P is unique, so $PP(u, u) = 0$ for $u \in [1, n]$. Therefore, we compute $P * P$ and then set diagonal elements of P^2 to be zeroes, the result is matrix PP. Then $\phi(u)$ can be calculated with the matrix P and PP.

At first, let's compute the local influence of a node u when there is no other seed in $N_2(u)$. At step 0, node u is active and local influence of node u is the expected number of nodes that get activated by u within two steps. Thus for each node $v \in V$, it has a probability to be activated by node u within two step. We first calculate the probability for each node v, and then add them up. The summation is the local influence of node u. For any node $v \neq u$, the probability that node v gets activated by node u within two steps is $1 - (1 - P(u, v))(1 - PP(u, v)) \approx P(u, v) + PP(u, v)$. For node u itself, the value is 1. Adding all the probabilities up, local influence of node u is shown as follows.

$$\phi(u) = 1 + \sum_{v=1}^{n} P(u, v) + PP(u, v). \tag{2}$$

Considering a more general situation, let's suppose there are already some nodes selected as seeds in $N_2(u)$ and S is the set that are already selected as seed nodes. In this case, local influence of node u decreases compared to the situation that there is no node in $N_2(u)$ selected as a seed. If node u gets activated by S

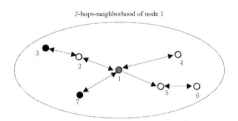

Fig. 2. An example of calculating local influence

within two steps, then selecting u as a seed makes no contribution to $\phi(u)$. This probability denoted as μ can be computed below.

$$\mu = 1 - \prod_{v}(1 - P(v, u))(1 - PP(v, u)) \approx \sum_{v} P(v, u) + PP(v, u), \quad (3)$$

where $v \in N_2(u) \bigcap S$. If node u is not activated by $v \in N_2(u) \bigcap S$ with the probability $(1-\mu)$, in this case, selecting u as a seed will activate additional nodes in $N_2(u)$. Node u has a probability to activate nodes through paths $\Gamma_2(u|S)$. $\Gamma_2(u|S)$ is the set of paths in $\Gamma_2(u)$ but excluding paths that have node belonging to S on it. The expectation of additional nodes get activated with the probability $(1 - \mu)$ is shown below.

$$\gamma = \phi(u) - \sum_{v} P(u, v)(1 + \varphi(v) - P(v, u)) + PP(u, v)$$

$$\approx \phi(u) - \sum_{v} P(u, v)(1 + \varphi(v)) + PP(u, v), \quad (4)$$

where $v \in N_2(u) \bigcap S$ and $\varphi(u) = \sum_{m=1}^{n} P(u, m)$. In conclusion, local influence of node u in this case is $(1 - \mu) * \gamma$ shown below.

$$\phi^{'}(u) = (1 - \sum_{v} P(v, u) + PP(v, u))*$$

$$(\phi(u) - \sum_{v} P(u, v)(1 + \varphi(v)) + PP(u, v)). \quad (5)$$

An example of calculating local influence in shown in Fig. 2. Node 3 and 7 are nodes already selected as seeds and propagation probability of each direct edge is set uniform p for simplicity. In this case, $\mu = 1 - (1 - p)(1 - p^2) \approx p + p^2$ and $\gamma = (1 + 4p + 2p^2) - p - p^2 = 1 + 3p + p^2$. Local influence of node 1 is calculated that $\phi(1) = (1 - \mu) * \gamma = (1 - p - p^2) * (1 + 3p + p^2)$.

Therefore, we propose an efficient heuristic algorithm called Local Influence Discount Heuristic (LIDH) which selects influential nodes with maximum local influence. The pseudo algorithm of LIDH is shown in Algorithm 2. In LIDH, $\phi(v)$ for each node is calculated at first according to equation (2) when there is no other seed selected. $dd(v)$, which is the updated local influence of node v, equals to $\phi(v)$ at the initial part. The algorithm selects the first node with the maximum $dd(v)$. After selecting the first node, local influence of nodes in $N_2(v)$ are updated according to equation (5) (update $dd(v)$). In similar ways, the second node is selected. The process goes on until all k seed nodes are selected.

Algorithm 2. LIDH algorithm

Input: $G = (V, E)$, k
Output: seed set S
 1: $S \leftarrow \emptyset$
 2: compute PP
 3: **for all** vertex $v \in V$ **do**
 4:    compute its local influence ϕ_v and φ_v
 5:    initialize μ_v and γ_v to 0
 6:    $dd_v = \phi_v$
 7: **end for**
 8: **while** $|S| < k$ **do**
 9:    Select $v = arg\ max_{u \in V \setminus S} dd_u$
10:    $S \leftarrow S \bigcup \{v\}$
11:    **for all** vertex $u \in N_2(v)$ **do**
12:        $\mu_u = \mu_u + P(v, u) + PP(v, u)$
13:        $\lambda_u = \lambda_u + P(u, v)(1 + \varphi_v) + PP(u, v)$
14:        $dd_u = (1 - \mu_u)(\phi_u - \lambda_u)$
15:    **end for**
16: **end while**
17: return S

---

5 Experiments

Experiments on two real-world datasets are conducted to evaluate both efficiency and effectiveness of different algorithms in the ICLPD. All the experiments are done on the same PC with Intel i5 2.5GHZ CPU, 4G memory and 750G disk. The directed and undirected graphs are both considered in the experiments. The first network denoted as NetComm originates from an online community for students at university of California [17]. In the online community, if student u is followed by student v, then an undirected edge (u, v) represents such a relationship. In NetComm, the propagation probability of each edge is set randomly from $0.01 \sim 0.05$. The second network, which is denoted as NetHEPT, is the scholar collaboration network coming from the section of high-energy physics theorem covering papers from January 1993 to April 2003 [18]. In NetHEPT, propagation probability of each edge is set uniform. The statistics of the two datasets is shown in Table 1.

We compare influence spreads and running time of the algorithms and heuristics below in the ICLPD.

Table 1. Statistics of the datasets

Datasets	Type	Node	Edge	Average Degree
NetComm	directed	1899	20296	21.4
NetHEPT	undirected	9877	51946	10.5

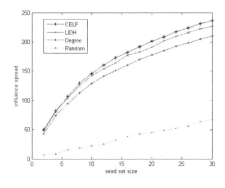 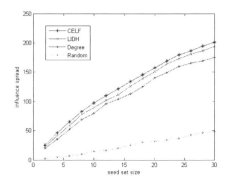

Fig. 3. Influence spreads on NetComm in ICLTD (all $\theta_u = 5$, $p \in [0.01, 0.05]$ randomly)

Fig. 4. Influence spreads on NetHEPT in ICLTD (all $\theta_u = 5$, $p = 0.05$ uniformly)

- CELF: The greedy algorithm with lazy-forward optimization [3]
- LIDH: The algorithm proposed in this paper (Algorithm 2)
- Degree: The algorithm of selecting k nodes with the largest degrees [1]
- Random: The algorithm of selecting k nodes randomly in the network, which it is a baseline heuristic algorithm [1]

We conduct experiments to evaluate the performance of different algorithms in the ICLPD. Experiments are done with all the limited propagation distance $\theta_u = 5$ for simplicity (in other cases, the results are similar).

At first, let's compare the influence spreads of different algorithms on both datasets. Fig. 3 and Fig. 4 show influence spreads of different algorithms on NetComm and NetHEPT in the ICLPD. As expected, the greedy algorithm CELF which exploits good properties of submodular functions outperforms all the other heuristics in terms of influence spread on both datasets. As is shown in the graph, the influence spread of the proposed LIDH is about 4.1% and 3.7% lower than CELF. Meanwhile, the degree algorithm performs even worse in terms of the influence spread that it is about 11.1% and 13.0% lower than the CELF.

Then we compare the running time of different algorithms for selecting 30 seed nodes. Fig. 5 depicts the running time (in sec on log scale) of different algorithms for selecting 30 nodes. As analyzed before, the CELF algorithm has low efficiency that it takes dozens of hours to select 30 seeds in a medium-size social network.

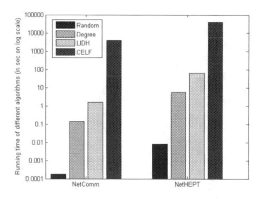

Fig. 5. Running time for selecting 30 seeds on two datasets (all $\theta_u = 5$)

However, the proposed LIDH only takes less than 100 seconds to get a result. In comparison with the execution time of the CELF and LIDH algorithms, it is obvious that the LIDH has extremely high efficiency compared to the greedy algorithm. The exception time for other compared heuristic algorithms is also shown in the graph. Though they run faster than LIDH, their influence spreads are significantly lower than that of LIDH.

Results show that LIDH is several orders of magnitude faster than CELF and its influence spread is close to that of CELF. Thus, LIDH can be applied to solve IM problem in ICLPD in large-scale networks with high efficiency and effectiveness.

6 Conclusions

In this paper, we establish a more sophisticated propagation model called Independent Cascade Model with Limited Propagation Distance (ICLPD) in which different information sources have different transmission capacities. It is proved that IM problem in the ICLPD is NP-hard and the influence spread function is submodular. Thus, a greedy algorithm can be applied to solve IM problem in the ICLPD which guarantees a $(1 - 1/e)$ approximation. In order to speed up the greedy algorithm, an efficient heuristic algorithm named LIDH is proposed. Experiments demonstrate that LIDH has extremely high efficiency compared to the greedy algorithm, while its effectiveness is close to the greedy algorithm. Furthermore, LIDH significantly outperforms other compared heuristics in terms of effectiveness in the ICLPD. For the future work, we plan to investigate our work to the Liner Threshold Model which is also a fundamental propagation model.

Acknowledgments. This work is sponsored by National Key Basic Research Program of China (2013CB329603), Research Fund for the Doctoral Program of Higher Education of China (20110071120037), and Program of Shanghai Subject Chief Scientist (13XD1425100).

References

1. Kempe, D., Kleinberg, J., Kleinber, J.: Maximizing the spread of influence through a social network. In: Knowledge Discovery and Data Mining (KDD), pp. 137–146 (2003)
2. Goldenberg, J., Libai, B., Muller, E.: Talk of the network: A complex systems look at the underlying process of word-of-mouth. Marketing Letters 12, 211–223 (2001)
3. Leskovec, J., Krause, A., Guestrin, C., Faloutsos, C., VanBriesen, J., Glance, N.: Cost-effective outbreak detection in networks. In: Knowledge Discovery and Data Mining (KDD), pp. 420–429 (2007)
4. Kempe, D., Kleinberg, J.M., Tardos, É.: Influential nodes in a diffusion model for social networks. In: Caires, L., Italiano, G.F., Monteiro, L., Palamidessi, C., Yung, M. (eds.) ICALP 2005. LNCS, vol. 3580, pp. 1127–1138. Springer, Heidelberg (2005)
5. Li, Y., Chen, W., Wang, Y.: Influence diffusion dynamics and influence maximization in social networks with friend and foe relationships. In: International Conference on Web Search and Data Mining (WSDM), pp. 657–666 (2013)
6. Li, H., Bhowmick, S.S., Sun, A.: Cinema: conformity-aware greedy algorithm for influence maximization in online social networks. In: International Conference on Extending Database Technology (EDBT), pp. 323–334 (2013)
7. Chen, W., Lu, W., Zhang, N.: Time-critical influence maximization in social networks with time-delayed diffusion process. In: Conference on Artificial Intelligence (AAAI), pp. 592–598 (2012)
8. Liu, B., Cong, G., Xu, D., Zeng, Y.: Time constrained influence maximization in social networks. In: International Conference on Data Mining (ICDM), pp. 439–448 (2012)
9. Domingos, P., Richardson, M.: Mining the network value of customers. In: Knowledge Discovery and Data Mining (KDD), pp. 57–66 (2001)
10. Chen, W., Wang, Y., Yang, S.: Efficient influence maximization in social networks. In: Knowledge Discovery and Data Mining (KDD), pp. 199–208 (2009)
11. Zhou, C., Zhang, P., Guo, J., Zhu, X.: Ublf: An upper bound based approach to discover influential nodes in social networks. In: International Conference on Data Mining (ICDM), pp. 907–916 (2013)
12. Estevez, P., Univ. de Chile, S., Vera, P., Saito, K.: Selecting the most influential nodes in social networks. In: International Joint Conference on Neural Networks (ICNN), pp. 2397–2402 (2007)
13. Chen, W., Wang, C., Guo, J., Wang, Y.: Scalable influence maximization for prevalent viral marketing in large-scale social networks. In: Knowledge Discovery and Data Mining (KDD), pp. 1029–1038 (2010)
14. Lu, W., Lakshmanan, L.V.S.: Simpath: An efficient algorithm for influence maximization under the linear threshold model. In: International Conference on Data Mining (ICDM), pp. 211–220 (2011)
15. Chen, Y.C., Peng, W.C., Lee, S.Y.: Efficient algorithms for influence maximization in social networks. Knowledge and Information Systems 33, 577–601 (2012)
16. Nemhauser, G.L., Wolsey, L.A., Fisher, M.L.: An analysis of approximations for maximizing submodular set functions. Mathematical Programming 14, 256–294 (1978)
17. NetComm, available at http://www.datatang.com/data/11979
18. NetHEPT, available at http://www.datatang.com/data/44859

Online Social Network Model Based on Local Preferential Attachment

Yang Yang, Shuyuan Jin, Yang Zuo, and Jin Xu*

Peking University, Beijing, 100871, P.R. China
{yyang1988,jinshuyuan,yangzuo,jxu}@pku.edu.cn

Abstract. Social network is a kind of complex networks which reflects social relations among people and is characterised by distinctive node degree correlation. In recent years, the node degree correlation of online social network (OSN) has been found to undergo a transition from assortativity to dissortativity. To reveal the mechanism of network evolution that leads to this transition, in this paper, an OSN model based on local preferential attachment (LPA) is proposed. The LPA model is proved to reproduce not only the transition of node degree correlation from assortative to dissortative, but also common characteristics of all social networks including "small world" phenomenon, "scale free" property and community structures.

Keywords: Local preferential attachment, Node degree correlation, assortativity.

1 Introduction

It is evident that the structural and functional properties of complex networks are dependent on their static or evolutionary network structures. In order to gain a better understanding of relationships between them, appropriate network models are needed for theoretical and experimental studies. For instance, the WS model [1] and NW model [2] were built to introduce randomness into regular networks by rewiring existing edges, which helps to explain that the "small world" networks are networks of intermediate randomness between regular networks and random networks with short average path length and high clustering coefficient. The BA model [3] creates networks with preferential attachment and proved that equivalent behaviors of webpage links or human interactions are the cause of pow-law distribution in corresponding real world networks.

The development of social network studies has partially stimulated the improvement of understanding complex networks. After thoroughly theoretical and experimental researches, a general belief exists that essential properties of social networks are short average path length, high degree clustering coefficient, scale-free node degree distribution, community structures and assortative node degree correlation (see Table 1). Therefore, different approaches have been taken

---

* Corresponding author.

W. Han et al. (Eds.): APWeb 2014 Workshops, LNCS 8710, pp. 35–46, 2014.

Table 1. Coefficient of node degree correlation for social networks, technological networks and biological networks

Network	size	r	Ref.
Physics coauthorship	52909	0.363	[4]
Biology coauthorship	1520251	0.127	[4]
Mathematics coauthorship	253339	0.120	[4]
File actor collaborations	449913	0.208	[4]
Company directors	7673	0.276	[4]
Power grid	4941	0.003	[5]
Internet	10697	-0.189	[4]
World Wide Web	269504	-0.065	[4]
Protein interaction	2115	-0.156	[4]
Neural network	307	-0.163	[4]
Marine food web	134	-0.247	[4]
Freshwater food web	92	-0.276	[4]

to build social network models that reproduce part or all of the essential features. In recent years, however, studies have found that OSNs are not assortative but mostly dissortative or inapparent assortative networks [6] (see Table 2). To make a further explanation, we firstly divide social networks into traditional social networks and OSNs according to the type of human interactions. In OSNs, social relationships are maintained by virtual web links, such as followships in twitter and acquaintanceships in facebook, while in traditional social networks, links are maintained by face-to-face contact, such as collaborations of actors or scientists. Thereby, an adjustment of the former conclusion may be made that traditional social networks are assortative networks and OSNs are dissortative ones. However, a recent study on a large social network called Wealink[1] with data sets over 27 months found that the network underwent a transition from the initial assortativity characteristic of traditional networks to subsequent dissortativity characteristic of many OSNs [6]. Time before this, no studies had ever been involved with the transition of node degree correlation of an OSN and there had been a general belief that social networks always belong to the same type of assortativity or dissortativity from time to time. In this paper, a novel model is proposed to reproduce the transition of node degree correlation in OSNs. Numerical simulations indicate that this network model also exhibits "small world" phenomenon, scale-free property and community structures.

The rest of the paper is organized as follows: in Section 2, we give a brief review of related works. Section 3 introduces the motivation of LPA model and

---

[1] http://www.wealink.com

Table 2. Node degree correlation coefficient of online social networks

Network	size	r	Ref.
Cyworld	12048186	-0.13	[7]
nioki	50259	-0.13	[8]
MySpace	100000	0.02	[7]
Orkut	100000	0.02	[7]
Xiaonei	396836	-0.0036	[9]
Flickr	1846198	0.202	[10]
LiveJournal	5284457	0.179	[10]
YouTube	1157827	-0.033	[10]
mixi	360802	0.1215	[10]

its algorithm. Simulation results are discussed in Section 4 and conclusions are made in the last section.

2   Related Works

It was argued in [11] that assortative mixing is an important feature that differentiates social network from other complex networks. This has been taken into consideration by many works in social network modeling [11–13]. However, there are still models not involve assortative mixing [14–17]. In this section, we briefly summarize the related work from the following two perspectives: social network models without assortative mixing, and social network models involve assortative mixing.

2.1   Models without Assortative Mixing

Social network models that does not include assortativity are the ones [15, 18] proposed before [11] or the ones that focus on aspects of other network properties [14, 16]. Taking geographical space into consideration, a spatial social network model was built in [14] by letting the edge probability between any two nodes dependent on their spatial distances. The model captures "small-world" properties, skewed degree distribution and community structures. Another model in [15] was built to reproduce short path length, high clustering, scale-free or exponential link distributions and it was based on the assumption that acquaintanceships tend to form between any two friends of a person. A community structure oriented model was built in [16] based on inner-community preferential attachment and inter-community preferential attachment mechanisms, its community structure and scale-free properties was proved by theoretical results and numerical simulations. In [18], a model which incorporates three features extracted from real human behavior was proposed and reproduces high level of clustering or network transitivity and strong community structures. Above all, all the models mentioned above do not incorporate with node degree correlation.

2.2   Models Involve Assortative Mixing

The assortative mixing pattern of social networks was firstly found in [4] and a model of assortatively mixed network was proposed. Within the model, networks were found to percolate more easily and more robust to vertex removal if they are assortative. In [11], a simple model was built to assist the explanation of latent relationship between assortative mixing and variation in the sizes of the communities. In [12], the social network model was built to produce efficiently very large networks to be used as platforms for studying sociodynamic phenomena. The model was built under a mixture of random attachment and implicit preferential attachment and is characterized by high value of clustering coefficient, assortativity and community structures. Inspired by individual's altruistic behavior in sociology, a model with "altruistic attachment" growth was developed in [13] and found that altruism leads to emergence of scaling and assortative mixing in social networks. The work in this paper extends the above models by proposing an evolutionary model that incorporates the transition of node degree correlation from assortative to dissortative during its growth. To the best of our knowledge, the model in this paper is the first one focusing on the transition of node degree correlation with other essential features including short average path length, high degree clustering coefficient, scale-free node degree distribution, community structures.

3   Model

3.1   Motivation for the Model

To explain the transition from assortativity to dissortativity in Wealink, a conjecture was proposed in [6] that in the early stage of network growth, the network structure of this OSN reflects traditional social networks, which means that users link to other ones who are their friends in real world and makes the OSN an identity mapping of traditional social network. As the OSN attracts more users, preferential attachments occur and the OSN gradually evolves into a dissortative one.

Consequently, we build LPA model based on following considerations: for a constantly growing OSN, there exists one tight-knit group at the initial stage. The group is an identity mapping of traditional social network and corresponds to the seed network of LPA model. Link establishment occurs when a new user join in the OSN. The user establishes relationships firstly with a pre-existing member chosen randomly and secondly with the a influential neighbor of the first chosen one. "Local preferential attachment (LPA)" comes from the second relationship formation process, during which the influential neighbor is chosen preferentially based on local information. Namely, the first chosen member with limited neighbors provides partial information for the new user and the uuser cannot get to know the globally most influential members. The influence of a user is measured under node degree centrality [19], i.e. , higher the degree, greater the influence.

The LPA model is characterised by three important features: (1) A complete seed network at initial stage. (2) Growing network size, as newcomers are introduced into the club. (3) Establishment of preferential attachment based on partial information, as newcomers attaches to influential neighbors of the first randomly chosen user.

3.2   Model Algorithm

The LPA model we build produces undirected and unweighted networks with nodes representing social network members and links representing relationship of acquaintances. The algorithm is described as follows:

Step 1. Initially, the target network N starts with a complete seed network N_s consisting of S_0 nodes.

Step 2. A new node n_a arrives at the network.

Step 3. A node n_b is chosen from N with probability P_r and a link is established between n_a and n_b.

Step 4. A node n_c is chosen from neighbors of node n_b with probability P_l and a link is established between node n_a and n_c.

Step 5. Repeat from step 3 to step 4 for $m(m < S_0)$ times.

Step 6. Repeat from step 2 to step 5 until the network size reaches S.

Probability P_r and P_l are defined as follows:

$$P_r = \frac{m}{S_0 + r}$$

$$P_l(n_j) = \frac{k_j}{\sum_i k_i}.$$

where r presents repetition times of step 6, ranging from 0 to $N - N_0 - 1$, k_i represents node degree of n_i.

4   Statistic Results

In this section we present the simulation results of LPA model. Statistical characteristics of the generated networks are proved to be evolutionary degree correlation, "rich club" phenomenon, "small world" phenomenon, "scale free" property and community structures. All results of statistical properties are averaged over 100 iterations.

4.1   Evolution of Node Degree Correlation

Node degree correlation is a comprehensive measurement in describing the pattern of links built between any node pairs. It is the Pearson Correlation coefficient of degree between pairs of linked nodes and defined by Newman [4] as follows:

$$r = \frac{M^{-1} \sum_i j_i k_i - [M^{-1} \sum_i \frac{1}{2}(j_i + k_i)]^2}{M^{-1} \sum_i \frac{1}{2}(j_i + k_i) - [M^{-1} \sum_i \frac{1}{2}(j_i + k_i)]^2}$$

where M is the total number of links in the network, j_i and k_i are the degrees of the two nodes at either end of the ith link, respectively. Networks with positive value of r are assortative networks and this indicates that nodes of similar degrees tend to link to each other. Networks with negative node degree correlation are dissortative networks and this indicates that nodes tend to link to others of dissimilar degrees.

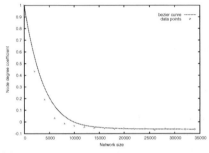 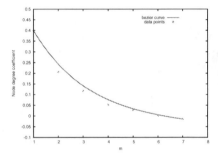

(a) Coefficient vs. network size. The network size ranges from 1000 to 35000 nodes with $m = 3$.

(b) Coefficient vs. m. The size of network is 5000 nodes with m ranging from 1 to 7.

Fig. 1. Evolution of node degree correlation

The value of node degree correlation not only manifests in its capability of differentiating social networks from other types of networks, but also it can be used to reveal the trend of acquaintanceship establishment between two people in social networks. As in physics coauthorship and film actor collaboration networks, both of these networks are assortative mixing and it can be inferred that partnerships are inclined to be established between people of same reputation or influence. While in OSNs, the dissortative mixing of some networks indicates a contrary followships establishment, and inconspicuous assortative mixing in some networks indicates random link formation. Particularly, from the evolutionary pattern of Wealink, it can be inferred that at the stage of website establishment, all of new registrants have few and similar links and the network is characterised by assortative mixing. With more registrants joining in, high influential ones emerge and followed by more and more less influential ones, leading the network evolves into a dissortative mixing. The LPA model is built to reveal this evolutionary pattern.

Fig. 1(a) shows the simulation results of node degree correlation coefficient. We can see that the coefficient drops dramatically when the network size is smaller than 10000 and it is followed by a smooth downtrend towards a steady negative value. At the end of evolutionary process, the correlation coefficient reaches a stable state. This process adequately reveals the characteristic of node degree correlation with the evolutionary process from assortative to dissortative and in the steady state, the network remains inconspicuous dissortative.

Fig. 1(b) shows another property of node degree evolution and node degree correlation coefficient decreases monotonically with the growth of m. It shows that if the newly added members are willing to make more acquaintances, the network will become less assortative. This can be used to explain the phenomenon that traditional social networks are assortative networks while OSNs are dissortative or inconspicuous assortative networks: the social activity in OSNs is maintained by web links, thus one can follow more people than he may be acquainted in real social environment.

From the simulation we can conclude that seed network has important impact on the evolutionary pattern of node degree correlation. In the initial stage, the network is a seed network of N_0 seed nodes and each node links to other $N_0 - 1$ nodes. Thus the network is assortative mixing with correlation coefficient equals 1. With new nodes of lower degrees added in, pre-existing nodes are more likely to be linked to and some of them grow into influential ones with high degree. While network size is small, the network correlation coefficient drops dramatically with more nodes of low degree attach to influential ones. Until the turning point around 0, the network grows into a larger size and the decline trend becomes smooth.

4.2   "Rich Club" Phenomenon

In human society, the "rich club" phenomenon refers to the fact that people who possess most of the social wealth tend to form tight social relationships. While in social network, the "rich people" correspond to influential nodes. Based on the measurement of node degree centrality, the "rich club" in social network are composed of high degree nodes, which are inclined to form links between each other.

Networks produced by LPA model are characterised by "rich club" phenomenon. During the evolutionary process in numerical simulation, one "rich club" forms up. The "club" is a sub-network with only nodes of high degrees linking to each other, while these nodes may be connected with low degree nodes which are not "rich club" members. The occurrence of "rich club" phenomenon can be explained as follows: during the evolutionary process, nodes of seed network are tightly connected with each other and their initial degrees are higher than any newly added member. Therefore, they are more likely to be linked to by newly added nodes and this advantage is amplified when the network size grows larger. Meanwhile, several newly added nodes happen to be linked by

Fig. 2. The "Rich club" phenomenon. This network consists of 1000 nodes with m equals 3. It is displayed using $Fruchterman - Reingold$ layout[20] scheme with nodes of higher degrees placed in the center. Seed nodes are painted in red, newly added nodes green and links between nodes of degrees higher than 100 yellow. In this figure, nodes linked to each other by yellow lines form a tight-knit group of high degree nodes with most of them are red nodes and few are green nodes.

other ones and become members of the "rich club". As a result, the "rich club" includes most of the seed nodes and a very small part of newly added nodes. This phenomenon is presented in Fig. 2 and we can see that a tightly connected "rich club" forms up at the end of the evolution process.

4.3 "Small-World" Phenomenon

Small world networks are highly transitive networks. These networks possess high proportion of cliques and quasi cliques, of which node pairs are mostly connected by short paths. A simple measurement of network transitivity is the average clustering coefficient, which is the number of closed triplets over the total number of triplets with an interesting saying: the probability of "the friends of one's friend is also his friend". The clustering coefficientC_i of any node i can be calculated as follows:

$$C_i = \frac{2e_i}{n_i(n_i - 1)}$$

Where n_i is the number of neighbors of node i and e_i is the number of links between these n_i neighbors. And the average clustering coefficient C is the average value of all nodes. It is calculated as follows:

$$C = \frac{1}{N} \sum_{i}^{N} C_i$$

Where N represents the network size. To verify the network of "small-world" phenomenon, a second measurement is the length of average shortest paths, which is the mean value of all shortest paths between node pairs.

Social networks are typically "small-world" networks. They are more transitive than random networks and are characterised by high average clustering coefficients and short average length of distances between node pairs.

Correlation between network evolutionary mechanism and transitivity is analysed in numerical simulations. The results are drawn in Fig. 3(a). In the figure, the network size grows from 0 to 50000 nodes with m equals 3 and network transitivity is calculated with every 1000 nodes added in. From the results, we

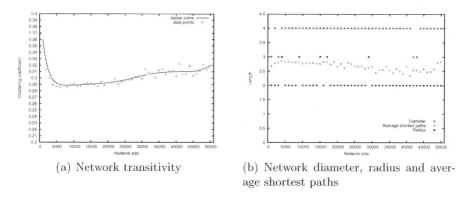

(a) Network transitivity

(b) Network diameter, radius and average shortest paths

Fig. 3. Evolution of network metrics

can see that there is a dramatic decline during the first stage and the descending trend stops when the network size reaches 5000, with the lowest point no less than 0.285. During the second stage, it rises gradually as the network size grows larger. At the equilibrium of network evolution, the clustering coefficient ranges around 0.310. It can be concluded that the networks produced by LPA model exhibit the transition from assortativity to dissortativity, which matches the one proposed by Hu et al [6] in analysing Wealink data sets. The value at the equilibrium of network evolution also matches those of Flickr and LiveJournal at 0.313 and 0.330. It can be explained as follows: initially, the seed network is a complete network with clustering coefficient equals 1. As new nodes added in, the network becomes more and more sparse with a downtrend of clustering coefficient. However, with newly added nodes repeat m iterations on linking to m randomly chosen nodes and their influential neighbors, a definite number of m triangles form and the lower limit of clustering coefficient is set when network grows into large scale.

Similarly, the simulation results of the diameter, radius and length of average shortest paths are shown in Fig. 3(b). In the figure, network size grows from 0 to 50000 nodes with m equals 3. Numerical values are calculated as every 1000 nodes add in. The length of diameter and radius are calculated under 1 iteration, while the length of average shortest paths is calculated under 50 iteration. The diameter of network falls between 3 and 4, while radius falls between 2 and 3. The length of average shortest paths ranges between 2 and 3. From the results, we can see that network diameter and radius are network size independent and it can be concluded that the strengths of association between any node pairs are steady and strong. This property matches those of traditional social networks [1] and online social networks [10].

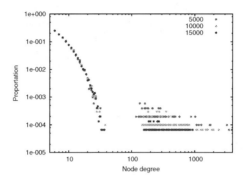

Fig. 4. Node degree distribution displayed in log scale. Networks with the size of 5000, 10000 and 15000 nodes displayed in red, green and blue colors.

4.4 "Scale Free" Property

"Scale free" networks are heterogeneous networks with power-law degree distribution. Simulations have been proceeded to verify that preferential attachment based on partial information produces "scale free" networks. Fig. 4 reveals the degree distribution of networks in different sizes. It apparently shows that all of them follow power law distribution and display apparent "big tails".

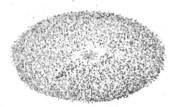

Fig. 5. Community division. A random choosen network of 50000 nodes is drawn in 15 colors. This network is displayed under DrL layout[21] with nodes belonging to the same community in one color. There are totally 15 communities of different size in this figure.

4.5 Community Structure

Community structure is one of the most important characteristics in social networks. Communities of different size and closeness are built based on common interests or organizational structure in our social interaction. Under fast algorithm of community detection [22] based on optimization method of modularity [23] or fast greedy algorithm [24] for community detection, simulation networks of different size are tested and different size of communities are found in these networks. To make a further explanation, a randomly chosen network of 5000 nodes is tested under fast greedy algorithm and 15 communities are found under modularity 0.394, which is represented in Fig. 5. The network is displayed under Drl layout [21]. As a result, social network model based on local preferential attachment produces networks with different size of communities, in accordance with real social networks.

5 Conclusions

In this paper, we propose a LPA model to characterize online social network evolution. This model is capable of reproducing networks with characteristics of evolutionary node degree correlation, "small world" phenomenon, "scale free" property and community structures. Unlike most existing models, the proposed model naturally produces networks with the same topological characteristics as real OSNs. From the statistical results we can see that the seed network has important influence on the evolution of node degree correlation, and the evolution process goes from strong assortativity to a steady state of weak dissortativity. Interesting phenomenon of the "rich club" phenomenon in human society is also found.

Acknowledgments. This work is supported by the State Key Development Program of Basic of Basic Research of China(973) under Grant Nos. 2013cd329601 and 2013cb329602.

References

1. Watts, D.J., Strogatz, S.H.: Collective dynamics of 'small-world' networks. Nature 393, 440–442 (1998)
2. Newman, M.E., Watts, D.J.: Renormalization group analysis of the small-world network model. Physics Letters A 263, 341–346 (1999)
3. Barabási, A.L., Albert, R.: Emergence of scaling in random networks. Science 286, 509–512 (1999)
4. Newman, M.E.: Assortative mixing in networks. Physical Review Letters 89, 208701 (2002)
5. Zhou, T., Lü, L., Zhang, Y.C.: Predicting missing links via local information. The European Physical Journal B 71, 623–630 (2009)
6. Hu, H., Wang, X.: Evolution of a large online social network. Physics Letters A 373, 1105–1110 (2009)
7. Ahn, Y.Y., Han, S., Kwak, H., Moon, S., Jeong, H.: Analysis of topological characteristics of huge online social networking services. In: Proceedings of the 16th International Conference on World Wide Web, pp. 835–844. ACM (2007)
8. Holme, P., Edling, C.R., Liljeros, F.: Structure and time evolution of an internet dating community. Social Networks 26, 155–174 (2004)
9. Fu, F., Liu, L., Wang, L.: Empirical analysis of online social networks in the age of web 2.0. Physica A: Statistical Mechanics and its Applications 387, 675–684 (2008)

10. Mislove, A., Marcon, M., Gummadi, K.P., Druschel, P., Bhattacharjee, B.: Measurement and analysis of online social networks. In: Proceedings of the 7th ACM SIGCOMM Conference on Internet Measurement, pp. 29–42. ACM (2007)
11. Newman, M.E., Park, J.: Why social networks are different from other types of networks. Physical Review E 68, 036122 (2003)
12. Toivonen, R., Onnela, J.P., Saramäki, J., Hyvönen, J., Kaski, K.: A model for social networks. Physica A: Statistical Mechanics and its Applications 371, 851–860 (2006)
13. Li, P., Zhang, J., Small, M.: Emergence of scaling and assortative mixing through altruism. Physica A: Statistical Mechanics and its Applications 390, 2192–2197 (2011)
14. Wong, L.H., Pattison, P., Robins, G.: A spatial model for social networks. Physica A: Statistical Mechanics and its Applications 360, 99–120 (2006)
15. Davidsen, J., Ebel, H., Bornholdt, S.: Emergence of a small world from local interactions: Modeling acquaintance networks. Physical Review Letters 88, 128701 (2002)
16. Li, C., Maini, P.K.: An evolving network model with community structure. Journal of Physics A: Mathematical and General 38, 9741 (2005)
17. Marsili, M., Vega-Redondo, F., Slanina, F.: The rise and fall of a networked society: A formal model. Proceedings of the National Academy of Sciences of the United States of America 101, 1439–1442 (2004)
18. Jin, E.M., Girvan, M., Newman, M.E.: Structure of growing social networks. Physical review E 64, 46132 (2001)
19. Freeman, L.C.: Centrality in social networks conceptual clarification. Social Networks 1, 215–239 (1979)
20. Fruchterman, T.M., Reingold, E.M.: Graph drawing by force-directed placement. Software: Practice and Experience 21, 1129–1164 (1991)
21. Martin, S., Brown, W., Klavans, R., Boyack, K.: Drl: Distributed recursive (graph) layout. SAND2008-2936J: Sandia National Laboratories (2008)
22. Blondel, V.D., Guillaume, J.L., Lambiotte, R., Lefebvre, E.: Fast unfolding of communities in large networks. Journal of Statistical Mechanics: Theory and Experiment 2008, P10008 (2008)
23. Newman, M.E.: Modularity and community structure in networks. Proceedings of the National Academy of Sciences 103, 8577–8582 (2006)
24. Clauset, A., Newman, M.E., Moore, C.: Finding community structure in very large networks. Physical Review E 70, 66111 (2004)

Temporal Curve Patterns Discovery of Information in BBS

Yanyu Yu[1,2], Yue Hu[2], and Ge Li[1,2]

[1] University of Science and Technology Beijing, Beijing, China
{yuyanyu,lige}@nelmail.iie.ac.cn
[2] Institute of Information Engineering, Chinese Academy of Sciences, Beijing, China
huyue@iie.ac.cn

Abstract. This paper study temporal curve patterns associated with BBS information and how the information's popularity grows and fades over time. We develop the Temporal-Peak clustering algorithm that accurately finds the curve pattern in BBS. According to the characteristics of BBS platform and applying sudden and durative describes temporal variation curve of information. The article demonstrates our approach on a massive dataset. The algorithm effectively avoids interference of random in the temporal node. Temporal-Peak accurately and succinctly finds temporal curve patterns in BBS.

Keywords: Social Networks, BBS, Temporal Curve Patterns.

1 Introduction

Bulletin Board System (BBS) is one kind of social networks that based on the shared space. Compare with other social networks' platform, BBS has following features:

- **Low Cost:** users doesn't need real-name authentication in BBS platform. A simple registration can publish the information.
- **Periodicity:** information in BBS has an obviously grows and fades over time every 24 hours.
- **Weak Real-Time:** BBS platform allows the users reply for the post no matter the information publisher or others. But nonsupport real-time remind of reply message. So there is strong stochastic behavior of the temporal node.
- **Less Frequency, Long Time:** Users' show less frequency log in and keep a long time on line in BBS. Users browse or reply the post which they are interested in.

These features make information temporal node has strong random and information curve has periodic. Temporal variation curve becomes more complicated.

There are two distinct lines of work related to the topics presented here: base on network node of information diffusion, and study on the general time series analysis.

Base on Network Node. Anderson and May confirmed epidemiological models like SI、SIR、SIS in theirs text-book, according the probability of node that be affected

W. Han et al. (Eds.): APWeb 2014 Workshops, LNCS 8710, pp. 47–57, 2014.
© Springer International Publishing Switzerland 2014

by the information, research the ability of the message diffusion [1]. Although the theory of diffusion of innovations. Rogers categorizes people into one of five categories [2]. And then, Cere Budak basis of how early/late a person adopts information and divide all network node to local node or goral node [3]. Pattern of human attention [4], the influence of neighborhoods and popularity [5, 6] have been extensively studied.

Time Series Analysis. Earlier work on temporal pattern is Riley Crane in 2008 put forward [7] there are two basic time-series model: emergency and un-emergency. Previous research on Youtube [8] data claims four temporal patterns exist in it. In paper of Yasuko Matsubara, they propose SpikeM [9], a concise yet flexible analytical model for the patterns of information diffusion. Time series clustering is used finding different classes of the information temporal curve. Jaewon Yang developed the K-SC algorithm [10] and found six main temporal shapes of information in online media.

However, BBS information exhibits rich temporal dynamics, and random node intensifies this process. Using existing methods can not accurately solve such problems. For example, Figure 1 illustrates the ambiguity of choosing a time series normalization method. Here we aim to group time series C1 and C2 in two clusters, where C1 (solid line) has two peaks and C2 (dash line) has only one sharp peak. K-SC aligns and scales time series by their peak volume and run the K-SC algorithm using Euclidean distance. (We choose this time series normalization method because it shows well in online media.) However, the K-SC algorithm identifies wrong puts C1, C2 into one cluster. This is because the peak scaling tends to focus on the global peak and ignores other time series node.

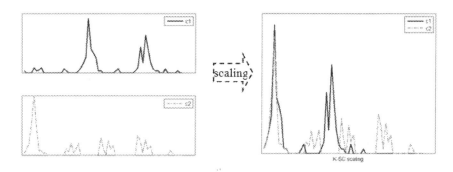

Fig. 1. C1 has two peaks and C2 has only one peak. Euclidean distance is nearly zero, in fact, there are two different clusters.

To tackle this problem this paper adopt a measure that using characteristic expression information temporal curve and applied a clustering algorithm, which gets rid of the interference factor from stochastic volatility in information curve. Two dimensional attributes: sudden and durative represent the complicated time series curve. And then, using clustering algorithm on the two dimensional characteristics.

The experiment results show that Temporal-Peak algorithm can effectively finds out the information temporal curve patterns in BBS.

2   Temporal-Peak Algorithm

Temporal variation curve refers to information's popularity grows and fades over time, such as the volume of click, the volume of reply or the volume of retweet every hour. The time series curve of BBS shows rich temporal dynamics. There are many interference factors, as periodicities, weak real-time, node stochastic volatility.

BBS information curve has cyclical change every 24 hours, peak is defined within one cycle, the maximum node is the summit of peak, and others node are the part of peak. By definition the factors of peak and the relationship of each other express the sudden and the durative of complicated information temporal curve.

2.1   Sudden

Sudden shows the explosive power of information, consists of three parts that the main peak time (Mpt), sum of peaks (Sp) and arduous of peak (Ap). Strong sudden means main peak often appear earlier in lifecycle of information, shorter time to reach a larger peak level, the higher degree of steep peaks, peaks will become more arduous.

Main Peak Time (Mpt): main peak time be defined as when the time of maximum peak appears from publish to the fall of whole lifecycle. Main peak time (Mpt) means the maximum peak time subtract mid-life of information's lift cycle. Mpt values as follows:

$$M\,pt = \; Max\,(\,T_{peak}\,) \; - \; T_{s \to f} \, / \, 2 \tag{1}$$

In the formula $Max\,(T_{peak})$: the time of maximum peak appear. $s \to f$: From grows to demise of whole lifecycle. $T_{s \to f}$:   the entire life cycle.

In addition, different values of Mpt:

$$Mpt = \begin{cases} > 0 \\ = 0 \\ < 0 \end{cases} \qquad \text{early>0, middle=0, late<0}$$

Sum of Peaks (Sp): total of peaks' value, the peaks are defined in this paper. Sp can avoid interference caused by the stochastic volatility temporal point. N is the number of all peaks. Sp defined as follows:

$$Sp = \sum_{i}^{n} NumPeak_i \tag{2}$$

The formula $NumPeak_i$: the number of one peak's values.

Arduous of Peak (Ap): the steep of peak is part of sudden, the more craggy the larger values of the sudden. Ap can indicates whether the information belongs to explosive and emergency information. Equation is as follows:

$$Ap = \sum_{i=1}^{n}\sum_{j=1}^{t}(NumPeak_i - Num_{i-j}) + \sum_{i=1}^{n}\sum_{j=1}^{t}(NumPeak_i - Num_{i+j}) \qquad (3)$$

The t means one cycle time. $NumPeak_i$ and Num_{i-j} : The arduous of peak on the left side.

We combine Eqs.1, 2 and 3. The sudden is defined as follows:

$$Sudden = \begin{cases} Mpt \mid \text{Max}(T_{peak}) - \lim_{s\to f} T/2 \\ Sp \mid \sum_{i=1}^{n} NumPeak_i \\ Ap \mid \sum_{i=1}^{n}\sum_{j=1}^{t}(NumPeak_i - Num_{i-j}) + \sum_{i=1}^{n}\sum_{j=1}^{t}(NumPeak_i - Num_{i+j}) \end{cases} \qquad (4)$$

Finally the equation is Sudden=F (α (β_1*|Mp|+β_2*Sp+β_3*Pa)). If the Mpt<0, α=-1, if the Mpt>0, α=1.Coefficientβ_1, β_2, β_3 are the normalization parameters.

2.2   Durative

Durative means the length of time of information diffusion. Number of peaks (Np), distance between two adjacent peaks (Pd), and reply rate of host (Rrh) compose the durative. More peaks and larger distance between two peaks said the stronger of the durative. The Rrh express the host wants information diffusion how long does it exist.

Number of peaks (Np): on curve likely appear a peak every 24hours. Every time the peak appeared the ability of durative will be continue. Np defined as follows:

$$Np = \sum_{s\to f}^{n} \text{def}\left(\frac{T}{peak}\right) \qquad (5)$$

Where $\text{def}\left(\dfrac{T}{peak}\right)$ conform to the definition of peak.

Peak Distance (Pd): the distance between each adjacent peak shows the ability of durative. Larger distance means the information silence for a period of time and then keeps continue.

$$Pd = \sum_{i=0}^{n} \left| T_{Peak_i} - T_{Peak_{i+1}} \right| \tag{6}$$

T_{Peak_i} and $T_{Peak_{i+1}}$ are two adjacent peak.

Reply Rate of Host (Rrh): Host is the publisher of the information. Reply of host is peculiar properties in BBS. BBS platform allows users reply to their self's post or other's reply. So it contribute to the host enhance the durative by reply to himself. Rrh is defined as follows:

$$Rrh = \frac{Num(Reply_{Host})}{Num(Reply)} \tag{7}$$

$Reply_{Host}$ is the number of reply from host, $Reply$ is the total number of reply.

We combine Eqs.5, 6 and 7 forms the durative. The expression of durative as follows:

$$Durative = \begin{cases} Np \mid \sum_{s \to f}^{n} \mathrm{def}\left(\underset{peak}{T} \right) \\ Pd \mid \sum_{i=0}^{n} \left| TPeak_i - TPeak_{i+1} \right| \\ Rrh \mid \dfrac{Num(Reply_{Host})}{Num(Reply)} \end{cases} \tag{8}$$

By definition Durative=F (β_4*Np+β_5*Pd+β_6*Rhr), the coefficient β_4, β_5, β_6 are the normalization parameters of Np, Pd and Rhr.

2.3   Algorithm Description

Algorithm Idea: Temporal-Peak according the various properties of the peak, the source data of BBS is converted to the peak curve, and then calculates the sudden value and the durative of information. Next, we use the classical K-means clustering algorithm that finds clusters of time series curve pattern. By definition C_i is number of clusters; U_i is the cluster center that the cluster centroids are then updated. At last it will output the clustering results. Calculation of different clusters finds distinct temporal pattern.

Algorithm Process: to deal with data, solving the sudden and durative attribute value of information uses clustering algorithm and develops different temporal curve pattern.   As shown in algorithm 1.

Algorithm 1. Temporal-Peak Algorithm

Input: information of post, the number of click/reply, posting time, reply time. Initial cluster C= {$C_1,C_2\cdots C_k$}

Output: cluster center U, U={$U*_1,U*_2\cdots U*_k$}

---

1. Posting time/Reply time translate into numberical(hour)
2. While(ReplyTime.list!=null){

    Split the ReplyTime.list, statistic the value every hours

        If (ReplyID==HostID) {

            HostReply++;}

   }

3. Random initial K cluster center $U_1,U_2\cdots U_k$
4. Repeat {

    Input time series j and assigns j to the cluster closest to it

    $$C_{i\_s} = \arg\min_j \left\| S_i[\partial_1 Mpt,\partial_2 Sp,\partial_3 Pa] - U_{j\_s} \right\|^2$$

    $$C_{i\_d} = \arg\min_j \left\| D_i[\partial_4 Np,\partial_5 Pd,\partial_6 Rrh] - U_{j\_d} \right\|^2$$

    Recalculate the cluster center U_i , which j join.

    $$U_{i\_s} = \frac{\sum_{i=1}^{n}\{c_{i\_s} = j\}s_i}{\sum_{i=1}^{n}\{c_{i\_s} = j\}} \quad , \quad U_{i\_d} = \frac{\sum_{i=1}^{n}\{c_{i\_d} = j\}d_i}{\sum_{i=1}^{n}\{c_{i\_d} = j\}}$$

    }

   Until

        Minimizes the sum of the squared distances between of the same clusters

5. Return cluster center U, U= {$U*_1$, $U*_2\cdots U*_k$}.

---

3   Experiments

3.1   Experimential Results

Data Set: data set from the world's largest Chinese BBS—Tian Ya BBS. In this paper randomly selected the 8 sections for data collection in more than 60 sections.

In experiments we used more than 100 million reply messages from 16216 posts. The time of data collection was from December 21 2013 to March 24 2014. In order to avoid noise, this paper truncates the length of the time series to 720 hours (reply within 30 days from the posting time).

Results: We apply the Temporal-Peak algorithm to the Tian Ya dataset, calculate the values of sudden and durative and execute clustering process. K is the number of clusters, requires to be specified in advance. We ran Temporal-Peak algorithm with different numbers clusters, experimented with K=3, K=4, K=5 and K=6. Figure 2 shows the results. In Figure2 each color represents a cluster.

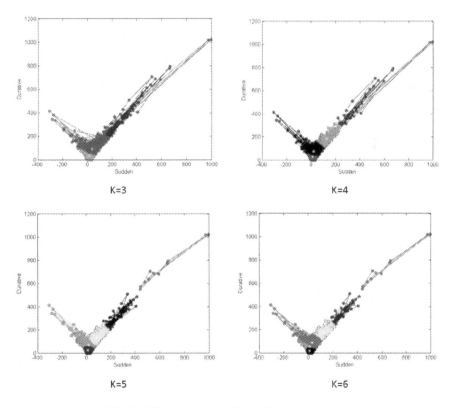

Fig. 2. Different values of K, the division of clusters

Measure Hartigan's index, Figure 5(Sec.3.2) shows that the value of K=4 gives better results. Then chose K=4 and draw information temporal curve for each cluster. Results as Figure 3 shows:

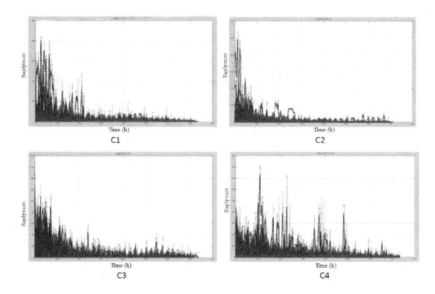

Fig. 3. K=4 and draw information temporal curve for each cluster, C= {C1, C2, C3, C4}

Finally, this paper fit the curve of the Figure 3 and develop there are four main temporal curve patterns of information in BBS. The result is shown in Figure 4.

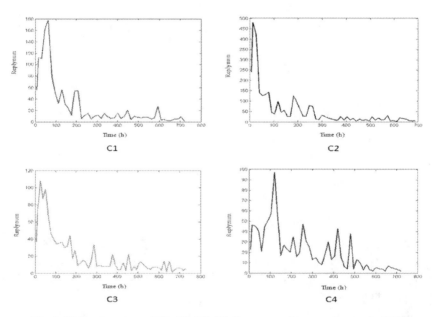

Fig. 4. Fitting the curve of C1, C2, C3, C4, four types of temporal curve in BBS

3.2   Experimential Analysis

In our experiments measured In-Group Proportion (IGP) to choose the most appropriate number of clusters. Figure 5 shows the value of the IGP. The higher the value the better the clustering. If the values are equal, the more categories the better.

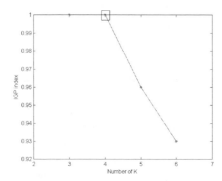

Fig. 5. K=3and K=4 are equal, the K=4 is better

Table 1. Statistics of the cluters from Figure 4. Sud/Dur: Sudden value/Duartive value, Per: Fraction of posts in the cluster, Vp: volume at main peak, Mpt: main peak appear time, DT: demise time of the information.

Cluster	C1	C2	C3	C4
Sud/Dur	50/48	382/505	184/200	[-152,82]/105
Per	56.40%	1.19%	5.08%	37.33%
Vp	178	475	105	140
Mpt	70	8	30	120
DT(h)	205	100	450	480

Figure 4 shows the cluster centers for K=4 clusters,and table 2 gives futher descriptive statistics for each of four clusters. Notice C1 and C4 are the top two, means there are 93.73% posts belong to C1 or C4. C2 is the smallest, only 1.19%.

The biggest cluster, C1, has the smallest value of sudden and durative. There are 56.4% posts fall into this cluster. Notice that C1 looks very much like the average of the four clusters. This is natural because in a large number of posts is likely to generate the average pattern. C2 has the biggest value of sudden and durative than other clusters. The volume at mian peak is the biggest, mian peak appear time and demise time of information is the earliest. C3 is characterize by a lowest volume at main peak, but the demise time can keep longer than C1 and C2. C4 is the only one cluster that contains the value of sudden is negative. The time of main peak appear at the latest, with the height of the main peak slowly declining.

In additon, it is found in experiments that 86.34% posts' mian peak appear time is early, means the value of sudden is postive. It is conform to the characteristic of BBS platform. This is natural because the new reply posts will replace the posts public recently in BBS. In Figure 4, all the curves are power law decay, this is consistent with the information diffusion rule.

4   Conclusion

In this paper, we studied the time series patterns discovery in BBS. First the characters of BBS platform were analyzed. We then developed Temporal-Peak, a novel algorithm for temporal curve patterns discovery that efficiently avoid the interference factors of BBS. Using Temporal-Peak, this paper analyzed real data of Tian Ya BBS. It was confirmed that Temporal-Peak algorithm accurately and succinctly finds out temporal curve patterns of information in this real dataset.

All in all, the article provides means to study temporal curve patterns of information in BBS, by identifying the patterns from massive amounts of real world data. We believe that our approach offers a useful starting point for information diffusion in BBS and how the temporal curve evolves over time.

Acknowledgements. This work was supported by the National 973 Project (No. 2013CB329605).

References

1. Aperjis, C., Huberman, B.A., Wu, F.: Harvesting collective intelligence: Temporal behavior in yahoo answers. ArXiv e-prints (January 2010)
2. Rogers, E.M.: Diffusion of innovations, 5th edn. Free Press (2003)
3. Budak, C., Agrawal, D.: Amr El Abbadi. Diffusion of Information in Social Networks: Is It All Local? Data Mining, 121–130 (2012)
4. Yardi, S., Golder, S.A., Brzozowski, M.J.: Blogging at work and the corporate attention economy. In: CHI 2009 (2009)
5. Wu, F., Huberman, B.A.: Novelty and collective attention. PNAS 104(45), 17599–17601 (2007)
6. Szabo, G., Huberman, B.A.: Predicting the popularity of online content. ArXiv e-prints (November 2008)
7. Crane, R., Sornette, D.: Robust dynamic classes revealed by measuring the response function of a social system. In: PNAS (2008)
8. Barabási, A.-L.: The origin of bursts and heavy tails in human dynamics. Nature 435, 207 (2005)
9. Matsubara, Y., Sakurai, Y., Aditya Prakash, B.: Rise and Fall Patterns of Information Diffusion:Model and Implications. In: KDD, pp. 6–14 (2012)
10. Yang, J., Leskovec, J.: Patterns of temporal variation in online media. In: WSDM, pp. 177–186 (2011)
11. Budak, C., Agrawal, D., El Abbadi, A.: Limiting the spread of misinformation in social networks. In: WWW, pp. 665–674 (2011)

12. Anagnostopoulos, A., Kumar, R., Mahdian, M.: Influence and correlation in social networks. In: KDD, pp. 7–15 (2008)
13. Goyal, A., Bonchi, F., Lakshmanan, L.V.S.: Learning influence probabilities in social networks. In: WSDM, pp. 241–250 (2010)
14. Backstrom, L., Kleinberg, J., Kumar, R.: Optimizing web traffic via the media scheduling problem. In: KDD 2009 (2009)
15. Szabo, G., Huberman, B.A.: Predicting the popularity ofonline content. ArXiv e-prints (November 2008)
16. Wu, F., Huberman, B.A.: Novelty and collective attention. PNAS 104(45), 17599–17601 (2007)
17. Kumar, R., Novak, J., Raghavan, P., Tomkins, A.: On the bursty evolution of blogspace. In: WWW 2002 (2003)
18. Li, L., Prakash, B.A.: Time series clustering: Complex is simpler! In: ICML (2011)

Understanding Popularity Evolution Patterns of Hot Topics Based on Time Series Features

Changjun Hu, Ying Hu, Wenwen Xu, Peng Shi, and Shushen Fu

School of Computer and Communication Engineering, University of Science and Technology Beijing, China
{huchangjunustb,huyingustb,xuwenwenustb}@163.com,
shipengustb@sina.com, likyflash@gmail.com

Abstract. Understanding popularity evolution patterns of hot topics is important for online recommendation systems and marketing services. Previous research has analyzed popularity evolution patterns based on the time series which only captures the feature of peaks. However, hot topics experience more complex popularity evolution patterns, not only peaks but also other time series features: level, trend and seasonality. In this paper, we present a method to model and understand popularity evolution patterns based on the three time series features for two types of hot topics: burst and non-burst. Our experimental results demonstrate that the seasonality of the time series is multiplicative, which means the size of the fluctuations in popularity evolution pattern of a hot topic varies with the change of trend. The level and trend of the time series are relatively unstable for burst topics compared with non-burst topics.

Keywords: Popularity Evolution Patterns, Hot Topics, Time Series.

1 Introduction

Thousands of topics diffuse rapidly through social networks every day. Only a few topics attract much attention [4, 5] and become very hot. Compared to ordinary topics, hot topics have more influence on people and they can affect trends in public opinion [1, 2] and information flow in groups [3]. It is useful to understand popularity evolution patterns for online recommendation systems and marketing services. Advertisers can also benefit from popularity analysis to make advertisements online better. What's more, it contributes to comprehending the human dynamics of consumption processes [9].

Hot topics can experience complex popularity evolution patterns. The popularity evolution pattern of topic "MH370" on Tianya BBS is shown in Fig. 1. It peaks around the 110th hour and 410th hour. Besides peaks, it exhibits increasing or decreasing trends. Also, the popularity fluctuates every 24 hours and we refer to this feature as seasonality.

There are already some studies on how to analyze the popularity of online content based on time series [8-11, 14, 15]. Those studies only took the feature of peaks into account, such as when it peaks, how long a peak lasts, how fast a peak rises and decays,

W. Han et al. (Eds.): APWeb 2014 Workshops, LNCS 8710, pp. 58–68, 2014.

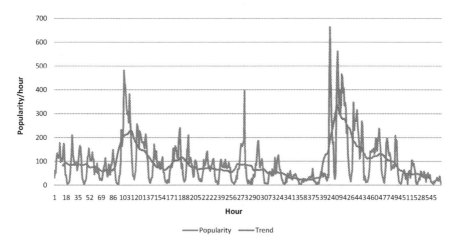

Fig. 1. Popularity evolution pattern of hot topic "MH370"

and what causes peaks. However, popularity evolution patterns of hot topics exhibit more time series features. As we discussed above, it exhibits not only peaks but also increasing or decreasing trends and seasonality. Hence, our research aims at understanding popularity evolution patterns hot topics based on the three time series features:

Level is a smoothed estimate of the data of several periods in a time series and indicates the average of the data of these periods.

Trend is a smoothed estimate of average growth of the data of several periods and indicates the increasing or decreasing trend during a period of time.

Seasonality is the tendency of time series data to exhibit behavior that repeats itself every several periods and indicates fluctuations in the time series.

By *popularity* in the following, we refer to the number of all posts and comments about a hot topic on Tianya BBS.

Hot topics are categorized into 2 types in this paper. One type is short-term and happens suddenly, for example "MH370", "H7N9". We define this type of hot topics as burst. Another type is long-term and happens periodically, for example "Christmas", "USA Election". We define this type of hot topics as non-burst. We crawled 4000 burst topics and 3000 non-burst topics on Tianya BBS as the dataset.

When giving the raw data of hot topics, we first choose proper data granularity for the two types of hot topics to get the time series. Then we use both MLR model (an additive model) and multiplicative seasonal Holt-Winters model (a multiplicative model) to model popularity evolution patterns and examine whether the seasonality of hot topics is additive or multiplicative. At last we analyze the features of trend and level according to the parameters of multiplicative seasonal Holt-Winters model. Our experiments demonstrate the size of the fluctuations in popularity evolution pattern of a hot topic varies with the change of trend, so a multiplicative model can perform better than an additive model when modeling popularity evolution patterns. The level and trend of the time series are relatively unstable for burst topics compared with non-burst topics, which means the popularity tends to be changeable and an increasing or decreasing trend only lasts for a short time for burst topics.

In Section 2 we discuss recent related work. Section 3 describes the dataset which was collected from Tianya BBS. Section 4 introduces some terminologies about time series and two analysis models we used in this paper. In section 5 we describe our analysis method and experimental results. At last, we will conclude the paper in section 6.

2   Related Work

Researchers have explored several methods for popularity analysis of UGC [16, 17] based on time series.   Crane and Sornette [8, 9] examined the time series of UGC and found the sudden peak and relatively rapid relaxation illustrates the typical signature of an "exogenous" burst of activity, and symmetric relaxation is characteristic of an "endogenous" burst of activity. Yang and Leskovec [15] studied how the content popularity grows and fades overtime based on the rate of rise and decay of a peak in the time series.   Figueriedo [10] found another type of popularity evolution patterns in which there are no peaks and UGC keeps attracting much popularity for a long time.   These work only considered the peaks in the time series of UGC. In this paper, we analyze more detailed time series features: level, trend and seasonality.

Much research was conducted for hot topic evolution modeling. Lin et al. [18] proposed a statistical method that models the popularity of topics over time, taking into consideration the burstiness of user interest, information diffusion on the network structure, and the evolution of textual topics. Romero et al. [19] studied how different kinds of topics spread by using hashtags and the subgraph structure of the initial adopters for different widely-adopted hashtags. Lehmann et al. [20] also studied the evolution pattern by using hashtags and found exogenous factors more important than endogenous factors to make a topic popular.   Ardon et al. [21] performed a rigorous temporal and spatial analysis, investigating the time-evolving properties of the subgraphs formed by the users discussing each topic and investigated the effect of initiators on the popularity of topics. They found that users with a high number of followers have a strong impact on topic popularity and topics become popular when disjoint clusters of users discussing them begin to merge and form one giant component that grows to cover a significant fraction of the network.

3   Dataset

The data were collected in two steps: first, 10 thousand hot topics were selected from Sina News [12] which is one of Sina channels. Everyday thousands of news is reported on Sina News. Second, we crawled all posts concerning hot topics gotten from step one and their comments with timestamps from Tianya BBS [13]. Tianya founded in 1999 is the largest BBS in China and has 85 million registered users, with 30 million visits every day. The website includes many sections, such as "Entertainment", "Politics", "Economy" and "Emotion" and so on.

Sina News has been collecting news since July 5, 2004. People can get news reported on someday by searching by date. 10 thousand hot topics were collected from "Entertainment", "Science", "Sports", "Economy" section. The topics were manually categorized into burst and non-burst types.

Then we took the hot topics collected from Sina News as keywords to search for posts on Tianya BBS. The website provides different searching criteria, such as "by relevance", "by posting time", "by full articles", "by title", among others. We chose to search by relevance and by title and crawled all posts and comments. In order to get qualified topics, we further selected 4000 ones from 6000 burst topics and 3000 ones from 4000 non-burst topics. For burst topics, the data were made by users between January 1, 2011 to December 31, 2013 and the popularity of each was more than 18000 in the first 15 days. For non-burst topics, the data were made between January 1, 2001 to December 31, 2012 and the final popularity was more than 15000.

4   Terminology and Time Series Models

This section introduces some basic terminologies about time series and two classic models. One is multiple linear regression model which is an additive model. The other is multiplicative seasonal Holt-Winters model which is a multiplicative model. We use the two models to examine whether the seasonality of hot topics is additive or multiplicative.

4.1   Terminology

The time series data are a sequence of observations and consist of 4 components which are level, trend, seasonality and noise. The seasonality can be classified into two types: additive and multiplicative. In the additive case, the series shows steady seasonal fluctuations, regardless of the trend of the series; in the multiplicative case, the size of the seasonal fluctuations varies, depending on the trend of the series [7]. An additive model should be chosen for additive seasonality and a multiplicative model should be chosen for multiplicative seasonality.

4.2   Analysis Models

Multiple Linear Regression Model. In this model, the time series is represented as follow:

$$Y_t = P_t + (a_1x_1 + a_2x_2 + \ldots + a_{l-1}x_{l-1}) + E_t \tag{1}$$

Where
t is a arbitrarily chosen point in time
Y_t is an actual value of popularity at time t

P_t is a polynomial function corresponding to t and represents the combination of trend and level component

l is the length of the season

E_t is the random error.

x_1, x_2, ..., x_{l-1} are dummy variables. There will be l-1 dummy variables if there are l periods in the season. The value of the period without a dummy variable can be taken as the reference value. A dummy variable takes the value 0 or 1. When variable t corresponds to a period, dummy of the period should take value 1 and the others should take value 0.

a_1, a_2, ..., a_{l-1} are coefficients of dummies.

Multiplicative Seasonal Holt-Winters Model. In this model, the time series is represented as follow:

$$Y_t = (L + tT) S_t + E_t \tag{2}$$

Where

t is a arbitrarily chosen point in time

Y_t is an actual value of popularity at time t

L_t is the level component

T_t is a linear trend component

S_t is a multiplicative seasonal factor

E_t is noise component.

$$L_t = \alpha y_t/S_{t-1} + (1-\alpha)(L_{t-1}+T_{t-1}) \tag{3}$$

$$T_t = \beta(S_t-S_{t-1})+(1-\beta)T_{t-1} \tag{4}$$

$$S_t = \gamma(y_t/L_t)+(1-\gamma)S_{t-1} \tag{5}$$

Where

y_t is an observation at time t.

α, β, and γ are called smoothing factors, must be estimated in such a way that the MSE(Mean square error) or MAPE(Mean absolute percentage error) is minimized. Values of the smoothing factors closer to one have a less smoothing effect and give greater weight to recent changes in the data, while values of the smoothing factors closer to zero have a greater smoothing effect and are less responsive to recent changes [6].

l is the length of the season.

The seasonal factors are defined so that they sum to the length of the season, i.e.

$$\sum_{1 \leq i \leq l} S_t = l \tag{6}$$

5　Analyzing Popularity Evolution Patterns of Hot Topics

We propose a framework to deeply study how popularity evolves for burst and non-burst hot topics. Multiplicative seasonality of hot topics is demonstrated which means the size of the seasonal fluctuations varies with the change of trend of the series. Also other characteristics of level and trend are also revealed in this section.

5.1   Framework of Popularity Evolution Pattern Analysis

When giving the raw data of the popularity of a hot topic, we first choose proper data granularity to get the time series. If the size in which data are subdivided is too fine, values of many periods will be close to zero, which should be avoided. Take non-burst topic "U.S. Presidential Elections" as an example; the popularity reaches the peak in November every four years but is at the bottom in the other months, shown in Fig. 3. If we take one day as the granularity, there will be many values close to zero in the time series. If the granularity is too coarse-- take one year as the granularity, there will be only 12 periods in the series, which is not adequate for analysis. So we should take one month as the granularity for non-burst topics and one hour for burst topics.

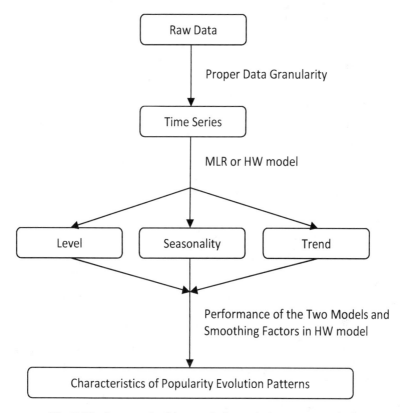

Fig. 2. The framework of the popularity evolution pattern analysis

We then use both MLR model and multiplicative seasonal HW model to best fit the time series we have already gotten above. The two models both can help us decompose the time series into level, trend, seasonality and noise component.

At last we compare the performance of the two models on our dataset to examine whether the seasonality of hot topics is additive or multiplicative. If multiplicative seasonal HW model performs better, the seasonality will be multiplicative, and vice versa. And smoothing factors in multiplicative seasonal HW model can reflect the stability of level and trend.

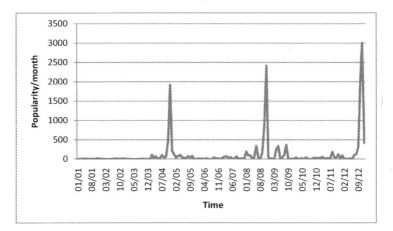

Fig. 3. Time series from Jan 2001 to Dec 2012 of the topic "U.S. Presidential Elections"

5.2   Experimental Results

In burst case, we take hot topic "Haze in Beijing" as an example. Fig. 4 shows its time series of popularity evolution pattern. We choose cubic polynomial instead of higher-order polynomial as the trend and level component for MLR model to avoid the problem of over fitting [6]. For multiplicative seasonal HW model, α, β, and γ are estimated by minimizing MAPE. The fitting results of the 5th to the 7th days for two models are shown in Fig. 5. In non-burst case, we take hot topic "Christmas" as an example and its time series of popularity pattern is shown in Fig. 6. Fitting results of the 6th and the 7th years are shown in Fig. 7.

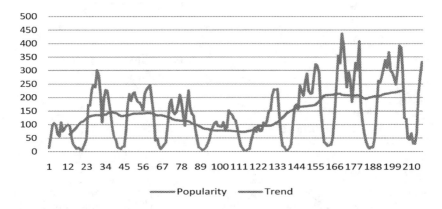

Fig. 4. The blue line indicates time series of the topic "Haze in Beijing" in first 9 days. And the red line indicates the trend of the time series and is plotted by moving averages method of which the window size is 24.

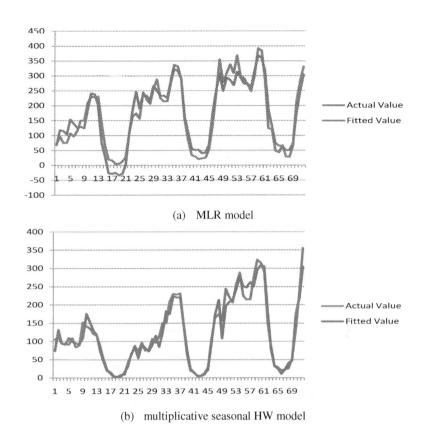

(a)   MLR model

(b)   multiplicative seasonal HW model

Fig. 5. Fitting results of the 5th to the 7th days for MLR model and multiplicative seasonal HW model on topic "Haze in Beijing"

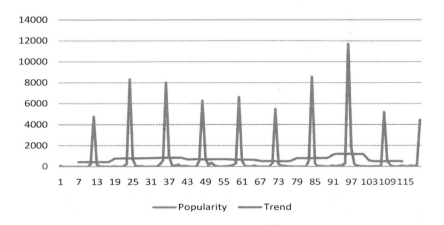

Fig. 6. The blue line indicates time series from Jan 2003 to Dec 2012 of the topic "Christmas". And the red line indicates the trend of the time series and is plotted by moving averages method of which the window size is 12. The popularity reaches the peak in December every year.

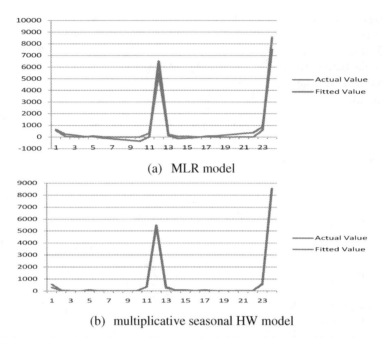

Fig. 7. Fitting results of the 6th to the 7th years for MLR model and multiplicative seasonal HW model on topic "Christmas"

We evaluate the efficiency of the two models by using average MAPE. For each hot topic, we have:

$$MAPE = \frac{\sum_1^m \frac{|\hat{y}-y|*100}{y}}{m} \tag{7}$$

where $\hat{y}$ is the fitted value of a period, y is the actual value of the period and m is the total number of periods in the series. For the average MAPE of all tested hot topics, we have:

$$\overline{MAPE} = \sum_1^n \frac{\sum_1^m \frac{|\hat{y}-y|*100}{y}}{m} \bigg/ n \tag{8}$$

where n is the number of all tested topics. The average MAPEs of MLR model and multiplicative seasonal HW model for burst and non-burst topics are given in table 1. We can conclude that multiplicative seasonal HW model (a multiplicative model) performs better MLR model (an additive model) for both types of hot topics. This result indicates that the size of the seasonal fluctuations varies and the seasonality of the time series of both types of hot topics is multiplicative. So, when modeling popularity evolution patterns or predicting future popularity, a multiplicative model is recommended rather than an additive model.

Interestingly, the smoothing factors of multiplicative seasonal HW model are usually in certain ranges: $\alpha<0.5$, $\beta<0.5$, $\gamma<0.07$ for non-burst topics, comparatively, $\alpha>0.55$, $\beta>0.5$, $\gamma<0.07$ for burst topics, which indicates that for burst topics, level and trend of the time series are relatively unstable. In other words, the popularity tends to be changeable and an increasing or decreasing trend only lasts for a short time for burst topics.

Table 1. The average MAPEs of MLR model and multiplicative seasonal HW model for two types of hot topics

	Burst	Non-Burst
MLR	39.84%	33.56%
HW	19.92%	14.73%

6   Conclusions and Future Work

This paper presented an intensive study on popularity evolution patterns of hot topics. We chose an additive and a multiplicative time series model to experiment on 4000 burst topics and 3000 non-burst topics. The experimental results show that multiplicative seasonal HW model make better performance than MLR model, which demonstrates that the seasonality of time series of hot topics is multiplicative and a multiplicative time series model will be a more advisable choice than an additive model when modeling popularity of hot topics. We have also found that level and trend of the time series are relatively unstable for burst topics.

In this work, we only analyzed the time series of popularity. There are also many factors (e.g. referrers, social influence, cascade, etc) affecting popularity. In future work, we will analyze the time series of those factors and help to shed light on popularity evolution patterns of hot topics.

Acknowledgement. This work was supported by the National 973 Project (No. 2013CB329605).

References

1. Domingos, P., Richardson, M.: Mining the Network Value of Customers. In: ACM SIGKDD International Conference on Knowledge Discovery and Data Mining, pp. 57–66. ACM Press, New York (2001)
2. Kempe, D., Kleinberg, J., Tardos, É.: Maximizing the Spread of Influence through a Social Network. In: ACM SIGKDD International Conference on Knowledge Discovery and Data Mining, pp. 137–146. ACM Press, New York (2003)
3. Wu, F., Huberman, B.A., Adamic, L.A., Tyler, J.R.: Information Flow in Social Groups. Physica A: Statistical Mechanics and its Applications 337(1), 327–335 (2004)
4. Wu, F., Huberman, B.A.: Novelty and Collective Attention. Proceedings of the National Academy of Sciences 104(45), 17599–17601 (2007)

5. Couronne, T., Stoica, A., Beuscart, J.S.: Online Social Network Popularity Evolution: an Additive Mixture Model. In: IEEE International Conference on Advances in Social Networks Analysis and Mining, pp. 346–350. IEEE Press, New York (2010)
6. Shmueli, G., Lee, H.: Time Series Forecasting. Tsing Hua University Publishing, Beijing (2012) (in Chinese)
7. Kalekar, P.S.: Time Series Forecasting Using Holt-Winters Exponential Smoothing. Kanwal Rekhi School of Information Technology 4329008, 1–13 (2004)
8. Crane, R., Sornette, D.: Viral, Quality, and Junk Videos on YouTube: Separating Content from Noise in an Information-Rich Environment. In: AAAI Spring Symposium: Social Information Processing, pp. 18–20. AAAI Press, Menlo Park (2008)
9. Crane, R., Sornette, D.: Robust Dynamic Classes Revealed by Measuring the Response Function of a Social System. Proceedings of the National Academy of Sciences 105(41), 15649–15653 (2008)
10. Figueiredo, F.: On the Prediction of Popularity of Trends and Hits for User Generated Videos. In: ACM International Conference on Web Search and Data Mining, pp. 741–746. ACM Press, New York (2013)
11. Ahmed, M., Spagna, S., Huici, F., Niccolini, S.: A Peek into the Future: Predicting the Evolution of Popularity in User Generated Content. In: ACM International Conference on Web Search and Data Mining, pp. 607–616. ACM Press, New York (2013)
12. Sina News, http://news.sina.com.cn/hotnews/
13. Tianya Search, http://search.tianya.cn/bbs/
14. Figueiredo, F., Benevenuto, F., Almeida, J.M.: The Tube over Time: Characterizing Popularity Growth of YouTube Videos. In: Proceedings of the 4th ACM International Conference on Web Search and Data Mining, pp. 745–754. ACM Press, New York (2011)
15. Yang, J., Leskovec, J.: Patterns of Temporal Variation in Online Media. In: Proceedings of the 4th ACM International Conference on Web Search and Data Mining, pp. 177–186. ACM Press, New York (2011)
16. Cha, M., Kwak, H., Rodriguez, P., Ahn, Y., Moon, S.: I Tube, You Tube, Everybody Tubes: Analyzing the World's Largest User Generated Content Video System. In: Proceedings of the 7th ACM SIGCOMM Conference on Internet Measurement, pp. 1–14. ACM Press, New York (2007)
17. Cha, M., Kwak, H., Rodriguez, P., Ahn, Y., Moon, S.: Analyzing the Video Popularity Characteristics of Large-scale User Generated Content Systems. IEEE/ACM Transactions on Networking 17(5), 1357–1370 (2009)
18. Lin, C.X., Zhao, B., Mei, Q., Han, J.: PET: A Statistical Model for Popular Events Tracking in Social Communities. In: Proceedings of the 16th ACM SIGKDD International Conference on Knowledge Discovery and Data Mining, pp. 929–938. ACM Press, New York (2010)
19. Romero, D.M., Meeder, B., Kleinberg, J.: Differences in the Mechanics of Information Diffusion across Topics: Idioms, Political Hashtags, and Complex Contagion on Twitter. In: Proceedings of the 20th International Conference on World Wide Web, pp. 695–704. ACM Press, New York (2011)
20. Lehmann, J., Gonçalves, B., Ramasco, J.J., Cattuto, C.: Dynamical Classes of Collective Attention in Twitter. In: Proceedings of the 21st International Conference on World Wide Web, pp. 251–260. ACM Press, New York (2012)
21. Ardon, S., Bagchi, A., Mahanti, A., et al.: Spatio-temporal and Events Based Analysis of Topic Popularity in Twitter. In: Proceedings of the 22nd ACM International Conference on Information and Knowledge Management, pp. 219–228. ACM Press, New York (2013)

VxBPEL_ODE: A Variability Enhanced Service Composition Engine

Chang-Ai Sun[1,*], Pan Wang[1], Xin Zhang[1], and Marco Aiello[2]

[1] School of Computer and Communication Engineering, University of Science and Technology Beijing, 100083 Beijing, China
casun@ustb.edu.cn
[2] Johann Bernoulli Institute, University of Groningen, 9747 AG, Groningen, The Netherlands
m.aiello@rug.nl

Abstract. Service compositions have become a powerful development paradigm to create distributed applications out of autonomous Web services. Since such applications are often deployed and executed in open and dynamic environments, variability management is a crucial enabling technique. To address the adaptation issue of service compositions, we proposed VxBPEL, an extension of BPEL for supporting variability, and a variability-based adaptive service composition approach which employs VxBPEL for variability implementation. In this paper, we present a VxBPEL engine for supporting the execution of VxBPEL service compositions. The engine is called VxBPEL_ODE and is implemented by extending a widely recognized open source BPEL engine, Apache ODE. We discuss key issues of developing VxBPEL_ODE, and three real-life service compositions are employed to evaluate and compare its performance with another VxBPEL engine we developed in our previous work. VxBPEL_ODE, together with analysis, design, and run-time management tools for VxBPEL, constitutes a comprehensive supporting platform for variability-based adaptive service compositions.

Keywords: Service Oriented Architecture, Service Compositions, Variability Management, VxBPEL.

1 Introduction

Service Oriented Architecture (SOA) is a mainstream paradigm for application development in the context of the Internet and highly-scalable systems [1]. Since individual services usually provide limited functionalities and are unable to meet in isolation the actual demand, service compositions are a powerful mechanism for integration of data and applications in the context of distributed, dynamic, heterogeneous environments. A service composition can be seen as a process in which multiple individual services participate and are coordinated to support complex business goals. Service compositions are able to support rapid business re-engineering and optimization. Business Process Execution Language (BPEL) is a process-oriented executable service composition language which can be used to construct loosely coupled systems by

W. Han et al. (Eds.): APWeb 2014 Workshops, LNCS 8710, pp. 69–81, 2014.

orchestrating a bundle of Web services [2]. Business processes implemented by such service compositions are expected to be flexible enough to cater for frequent and rapidly-changing requirements [3]. For instance, a service under composition may become unavailable or fail to provide the expected functionalities. Consequently, flexibility of service compositions is highly desirable. Unfortunately, the standard version of BPEL is limited in supporting such an expectation [4].

To address the adaptation issue of service compositions, we have proposed a variability-based adaptive service composition approach [5]. In the proposed approach, the changes are treated as the first-class objects, and VxBPEL [6], an extension to BPEL, is used for specifying variation. VxBPEL provides a set of constructs for supporting variation design, such as variation points and variants [6][7]. Since these constructs are not of standard BPEL elements and cannot be executed by standard BPEL engines, we developed a VxBPEL engine called VxBPEL_ActiveBPEL [8] by extending ActiveBPEL [10], a well recognized BPEL engine. To support the analysis and maintenance of variation, we developed a tool, ValySec [9], which can be used to automatically extract variation definitions from VxBPEL service compositions and visualize variation configurations. With the proposed approach and tools, designers can implement adaptive service compositions by identifying possible changes within service compositions and specifying them with alternatives.

In this paper, we present a VxBPEL engine, VxBPEL_ODE, to further provide an integrated supporting platform for variability-based adaptive service compositions. As a mainstream application development platform, Eclipse provides two open source plug-ins for BPEL, namely BPEL Designer and Apache ODE [11]. In our previous work [5], we developed a visual VxBPEL design tool called VxBPEL Designer by extending BPEL Designer. Since both VxBPEL_ODE and VxBPEL Designer are implemented as Eclipse plug-ins, VxBPEL_ODE can be seamlessly integrated with VxBPEL Designer, which facilitates the provision of an integrated supporting platform for VxBPEL-based adaptive service compositions. Furthermore, VxBPEL_ODE is significantly different in architecture from VxBPEL_ActiveBPEL that relies on ActiveBPEL, which is no longer available as open source. In this paper, we examine key issues relevant to development of VxBPEL_ODE and compare the performance of VxBPEL_ODE with that of VxBPEL_ActiveBPEL using three VxBPEL service composition cases.

The rest of the paper is organized as follows. Section 2 introduces the underlying concepts of VxBPEL. Section 3 discusses the design and implementation of VxBPEL_ODE. Section 4 validates VxBPEL_ODE and compares its performance with VxBPEL_ActiveBPEL through case studies. Section 5 describes related work. The conclusion is reported in Section 6.

2   Background

BPEL [2] is a process-oriented, executable composition language which describes control flows between activities in a business process. Activities are basic interaction units, and divided into basic and structural activities. Basic activities are an atomic execution step, such as *assign* and *invoke*. Structural activities are compositions of basic activities and/or structural activities. A standard version of BPEL process specification is fixed, which means that all activities and their relationships must be exactly defined before deployment and run-time changes are not allowed after deployment.

Variability is the ability of a software system to extend, change, customize or configure itself in any particular environment [4]. The core concepts of the variability include *variation points*, *variants*, and *dependency*. The *variation point* is the part of the software system that may change. Typically, a *variation point* occurs when multiple design options are to be chosen. After one option at a variation point is selected, a so called *variant* is obtained; the collection of choices at each variation point is referred to as a variation *configuration*. *Dependency* specifies constraints between the different variations corresponding to the variant.

In order to introduce variability into service compositions, we proposed VxBPEL which extends BPEL for modeling variability [4]. VxBPEL employs the COVAMOF variability framework [12] and provides constructs such as *variants*, *variation points*, and their *configurations* and *constraints*. A *variant* is defined using the tag <vxbpel:Variant>. The associated name is used to identify a variant. The element enclosed by the tag <vxbpel:VPBPELCode> is used to specify the choice contained within the variant, which corresponds to original BPEL activity, such as an *invoke* activity. A *variation point* is defined using the tag <vxbpel:VariationPoint>. It is identified by its unique name and may contain one or more variants enclosed by the tag <vxbpel:Variants>. The tag <vxbepx:ConfigurableVairationPoint> is used to specify the selection of variants associated with variation points, and the tag <vxbpel:RequiredConfiguration> specifies the realization between the higher variation point and the lower variants. *Constraints* defined in the tag <vxbpex:Constraint> provide a powerful mechanism for defining variant dependencies. To further facilitate its adoption, an integrated supporting platform is expected. VxBPEL_ODE, a core component of such a platform, is our proposal described next.

3   Design and Implementation

The variability-based adaptive service composition approach consists of two steps: (1) BPEL is first used to specify service compositions, and (2) variability constructs defined in VxBPEL are then used to define and configure variations of the BPEL service compositions. The resulting service compositions are referred to as *VxBPEL processes*. Obviously, VxBPEL processes cannot be executed by the existing BPEL engines at run-time, since they contain non-standard BPEL elements. We next discuss how to develop VxBPEL_ODE by extending Apache ODE.

3.1   Design

The key issue of developing a VxBPEL engine is how to treat newly introduced variability elements in a BPEL process (i.e. a VxBPEL interpreter is needed) and how to enforce them with the existing BPEL engine (i.e. the interaction between the BPEL engine and the VxBPEL interpreter). A general rule of thumb we followed for designing VxBPEL_ODE is to leverage the implementation of Apache ODE to the maximum without significantly modifying its architecture. Fig. 1 shows the architecture of VxBPEL_ODE. *VxBPEL Compiler* and *Configuration Management* are newly introduced components in order to enable the execution of VxBPEL elements. VxBPEL_ODE first invokes *ODE BPEL Compiler* and *VxBPEL Compiler* to compile the VxBPEL process. The former is in charge of the compilation of all standard BPEL elements, while the latter is in charge of the compilation of VxBPEL elements, mapping variants to BPEL activities, and storing association of variation points and variants. *Configuration Management* is responsible for the interactions between *VxBPEL Complier* and *ODE BPEL Compiler* and selecting the corresponding variants for the serialization. The compiled VxBPEL representation (i.e. *Compiled Process Definitions*) is an object model similar in structure to the underlying BPEL process document, and all variability elements are resolved after the compilation process. Finally, *ODE BPEL Runtime* is used for the execution of compiled processes. We next discuss each major component individually.

- *VxBPEL Compiler* is responsible for preprocessing VxBPEL-specific elements. First, it creates an object model for all VxBPEL elements similar in structure to the object model for BPEL elements. Second, it maps all variants of a variation point to BPEL activities. Third, it stores the association of variation points and variants.

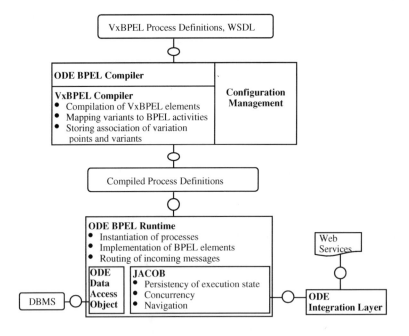

Fig. 1. Architecture of VxBPEL_ODE

- *ODE BPEL Compiler* is responsible for the conversion of standard BPEL elements into a compiled representation.
- *Configuration Management* is used to manage variants associated with a variation point. It analyzes variation configurations specified in ConfigurableVairationPoint and provides the operations for changing the variability configuration at run-time. During the serialization, a specific variant of a variation point is selected depending on the variability configuration and its corresponding object is created. Through the treatment, all VxBPEL-specific elements are resolved. The compiled representation has resolved the various named references present in the VxBPEL, internalized the required WSDL and type information, and generated various constructs.
- *ODE BPEL Runtime* is used to interpret the compiled process definitions, including creating a new process instance, implementing the various BPEL constructs, and delivering an incoming message to the appropriate process instance.
- *ODE Data Access Objects (ODE DAOs)* mediates the interaction between the ODE BPEL Runtime and an underlying database management system DBMS. ODE DAOs normally provides interfaces for tracking active instances, routing message, referring to the values of BPEL variables for each instance, referring to the values of BPEL partner links for each instance, and serializing process execution states.
- *JACOB* provides an application concurrency mechanism, including a transparent treatment of process interrupt and persistency of execution state.
- *ODE Integration Layer* provides an execution environment by providing communication channels to interact with Web services, thread scheduling mechanisms, and the lifecycle management for ODE BPEL Runtime.

We next examine interactions among components of the engine using a UML collaboration diagram, as shown in Fig. 2. Note that the name of each component is described by texts above the box. The execution process is described as follows.

Phase 1. When the engine is started, it first configures the process deployment directory through the container in which the engine runs, and then activates *Deployment-Poller* in ODE Integration Layer to check whether a VxBPEL process is deployed (i.e. Step 1). If detected, an empty file will be created. After that, the engine employs a utility tool to monitor this directory. If a file in this directory is updated, *DeploymentPoller* will re-deploy the process; if the deployed process is removed, *DeploymentPoller* is responsible for removing this process, accordingly.

Phase 2. The engine invokes the *"deploy ( )"* method provided by *ProcessStoreImpl* in ODE Data Access Objects to deploy the VxBPEL process (i.e. Step 2). During the deployment, major activities include parsing the VxBPEL file, creating an object model for the execution, and serializing all relevant objects in a binary file. These activities are implemented by *DeploymentUnitDir* in ODE Data Access Objects, which encapsulates the *"Compile (File bpelFile)"* method provided by *BpelC* in ODE BPEL Compiler. *DeploymentUnitDir* is in charge of a deployment unit, which corresponds to a process directory. The *"Compile (File bpelFile)"* method is responsible for the following tasks (i.e. Steps 3~11):

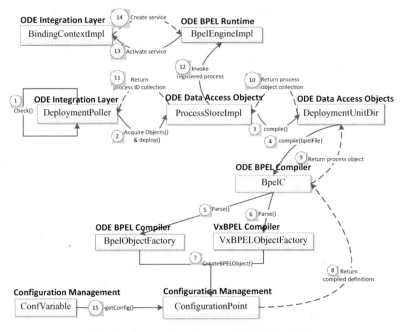

Fig. 2. The running process of VxBPEL_ODE

(1) *Parsing the VxBPEL file (i.e. Steps 3~6)*: This is implemented by the "*parse()*" methods provided by *BpelObjectFactory* in ODE BPEL Compiler and *VxBPELObjectFactory* in VxBPEL Compiler. This method parses the XML file and generates Java objects for process elements. During the paring phase, it generates objects for standard BPEL elements, and records the information about variation definitions and configurations. *ConfigurationPoint* in Configuration Management is responsible for parsing the VxBPEL configuration elements and providing interfaces for the configuration modifications.

(2) *Creating the binary file* (i.e. Steps 7~11): Objects generated during the parsing phase contains process and variation attributes. When the variation configuration is met for the first time, a variant specified by the *defaultVariant* configuration is selected. When the variation configuration is switched to a variation configuration scheme, the engine will invoke a method provided by *ConfigurationPoint* to select the specified variants. This variant selection process repeats until all variation points are handled. After the treatment, only BPEL process objects remain because variants associated with a variation point are of standard BPEL elements. The engine allows for changing the variation configurations at run-time, which is implemented by methods provided by *ConfigurationPoint* and *BpelC*. The variation configuration parameter is transferred to *ConfVariable* in Configuration Management, which maintains the current variation configuration (i.e. Step 15). Finally, all process relevant objects are serialized into a binary file.

Phase 3. The *BpelEngineImpl* in ODE BPEL Runtime executes the registered process (i.e. Steps 12~14). If an object is involved in the registered process, its relevant information is extracted from the binary file. For instance, when a Web service is invoked, the engine will call interfaces provided by *BpelBindContexImpl* in ODE Integration Layer to create the bound service for execution.

3.2   Implementation

We have implemented VxBPEL_ODE using the Java language. Two key issues related to the implementation are described next.

(a) *Compilation and deployment of VxBPEL elements*: As mentioned above, the engine first needs to deploy a process before its execution. An important task of the deployment is to compile the VxBPEL file and create an object model suitable for execution. This compilation process involves BPEL and VxBPEL elements that are distinguished by their name spaces. The former are handled by *ODE BPEL Compiler*, while the latter are handled by *VxBPEL Compiler*. To deliver an isomorphic implementation of the engine, we referred to the implementation of *ODE BPEL Compiler* during the implementation of *VxBPEL Compiler*.

(b) *Implementation of run-time variation configuration*: To support the implementation of variability, the engine first stores the variation definitions, and then selects variants based on the variation configuration during the serialization of process activities. The variation configuration is implemented by the component *ConfigurationPoint*. It records the name of the current VxBPEL process and the information about variation configurations, and selects the variants for execution at the serialization phase. Furthermore, variation configuration of the VxBPEL process can be changed at run-time via a method provided by *ConfigurationPoint*.

4   Evaluation

We report on an evaluation of VxBPEL_ODE using three VxBPEL processes. The goal of the evaluation is three-fold: (i) testing VxBPEL_ODE; (ii) evaluating the overhead of variability management by comparing BPEL and VxBPEL implementations for the same business scenarios, and (iii) comparing the performance of VxBPEL_ODE with that of VxBPEL_ActiveBPEL using the same VxBPEL processes for the same business scenarios.

4.1   Subject Programs

The *travel agency* process is a typical scenario describing how a travel request is processed [5]. When constructed from distributed services, the process often involves the composition of *travel agency*, *hotel*, *flight*, and *banking services*. These services can be provided by independent organizations. Once the system receives a request, the customer first chooses a travel agency, then selects the desired flights as well as hotels, the system asks the consumer to pay the package via an online banking service.

The *car estimation* process [5] describes a common scenario how a car repair is negotiated depending on the situation of the car and its repair cost. This process consists of five services, namely *Initial Estimate, Exterior Estimate, Interior Estimate, Powertrain Estimate*, and *Final Estimate*. Before the repair, a company needs to know the situation of the car, and the customer then decides to accept the reparation or not after being informed of the costs. Once the customer submits a request for estimation, the process first invokes an *Initial Estimate* service. In case of a simple estimation, the process invokes *Final Estimate* and notifies the cost to the customer; in case of a complex situation, the request is transferred to *Exterior Estimate, Interior Estimate*, and *Powertrain Estimate* services for a deep estimation, and the process then invokes the *Final Estimate* service and returns the cost.

The *smart shelf* process represents a complex shopping situation of shelf management [5], where one monitors quantity as well as quality of goods. Consumers first send a shopping request which includes the name and quantity of goods to be purchased. After receiving the request, the system first creates a formal request which is composed of *service time, quantity of goods, category of goods* and *period of goods* and then checks the *quantity, location* and *period* of goods. Next, the system checks the quantity, location, and period of goods, and executes various routine procedures based on the checking results. Finally, the system responds to the consumer with a confirmation message if *quantity, location*, and *period* are all qualified; otherwise, the process sends a failure message to the consumers and cancels the ordering list.

4.2   Evaluation Procedure

The following procedure is used for evaluation.

1. *Implementing business processes using VxBPEL*: The variability-based adaptive service composition method proposed in [5] guided us implement three subject programs using VxBPEL. We treat the locations that changes may happen as variation points, and possible choices are identified as variants. Furthermore, constraints are used to specify variation dependencies, which result in consistent and valid process variants (i.e. business scenarios) at run-time. Three VxBPEL programs are derived, as summarized in Table 1.

Table 1. A summary of subject programs

Name	VxBPEL Lines of Code[a]	Number of Web Services
travel agency	603	8
car estimation	607	7
smart shelf	1146	13

[a.] Measured in lines of XML code.

2. *Deploying and managing VxBPEL processes using MX4B*: Variation definitions and configurations defined in the VxBPEL process improve the adaptation of service compositions. Variation configuration switching is very complex and time-consuming. To facilitate this process, MX4B was developed to provide an efficient way for the deployment, configuration switching, and maintenance of VxBPEL

service compositions [5]. In this experiment, we employed MX4B for the following tasks: (i) deploying VxBPEL process artifacts, (ii) managing VxBPEL
processes, and (iii) switching variation configurations at run-time.

3. *Evaluating and comparing performance*: We evaluate two aspects of the performance: (1) for the same business scenarios, we compare the execution time of the
BPEL process running on Apache ODE with that of the VxBPEL process running
on VxBPEL_ODE; (2) for the same VxBPEL process, we compare overhead of
VxBPEL_ODE with that of VxBPEL_ActiveBPEL. The experimental setting is
described in Table 2.

Table 2. Experimental setting

CPU	2.40*4 GHz
Memory	8 GBytes
Hard Disk	750GBytes
Operating System	Win7 with 64bit

4.3   Results and Analysis

We summarize the evaluation results and provide an analysis of comparative comparisons next.

(a) *Feasibility*: Through the above evaluation procedure, we observe that
VxBPEL_ODE is able to provide an executable context for VxBPEL service compositions. Furthermore, the engine showed a good scalability, namely it is able to
process a simple service composition of one Web service, as well as a complex one
of 13 Web services, as shown in the case of the *smart shelf* process.

(b) *Performance*: Tables 3, 4 and 5 summarize the performance evaluation results in
ms of Apache ODE, VxBPEL_ODE, and VxBPEL_ActiveBPEL for three subject
programs, respectively. Column "Scenario Name" lists the name of business scenarios implemented using BPEL or VxBPEL; Column "Number of WS" shows the
number of Web service involved in the current scenario; Column "A: BPEL by
Apache ODE" shows the time cost of BPEL processes using Apache ODE; Column
"B: VxBPEL by VxBPEL_ODE" shows the time cost of VxBPEL processes using
VxBPEL_ODE; Column "C: VxBPEL by VxBPEL_ActiveBPEL" shows the time
cost of VxBPEL processes using VxBPEL_ActiveBPEL; Column "B/A (%)"
shows the ratio of Column "B: VxBPEL by VxBPEL_ODE" against Column "A:
BPEL by Apache ODE"; Column "B/C (%)" shows the ratio of Column "B:
VxBPEL by VxBPEL_ODE" against Column "C: VxBPEL by
VxBPEL_ActiveBPEL". Note that the time cost in Column "A: BPEL by Apache
ODE" is the sum of deployment time and execution time of BPEL processes, while
the time cost in Columns "B: VxBPEL by VxBPEL_ODE" and "C: VxBPEL by
VxBPEL_ActiveBPEL" is the sum of variation switching time and execution time.
From Tables 3-5, we have the following observations:

• For all scenarios in three subjects, the performance of VxBPEL processes is very
close to that of BPEL processes. In the case of *travel agency* and *car estimation*, the
performance of all process scenarios implemented using BPEL is slightly higher

than that implemented using VxBPEL; In case of *smart shelf*, the performance of most process scenarios except the "insufficient" scenario implemented using BPEL is slightly higher than that implemented using VxBPEL. This indicates that variability management does not introduce extra performance overhead. Furthermore, VxBPEL_ODE does not evidently decrease the performance of Apache ODE although some extensions are added. An insight investigation shows that most of time overhead of VxBPEL_ODE is dedicated to processing of standard BPEL elements.

- The performance of VxBPEL_ODE is slightly lower than that of VxBPEL_ActiveBPEL. In the case of *travel agency*, the performance of the former is comparable to that of the latter; In the case of *car estimation*, the performance of the former is higher than the one of the latter for the *simple* scenario, while lower for the *normal* or *expert* scenarios; in the case of *smart shelf*, the performance of the former is lower than that of the latter by around 50%. More interestingly, when the number of involved Web services in a scenario is small, the performance of the former is higher than that of the latter, while the situations change with the increasing number of involved Web services. This performance difference is mainly due to the fact that Apache ODE and ActiveBPEL have different architectures. Apache ODE introduces the serialization of the object models, and integrates with open source components, while ActiveBPEL adopts a coherent architecture and design patterns that significantly improves its performance.

Table 3. Performance evaluation results for the travel agency program

Scenario Name	Number of WS	A:BPEL by Apache ODE (ms)	B: VxBPEL by VxBPEL_ODE (ms)	B/A (%)	C: VxBPEL by VxBPEL_Active BPEL (ms)	B/C (%)
A	3	1508	1611	106.83	1435	112.26
B	3	1548	1608	103.88	1547	103.94
C	3	1564	1577	100.83	1660	95.00
D	3	1501	1607	107.06	1604	100.19
E	3	1475	1584	107.39	1440	110.00
F	3	1534	1569	102.28	1563	100.38

5   Related Work

One category of approaches has turned to Aspect Oriented Programming (AOP) technique [13]. AdaptiveBPEL [14] is a service composition framework which leverages AOP technique to handle various concerns that are separately specified in BPEL processes. The adaptation process is driven by policies, namely a policy mediator is used to negotiate a composite policy and oversee the aspects weaving to enforce the negotiated policy, and a run-time aspect weaving middleware is integrated on top of a BPEL engine. AOBPEL [15] is an aspect-oriented extension to BPEL, which provides a solution for the modularization of crosscutting concerns and supporting dynamic changes in BPEL. AOBPEL specifies extra concerns associated with business processes as aspects, and provides generic aspect constructs. These aspect-oriented approaches can enhance the adaptation of BPEL processes via aspects. However, aspects split up the process logic over different files, which may cause it a difficult task to comprehend the variation, especially when service compositions are complex.

Table 4. Performance evaluation results for the car estimation program

Scenario Name	Num-ber of WS	A: BPEL by Apache ODE (ms)	B: VxBPEL by VxBPEL_ODE (ms)	B/A (%)	C: VxBPEL by VxBPEL_Active BPEL (ms)	B/C (%)
simple	1	441	464	105.21	1354	34.27
normal	5	2433	2514	103.33	1610	156.15
expert	5	2505	2524	100.76	1561	161.69

Table 5. Performance evaluation results for the smart shelf program

Scenario Name	Num ber of WS	A: BPEL by Apache ODE (ms)	B: VxBPEL by VxBPEL_ODE (ms)	B/A (%)	C: VxBPEL by VxBPEL_Active BPEL (ms)	B/C (%)
default	6	2539	2735	107.71	1850	147.83
location	7	2753	2803	101.81	1845	151.92
status	7	2791	2843	101.86	1995	142.50
locationstatus	8	2768	2897	104.66	1844	157.10
sufficient	9	2806	2836	101.07	2100	135.05
insufficient	9	2710	2464	90.92	1753	140.56
warelocation	10	2752	2777	100.90	1872	148.34
warestatus	10	2765	2929	105.93	1983	147.71
warelocation status	11	2866	2932	102.30	1980	148.08

The other category of approaches is based on the proxy (or broker) mechanism. Trap/BPEL and its predecessors [16] are a family of extensions to BPEL for enhancing robust service compositions through static, dynamic, and generic proxies, respectively. Events such as faults and timeouts during the invocation of partner Web services at run-time are monitored, and the adapted process is augmented with a proxy that replaces failed services with predefined or newly discovered alternatives. wsBus [17] is a framework which is capable of realizing QoS adaptation of service compositions by means of the concept of virtual endpoints. A virtual endpoint is used to select appropriate services based on the attached policy for execution at run-time. All requests are sent to this virtual endpoint and redirected to the real service. The selection of services based on the monitoring data and QoS metrics. SCENE [18] is a service composition execution environment that supports dynamic changes disciplined through rules. The implementation of adaptation is based on the proxy mechanism, which is used to bind the discovered services to the proxy associated with each activity in the BPEL specifications. These proxy-based approaches implicitly implement the adaptation of processes at the messaging/event layer. Since the changes are not treated as first-class objects and variation dependencies are not clearly handled, it may result in variation configuration and maintenance a difficult task.

Unlike the existing efforts, our efforts have been made to achieve the adaptation of BPEL processes through variability management [4-9]. We extended BPEL to provide a set of variability constructs for explicitly specifying variation of service compositions, and the engine presented in this paper provides an executable environment for service composition with variability design. This engine, together with analysis, design, and run-time management tools for VxBPEL, forms an integrated and comprehensive platform, which not only makes the variability-based adaptive service composition approach viable, but also improves its efficiency.

6   Conclusion and Future Work

We have presented a VxBPEL engine, VxBPEL_ODE, by extending an open source BPEL engine, Apache ODE, to enable the execution of VxBPEL processes. VxBPEL_ODE is a core component of an integrated supporting platform which facilitates the adoption of the variability-based adaptive service composition approach. In this paper, we examined the key issues of the design and implementation of such an engine, and validated its effectiveness through case studies. Furthermore, we evaluated the performance of VxBPEL_ODE and compared it with that of VxBPEL_ActiveBPEL. From the experimental results, we observe that VxBPEL_ODE shows a comparable performance of VxBPEL_ActiveBPEL while benefits an integrated design and execution environment for VxBPEL processes.

For future work, we plan to extend VxBPEL to support unplanned changes of service compositions at run-time via the dynamic binding technique, which will accordingly require the further extension of VxBPEL_ODE for binding abstract services with concrete services searched at run-time. We are also interested in examining the variability-based adaptive service composition approach in the development of social network analysis tools that desire the adaptation ability.

Acknowledgment. This research is supported by the National Natural Science Foundation of China (Grant No. 61370061), the Beijing Natural Science Foundation of China (Grant No. 4112037), the Fundamental Research Funds for the Central Universities (Grant No. FRF-SD-12-015A), and the Beijing Municipal Training Program for Excellent Talents (Grant No. 2012D009006000002).

References

1. Papazoglou, M., Traverso, P., Dustdar, S., Leymann, F.: Service-oriented computing: a research roadmap. International Journal on Cooperative Information Systems 17(2), 223–255 (2008)
2. OASIS. Web services business process execution language version 2.0 (2007), http://docs.oasis-open.org/wsbpel/2.0/OS/wsbpel-v2.0-OS.html
3. Aiello, M., Bulanov, P., Groefsema, H.: Requirements and Tools for Variability Management. In: Proceedings of REFS 2010, pp. 245–250. IEEE Computer Society (2010)
4. Koning, M., Sun, C., Sinnema, M., Avgeriou, P.: VxBPEL: Supporting variability for Web services in BPEL. Information and Software Technology 51(2), 258–269 (2009)
5. Sun, C., Wang, K., Xue, T., Aiello, M.: Variability-Based Adaptive Service Compositions (submitted for publication, 2014)

6. Sun, C., Aiello, M.: Towards variable service compositions using VxBPEL. In: Mei, H. (ed.) ICSR 2008. LNCS, vol. 5030, pp. 257–261. Springer, Heidelberg (2008)

7. Sun, C., Rossing, R., Sinnema, M., Aiello, M.: Modeling and managing the variability of Web service-based systems. Journal of Systems and Software 83(3), 502–516 (2010)

8. Sun, C., Xue, T., Hu, C.: Vxbpelengine: A change-driven adaptive service composition engine. Chinese Journal of Computers 36(12), 2441–2454 (2013)

9. Sun, C., Xue, T., Aiello, M.: ValySeC: A Variability Analysis Tool for Service Compositions Using VxBPEL. In: Proceedings of APSCC 2010, pp. 307–314 (2010)

10. ActiveBPEL, Active Endpoints (2007), http://www.activebpel.org

11. Apache, Apache ODE (2006), http://ode.apache.org/

12. Sinnema, M., Deelstra, S., Nijhuis, J., Dannenberg, R.B.: COVAMOF: a framework for modeling variability in software product families. In: Nord, R.L. (ed.) SPLC 2004. LNCS, vol. 3154, pp. 197–213. Springer, Heidelberg (2004)

13. Kiczales, G., Lamping, J., Mendhekar, A., Maeda, C., Lopes, C., Loingtier, J., Irwin, J.: Aspect-Oriented Programming. In: Akşit, M., Matsuoka, S. (eds.) ECOOP 1997. LNCS, vol. 1241, pp. 220–242. Springer, Heidelberg (1997)

14. Erradi, A., Maheshwari, P.: AdaptiveBPEL: a Policy-Driven Middleware for Flexible Web Services Compositions. In: Proceedings of International Workshop on Middleware for Web Services (MWS 2005), pp. 5–12 (2005)

15. Charfi, A., Mezini, M.: AO4BPEL: An Aspect-Oriented Extension to BPEL. World Wide Web Journal 10(3), 309–344 (2007)

16. Ezenwoye, O., Sadjadi, S.M.: TRAP/BPEL-A Framework for Dynamic Adaptation of Composite Services. Proceedings of WEBIST (1), 216–221 (2007)

17. Erradi, A., Maheshwari, P.: wsBus: QoS-aware middleware for reliable web services interaction. In: Proceedings of EEE 2005, pp. 634–639. IEEE Computer Society (2005)

18. Colombo, M., Di Nitto, E., Mauri, M.: SCENE: a service composition execution environment supporting dynamic changes disciplined through rules. In: Dan, A., Lamersdorf, W. (eds.) ICSOC 2006. LNCS, vol. 4294, pp. 191–202. Springer, Heidelberg (2006)

Predicting User Likes in Online Media
Based on Conceptualized Social Network Profiles

Qiang Liu[1,2], Yuanzhuo Wang[1], Jingyuan Li[1], Yantao Jia[1], and Yan Ren[3]

[1] Institute of Computing Technology, Chinese Academy of Sciences, Beijing 100190
[2] University of Chinese Academy of Sciences, Beijing 100049
[3] National Computer Network Emergency Response Technical Team Coordination
Center of China, Beijing 100029

Abstract. Predicting user likes in online media and recommending related products to the user would bring great profits to certain service providers. Therefore, prediction approaches have become a popular research topic in both industry and academia over the past decades. However, data sparsity makes many well-known prediction algorithms perform poorly in cold start situation. In this paper, we apply attributes from user profile in social network sites to help recommending user likes in a video sharing site and propose a model to conceptualize unstructured words in the profile attributes into interests vector by knowledge base. Based on the model, we designed a recommendation framework to predict user clicks. Experiment results on dataset show that our approach is an efficient one.

1 Introduction

In the past decades, prediction approaches have become one of the most popular research topics in the field of information retrieval, where both service providers and consumers would benefit much from its progress. For example in YouTube, the number of clicks is very likely to be proportional to the profits the service providers might gain. Therefore, predicting a user's likes and recommending the related products to the user is very profitable. Moreover, with the explosion of multimedia contents, users can not consume the information effectively, so helping users to get his or her interest information would help the user to solve the problem of information overload [1].

Because of its usefulness, prediction approaches have been studied extensively, and many famous models have been proposed, such as collaborative filtering. However, a key challenge for prediction approaches is to provide sufficient high quality user information. For example, prediction approaches are not very helpful in a cold-start situation [2], where there is a serious lack of information on historical behaviors.

To solve the cold-start problem, many researchers proposed hybrid models by using information from auxiliary networks. For example, with the development of social networks, many large service providers build social network sites for people to communicate with each other and share their status. Google provides Google+ for users to connect and communicate with their friends, to watch and share videos from You-

W. Han et al. (Eds.): APWeb 2014 Workshops, LNCS 8710, pp. 82–92, 2014.
© Springer International Publishing Switzerland 2014

Tube, and to buy things from Google shopping. The aggregated online services become a powerful tool to obtain users full portraits, such as user interests.

Although user's general public information can be collected from social network sites, there are several issues that prevent service providers from leveraging collected information to improve the prediction accuracy. First, it is difficult to extract key features from a heterogeneous media, and conceptualize those features to describe the user with a proper explanation. Second, the actions of a user in social network sites may not comply with those actions in video sites of the very same user, because different on-line services have different characteristics [3]. A user might be a gentleman in a social network site because every action will be seen by his or her parents or the close friends, while he or she might prefer to watch vulgar video clips in a media site because that is what he or she really is.

In this paper, we study the aforementioned issues and test our solutions under datasets collected from YouTube and Google+. The main contributions are as follows.

- We extract key features from user social profiles, and convert user descriptions into interest vectors by knowledge base.
- We propose several-prediction strategies based on user's demographic difference and interest vectors, and conduct experiment under datasets from YouTube and Google+, the top video service site, and the top social network site, respectively.

The rest of this paper is organized as follows. In section 2, we review some related work in prediction approaches. In section 3, we focus on understanding the key features for users in social network, especially user descriptions, and the conversion procedure from descriptions to interest vectors by knowledge base. Section 4 introduces different prediction approaches based on different features. Experiments and results are presented and discussed in section 5, and section 6 concludes the paper.

2   Related Work

Predicting user's likes have received increasing attention and have been thoroughly studied over the past decades. The methodologies are two folds: collaborative filtering and content-based methods. Collaborative filtering assumes that users who carry similar characteristics tend to like similar products [4], including user-based and item-based approaches. User-based collaborative filtering predicts items to a target user using collected information from similar users, while users are typically represented in a vector space which summarizes their characteristics. Similarly, item-based [5] collaborative filtering methods take advantage of rating information of similar items which are reviewed by the target user in the past. In contrast to collaborative filtering, content-based methods often utilize the vast overload of information on the media [6], such as product reviews, customer opinions, and social media to directly make product recommendation.

Although those approaches are studied extensively, data sparsity makes many well-known prediction approaches perform poorly, such as in a cold-start situation. Researchers proposed hybrid approaches to incorporate both user-item rating dataset as well as other contextual information in different scenarios, including social network information, time information [7], etc. For instance, social trust or friend aware recommender approaches model trustworthiness or similarities of users. Jovian [8] combined the feature of Twitter-followers, and generated a much more accurate estimation of how likely a target user would like an App. Zhang [9] showed that there are significant correlations between social network information and online purchases, and presented a system that use Facebook likes for solving the task of products recommendation. However, these research findings focused on the heterogeneous entity relationships [10], and could not fully take advantage of the rich features in user social profiles like descriptions, which is poorly structured and difficult for machine to understand.

Although a large amount of users preferred to introduce themselves comprehensive through the description section of the social network sites, the semantic analysis is difficult. Many studies have been devoted to the core problem of language understanding. Latent Dirichlet Allocation (LDA) and Latent Semantic Analysis (LSA) first proposed to discover latent topics or concepts from large collection of documents in the early time, which are however not useful for short texts due to topic sparseness [11]. To improve understanding relevant information in short texts, ontology development was considered as backbone to improve understanding concepts in semantic web [12]. However, ontology is complex to build, and efforts would be wasted in unnecessary labels. Therefore, to more generally understand the distribution of short messages, we try to utilize knowledge bases to detect concepts of the user's descriptions.

Motivated by previous work, we propose a hybrid prediction framework, which leverages user social profiles to predict user behaviors. Moreover, we focus on model user features generated from social profiles by knowledge base, which have been proved to be better in understanding the target users and can achieve more promising performances.

3   Extracting Key Features from Social Profiles

In this section, we extract key features from user social profiles. We then tried to explain users' interests based on knowledge base.

3.1   Social Profiles

As we all know, when a user enters a social network, he or she needs to fill in some personal information, like name, gender, and age, and most of them write a description of him or her in a short paragraph. Many research findings proved that the description could describe the user sufficiently. We are to exploit the demographic and text-content difference to facilitate prediction.

Demographic Difference: A large amount of work has studied demographic distinctions in personalized services. Shepstone [13] showed that ratings for advertisements recommended using the age and gender data were significantly higher than rating from randomly selected advertisements. Kau [14] indicated that "on-off shoppers" (people who inquire online but buy offline) are prevalently teenagers; "comparative shoppers" (people comparing product features before buying) tend to be males in their twenties; offline shoppers are mostly the people over 40. Based on previous work, we take users' gender, age into consideration.

Text-Content Difference: Formally, social network user offers a self-description for self-introduction, which can be considered as a small window to make a big impression. Usually, the description is in just a few sentences, or a few vocabularies, but it is a comprehensively detailed information about the user. Alexandre [15] explored text-mining methods to improve classical collaborative filtering methods, and showed that the result was quite encouraging. In order to better understand users' interest, we first convert the description into a bag-of-words vector, and try to conceptualize those words by knowledge base.

3.2   Conceptualized User Interest by Knowledge Base

Many machine learning algorithms require the input to be represented as a fixed-length feature vector. When it comes to texts, one of the most common fixed-length features is bag-of-words. Bag-of-words features have two major weaknesses. On the one hand, they lose the ordering of the words, therefore "the big bang" and "the bang big" would have the same representation. On the other hand, they ignore semantics of the words. For example, "football" and "soccer" is considered as two different objects while essentially they are the same thing. This is why we try to describe user interests by knowledge base called Freebase. We name the process of converting unstructured user descriptions into structured interest vectors as the conceptualized process.

Freebase is a large collaborative knowledge consisting of entities and relationships composed mainly by its community members. It now contains 2.3 billion instances of relationships between 43 million entities. Freebase is widely used in academia and industry in many research problems and applications. Each entity in Freebase is associated with multi-domains that are related to this entity, each domain is with a weight to the entity.

To resolve the weakness of bag-of-words referred above, we divided the understanding user interest into three distinct steps.

Step 1, to understand semantics of each word, as fig. 1 shows, we first segment a sentence into sequential words vector $W_{single} = \{w_1, w_2, w_3 ...\}$, and try to search the most related topics $T_i = \{t_{i_1}, t_{i_2}, ..., t_{i_k}\}$ limited by k for each word w_i by Freebase. t_{i_j} is a topic name $n_{i_{j_k}}$ with weight associated to the word $d_{i_{j_k}}$, which can be represented as $t_{i_j} = \{n_{i_{j_k}} : d_{i_{j_k}}\}$. We simply join the topics to $T_{Single} = \{t_1, t_2, t_3 ...\}$, each topic T_i is normalized by weight. When the same topic occurs, we combine the topic together with the sum of weights.

Step 2, contrast to step 1, we segment sentences to double words represented as $W_{double} = \{w_1 w_2, w_2 w_3, ...\}$. Similarly with step 1, we get $T_{double} = \{t_1, t_2, t_3 ...\}$. Generally speaking, an entity phrase name is less than 4 words. So we continue to calculate T_{triple} and $T_{quadruple}$.

Step 3, we combine those topics to represent the user interests $T = \theta_1 T_{Single} + \theta_2 T_{double} + \theta_3 T_{triple} + \theta_4 T_{quadruple}$, while θ is the weight for each topic with $\theta_1 + \theta_2 + \theta_3 + \theta_4 = 1$.

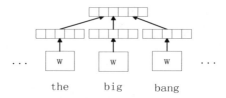

Fig. 1. Conceptualized user interest vector by knowledge base

4   Prediction Framework

In section 3, we have discussed the features in social profiles. In this section, we focus on training those features to predict user likes in YouTube. First, we will give the formal definition of the problem. Then we try to predict user interested videos by social profiles.

4.1   Problem Definition

Suppose we are given a social network G^S and an online media like YouTube G^Y. Suppose the social network G^S have comprehensively detailed user profiles, while the online media G^Y is without any information but user historical behaviors. The social network $G^S = (U, F^S)$ contains user profiles, where $U = \{u_1, u_2, ..., u_n\}$ is the set of users, and $F^S = D^S \cup T^S$ is the features in the social network. $D^S = \{d_1, d_2, ..., d_k\}$ represents a set of demographic difference like age, gender, locations etc. T^S represents the text-content difference for user descriptions, which we can simply use bag-of-words methods to describe, or use conceptualized interest vector by knowledge base. $G^Y = (U, C^Y)$, is with the same users U, while $C^Y = \{c_1, c_2, ..., c_m\}$ is the set of categories user likes in YouTube.

Prediction Problem: suppose we have a social network G^S and an online media G^Y, the task is to predict the categories that a user without any historical behaviors would be like in G^Y with profiles from G^S.

4.2   Prediction by Social Profiles

Prediction by User Demographic Difference
As we have mentioned before, users with different demographic features like age, gender usually interest in different video clips. Since the demographic information is abundant in Google+, we take it as a key feature to model user similarity [16]. To represent the demographic information of a user, we collect all tags in the registration information and build a tag space with dimension d. The tags of a user are converted

into a demographic vector. The user u_i can be represented by a vector $x_i \in R^d$. The normalized linear kernel to measure the user similarity is denoted as:

$$Sim(u_i, u_j) = \frac{x_i{}^T x_j}{\sqrt{x_i{}^T x_i} \sqrt{x_j{}^T x_j}}$$

To evaluate the degree of a user like a category in YouTube, we just calculate whether the similar users like the category as follows:

$$p_D(c_i | u) = \sum_{k=1}^{n} Sim(u, u_k) \frac{like(u_k, c_i)}{\sum_{c \in C} like(u_k, c)}$$

where $like(u, c)$ is the number of liked videos in category c by user u, and C is the set of all 18 YouTube categories. For example, if a user likes 3 videos from category Music, and 2 videos from category Sports, we have $\frac{like(u, Music)}{\sum_{c \in C} like(u, c)} = 0.6$ and $\frac{like(u, Sports)}{\sum_{c \in C} like(u_k, c)} = 0.4$.

Prediction by Text-Content Difference

A user use a short paragraph to introduce himself. Through the description is very succinct, it fully expresses the user's individuality. To predict a user's interest videos, we cannot just find the similar phrase used by users, because the words, like "favorite", are meaningless but frequently used, while noteworthy words like "football" are sparse. Therefore, we do not collaborative filter by similar phrase, but set each word as unit, and calculate the videos liked by each word, as follows:

$$P_W(w, c_i) = \sum_{k=1}^{n} B(u_k, w) \frac{like(u_k, c_i)}{\sum_{c \in C} like(u_k, c)}$$

where $B(u, w)$ equals 1 if the word w in user u description, and 0 otherwise. To evaluate the degree of a user like a category in YouTube, we calculate the degree of a user like videos as follows:

$$p_{TW}(c_i | u) = \sum_{w}^{W} tfidf(w) P_W(w, c_i)$$

were W is the words that the user u uses in his descriptions, $tfidf(w)$ is the normal tf-idf weight of the word w. The above formulation uses bag-of-words to represent user's descriptions. However, as we have explained, it is better to use conceptualized interest vector calculated by knowledge base rather than bag-of-words to represents the user. So, when it comes to interest vector, we first calculate as follows:

$$P_I(i, c_j) = \sum_{k=1}^{n} B(u_k, i) \frac{like(u_k, c_j)}{\sum_{c \in C} like(u_k, c)}$$

where $B(u, i)$ equals 1 if the interest i in user u conceptualized interest vector, and 0 otherwise. Therefore, the degree of a user liking category in YouTube can be calculated as follows(I is the interest vector of user u):

$$p_{TI}(c_j | u) = \sum_{i}^{I} tfidf(i) P_I(i, c_j)$$

4.3   Prediction by Online Media Characteristic

Different network sites have different characteristic. For example, social network site provide methods for user to intercommunicate with friends, like updating messages, or sharing pictures. Video websites on the other hand provide ways to share video clips to anyone for entertainment. Since we predict users' like videos in YouTube, we must take YouTube characteristics into consideration. As we can see from fig. 4, the number of videos is unevenly distributed in different categories. Therefore, a reasonable system predicts categories according to their popularity as follows.

$$p_P(c_i|u) = \sum_{k=0}^{n} \frac{like(u_k, c_i)}{\sum_{c}^{C} like(u_k, c)}$$

4.4   Prediction Framework

Section 4.2 and section 4.3 have presented the prediction methods of user likes in video sites, based on social network user profiles and site characteristic, respectively, we then presented combined strategy as follows:

$$p(c_i|u) = \alpha\, p_P(c_i|u) + (1 - \alpha)(\beta\, p_D(c_i|u) + (1 - \beta)p_{T?}(c_i|u))$$

where α and β is the weight, and $p_{T?}(c_i|u)$ is either $p_{TW}(c_i|u)$ or $p_{TI}(c_i|u)$.

5   Experiments

In this section, we first outline the real-world data set we have collected, and then propose the predication strategies. In the end, we compare the effectiveness of different strategies and try to analysis the result reasonably.

5.1   Data Set

To evaluate the performance of prediction approach, we use YouTube as the target platform to predict likes, and Google+ as the auxiliary platform where social profiles are used. YouTube, the largest video-sharing website in the world, provides interfaces for its users upload, view and share videos. Once a user likes a video, the behavior will be recorded and showed on the user's homepage. Google+ is the second largest social networking site in the world that is also owned by Google Inc. Google has described Google+ as a "social layer" that enhances many of its online properties. Many users in Google+ share the link of their YouTube homepage, so we can build the relationship of user between YouTube and Google+ as the ground truth of our proposed strategies.

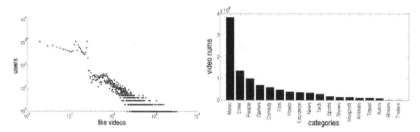

Fig. 2. Distribution of like videos for users   **Fig. 3.** Number of videos liked categories

We collected 20,956 users from Google+ with detailed social profiles and who have liked at least one video in YouTube. The social profile is described as Table 1. The percentage of male user is 60.34%, while female is 39.34%, which is consistent to the gender distribution published by Google. Users in this dataset liked 1,169,913 videos, where fig. 3 displays the power-law distribution of user liked videos. We tagged the videos into 18 categories by YouTube categories. As fig. 4 shows, due to the characteristic of video-sharing site, the Music category is the most popular of all 18 kinds.

In the experiments, we split the data into 11 sets, where 10 sets are used as train sets, while the other one set is used as test set.

5.2   Prediction Strategies

We addressed the personalized video prediction problem by introducing cross-platform user modeling, and tried to conceptualize those social profiles by knowledge base. In our experiment, we set three strategies for prediction as follows.

1) Predicting only by YouTube popularity characteristic.(**S1**)
$$\alpha = 1$$
2) Predicting by YouTube popularity characteristics and social profiles from Google+, where user description was represented as bag-of-words.(**S2**)
$$p_{T?}(c_i|u) = p_{TW}(c_i|u)$$
3) Predicting by YouTube popularity characteristics and social profiles from Google+, where user description was represented by interest vector conceptualized by Freebase.(**S3**)
$$p_{T?}(c_i|u) = p_{TI}(c_i|u)$$

5.3   Result Analyses

In the experiment, we tried to adjust the weight factors α and β for acquiring the best performance, where the minimum difference of α and β was 0.01 and the range was within [0, 1]. The result shows that the prediction by popularity has a significant impact, especially in **S2**, α accounts for 50%. The result means that the characteristic of an online media is worth to take into consideration. We can see that from fig. 3, the proportion of category Music is 38%, for the reason that music video clips are generally short but have high appreciating rates, which is suitable for video sharing media. On the other hand, the weight factor β is considerably small, which is not surprising because the unstructured user description provide a much richer and more relevant information with respect to demographic difference like age and gender. Moreover, the weight factor α and β in **S3** are smaller than **S2**, which shows that the proportion of user description is of greater importance after being conceptualized by knowledge base, which means the conceptualization makes a better performance.

Applying the best performance of the above weight factors, we claulated the average precision and NDCG through the number of recommended categories from 1 to 13 by the strategies. As we can see from fig. 4, through the characteristic of YouTube makes the popularity prediction perform well, the social profile promoting the result, which demonstrates the heterogeneous information is effective in improving prediction performance. Furthermore, after the descriptions conceptualized by knowledge base, the result is much more encouraging.

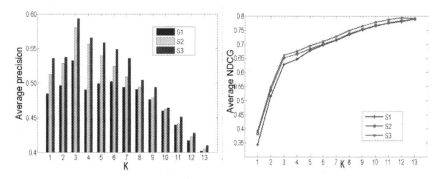

Fig. 3. Performance of different strategies

More than 51% of the users liked video clips in less than 5 categories, therefore the more invalid category we ever predicted, the less precision it would be, with the raise of recall rate. Therefore, we chose the average presicison, rather than the F-score to demonstrate the result. Due to the uneven distribution of video clips in YouTube, users are very liked to like video clips those top popular categories. However, as user cast his or her likes with more clips, the personality will be reflect, that is why the strategy of popularity is not stable, and perform poorly when k is around 5. While with the aid of information from social network, the prediction system became personalized and stable. Moreover, as we have explained before, bag-of-words model cannot describe the user comprehensively. When the words were conceptualized by knowledge base in our experiment, the unstructured descriptions were converted to interest vectors, and indeed described the users better.

6   Conclusion

In this paper, we proposed a novel approach to predict user interest video categories by utilizing user profiles from social network sites. By analyzing the features of user social profiles, we tried to describe a user through certain characteristics, especially user description, which had been conceptualized from unstructured word into interest vectors by knowledge base. Based on the differences of user features, we presented approaches to predict user interest video clips, and several strategies by combining those approaches together. We applied our strategies on a dataset of 20,956 selected Google+ users, and the result demonstrated that heterogeneous social profile is effective in improving prediction performance, especially with conceptualization of user descriptions by knowledge base, which can obtain 8% average improvement in precision.

Acknowledgements. This work was supported by National Grand Fundamental Research 973 Program of China (No. 2013CB329602, 2014CB340401), National Natural Science Foundation of China (No. 61173008, 61232010, 61303244, 61370132), the 242 Projects (No. 2012F86, 2013F97), and Beijing Nova Program (No. Z121101002512063).

References

1. Davidson, J., Liebald, B.: The YouTube video recommendation system. In: RecSys 2010, pp. 293–296. ACM (2010)
2. Chen, C.C., Wan, Y.H., Chung, M.C., Sun, Y.C.: An effective recommendation method for cold start new users using trust and distrust networks. In: Information Sciences, pp. 19–36. Elsevier (2013)
3. Yu, J.Y., Wang, Y.Z., Jin, X.L.: Evolutionary analysis of online Social Network through Social Evolutionary Game. In: WWW 2014 (2014)
4. Chen, K., Song, G.H., Sun, S.T.: Collaborative personalized tweet recommendation. In: SIGIR 2012, pp. 661–670. ACM (2012)
5. Pirasteh, P., Jung, J.J., Hwang, D.: Item-based collaborative filtering with attribute correlation: a case study on movie recommendation. In: Nguyen, N.T., Attachoo, B., Trawiński, B., Somboonviwat, K. (eds.) ACIIDS 2014, Part II. LNCS, vol. 8398, pp. 245–252. Springer, Heidelberg (2014)
6. Liu, D.W., Wang, Y.Z., Jia, Y.T.: LSDH: a Hashing Approach for Large-Scale Link Prediction in Microblogs. In: AAAI 2014 (2014)
7. Jia, Y.T., Wang, Y.Z., Jin, X.L., Cheng, X.Q.: TSBM: The Temporal-Spatial Bayesian Model for Location Prediction in Social Networks. In: WI 2014 (2014)
8. Lin, J., Sugiyama, K., Kan, M.Y., Chua, T.S.: Addressing cold-start in App recommendations: latent user models constructed from Twitter followers. In: SIGIR 2013, pp. 283–292. ACM (2013)
9. Zhang, Y.Z., Pennacchiotti, M.: Predicting purchase behaviors from social media. In: WWW 2013, pp. 1521–1531. ACM (2013)
10. Jia, Y.T., Wang, Y.Z., Li, J.Y.: Structural-Interaction Link Prediction in Microblogs. In: 22nd International World Wide Web Conference (WWW), May 13-17 (2013)
11. Yu, J.Y., Wang, Y.Z.: Identifying Interaction Groups in Social Network Using A Game-Theoretic Approach. In: WI 2014 (2014)
12. Saruladha, K., Aghila, G.: COSS: cross ontology semantic similarity measure-an information content based approach. In: IEEE, pp. 485–490. IEEE (2011)

13. Shepstone, S.E., Tan, Z.H.: Demographic recommendation by means of group profile using speaker age and gender recognition. In: Proceedings of the 14th Annual Conference of the International Speech Communication Association (2013)
14. Kau, A.K., Tang, Y.E., Ghose, S.: Typology of online shoppers. Journal of Consumer Marketing 20, 139–156 (2003)
15. Alexandre, S., Desmarais, M.C.: Combing collaborative filtering and text similarity for expert profile recommendation in social websites. In: User Modeling, Adaption, and Personalization, pp. 178–189. Springer (2013)
16. Liu, D.W., Wang, Y.Z., Jia, Y.T.: From Strangers to Neighbors: Link Prediction in Microblogs using Social Distance Game. In: WSDM 2014. Workshop on Diffusion Networks and Cascade Analytics (2014)

Demographic Prediction of Online Social Network Based on Epidemic Model

Xiang Zhu*, Yuanping Nie, and Aiping Li

College of Computer, National University of Defense Technology, China
zhuxiang@nudt.edu.cn,
{yuanpingnie,apli1974}@gmail.com

Abstract. With the development of Online Social Network, more and more people are inclined to use OSNs to publish information due to its strong interpersonal interaction. In this paper we apply epidemic model to demonstrate user adoption and abandonment procedure in OSNs, where adoption is analogous to the infective and abandonment is analogous to the removal. We modified the traditional SIRS model by taking infective-remoal theory into consideration such that the population of the removal will influence the infective. The modified irSIRS model is verified by real data crawled from Renren Network and Sina Weibo, and the best fit curve exhibit the infective population increase rapidly and decline slowly to an proportion in future time. Through experiments irSIRS model is proved to predict demographic evolution well.

Keywords: epidemic model, social network, dynamic analysis.

1 Introduction

Online social networks(OSNs) are the most popular media today due to its strong interpersonal interaction. With the development of OSNs, a lot of companies like Renren Network, Sina, Tencent have established their own OSNs. It is more obvious when the six degrees [1] of separation concept applied to OSNs, people are connected by social ties or common interests and hobbies, so people from all over the world may have a connection through OSN. Nowadays, some OSNs as Sina Weibo and Renren Network have achieved great success, both of which is worthy of high valuations based on their large amount of users and expectations of growth in future. In spite of the success of those two OSNs, some OSNs have risen and fallen in popularity during the last two decades, most representatively the qzone created by Tencent. The active users in qzone is declining as well as the value of qzone. The demographic dynamics governing the rises and falls in OSN are therefore not only an academic problem but also a financial problem.

* Sponsored by National Key Technology Research and Development Program No.2012BAH38B04, National Key fundamental Research and Development Program No.2013CB329601.

W. Han et al. (Eds.): APWeb 2014 Workshops, LNCS 8710, pp. 93–103, 2014.

In this paper we analyze the demographic dynamics of OSNs by likening to epidemic model, which is used to depict the infectious disease spread. The application of epidemic model to OSNs demographic dynamics is intuitive, since users usually join OSNs because their friends have already joined, which is analogous to the procedure of spreading of infective disease. The abandonment procedure is also similar to immunity, when users will leave OSNs for lack of interest, which is analogous to getting immunity. Users may return to OSNs after a period of time as the recover individual will lose immunity. From the above, we can apply epidemic model to demographic prediction intuitively. In section 2, we summarize the related work for epidemic model and dynamic demographic analysis. In section 3, an improved epidemic model is proposed to depict dynamic demographic prediction based on traditional SIRS model. Then some experiments are shown in section 4 to verified the improved model. At last in section 5 we summarize our work and make a conclusion about application of epidemic model to OSNs.

2   Related Work

The outbreak and spread of infectious disease has been questioned and studied for many years. The ability to make prediction to infectious disease spread is important to control the mortality rate of a particular epidemic. In 1927, Kermack and McKendrick created a model [2] in which they considered a fixed population with only three compartments:susceptible, infective and removal. They use those compartment to depict the dynamics in epidemiology. In 1990s, a SIS model [3] depicting computers and epidemiology is proposed by Kephart and White, which helps people to realize how computer virus spread among computers. Moreover, Pastor-Satorras proposed more general case the removal will lose immunity at a certain ratio and can be infected again [4].

The OSNs can be depicted as a society infected by an epidemic disease as the procedure of adoption and abandonment is similar to the procedure of infection and recovery. A model of epidemic spreading in a population with a hierarchical structure of interpersonal interactions is described and investigated numerically by Grabowski and Kosiski [5], in which the structure of interpersonal connections is based on a scale-free network. Social network theory can be also help with understanding of process within animal populations such as disease transmission and information transfer [6].

As OSNs and epidemic model have internal relations intuitively, it is reasonable to apply epidemic model to predict demographic dynamics.

3   Epidemic Model

3.1   Traditional SIRS Model

The SIRS Model is a modified model of SIR model in which they considered a fixed population with only three compartments: Susceptible, $S(t)$; Infective, $I(t)$; Removal, $R(t)$. The compartments used for this model are composed of three categories. $S(t)$ is used to stand for the number of people not yet infected with the disease at time t, or those susceptible to the disease. $I(t)$ represents the number of individuals who have been infected with the disease and capable of spreading the disease to individuals in the susceptible category. $R(t)$ denotes those individuals who have been infected and then removed from the disease, either due to immunization or due to death.

SIRS model have the same compartments with SIR model, however, the only difference between them is $R(t)$. In SIR model, individuals in $R(t)$ category cannot be infected again or transmit the infection to others. The flow of this model can be depicted as follows: $S{\rightarrow}I{\rightarrow}R$. In SIRS model, $R(t)$ category allows members of the removals to be free of immunization and rejoin the susceptible class. That model is simply an extension of SIR model and we can see from its construction: $S{\rightarrow}I{\rightarrow}R{\rightarrow}S$.

The classical SIRS Model for the spread of infectious disease is demonstrated in Equations 1 - 3. The variable names are summarized in Table 1, which compares the epidemic interpretation of the variables to equivalent OSN demographic interpretation of the variables. In general, the SIRS model is described by three ordinary differential equations that depict how the rate of three compartments of the whole population N(susceptible S, infective I, removal R) evolve in the period of a disease outbreak. We can learn from ODEs that Equations 1 - 3 sum up to 0. Mathematically, $S+I+R = N$, which means we assume the population remains constant during outbreak of disease. That assumption is reasonable because the period of disease outbreak is short compared to the lifespan of all the population.

$$\frac{dS}{dt} = -\beta SI + \alpha R \tag{1}$$

$$\frac{dI}{dt} = \beta SI - \gamma I \tag{2}$$

$$\frac{dR}{dt} = \gamma I - \alpha R \tag{3}$$

Table 1. Terminology of Epidemic Model and OSN Demographic Prediction

Symbol	Units	Epidemic Model Parameter	OSN demographic Parameter
S	People	Susceptible	Latent OSN users
I	People	Infective	OSN users
R	People	Removal with immunity	OSN abandoner
β	$Time^{-1}$	Infection rate	Rate of latent users joining OSN
γ	$Time^{-1}$	Recovery rate	Rate of OSN users abandoning OSN
α	$Time^{-1}$	Average loss of immunity rate	Rate of users becoming latent users
μ	$Time^{-1}$	Average death rate	Rate of individuals leaving all categories
B	$Time^{-1}$	Average birth rate	Rate of newcomer joining latent users

Equation 1 shows that the rate at which the susceptible transfer to the infective in proportion to the infection rate β, the infective population and the susceptible population. This indicates that an individual in the population must be considered as having an equal probability as every other individual of contracting the disease with a rate of β. Therefore, an infected individual makes contact and is able to transmit the disease with βN to others per unit time and the fraction of contacts by the susceptible is S/N, so the rate of new infection, or the people leaving susceptible category, is $\beta N(S/N)I = \beta SI$[7]. The removal lost their immunity at a rate of α and then rejoin the susceptible class, so the rate of new susceptible is αR. For the second Equation, βSI reveals the population leaving susceptible class is equal to the population entering infective category. However, a part of infection are leaving this class at a rate of γ to the removal. γ represents the mean recovery rate or $1/\gamma$ stands for the mean infective period. In the last Equation, new removal is proportionate to recovery rate γ and the removal lost their immunity at a rate of α. Those processes which occur simultaneously are referred to as the Law of Mass Action, a widely accepted idea that the rate of contact between two groups in a population is proportional to the size of each of the groups concerned[8]. Finally, it is assumed that the rate of infection and recovery is faster compared with birth and death rate, which are ignored in this model.

There is no analytical solution to SIRS model, but the ODEs can be calculated by numerical method with initial conditions for each of the population compartment: $S(0) = S_0$, $I(0) = I_0$ and $R(0) = R_0$. S_0 stands for initial population that is susceptible to contracting disease, and generally represents the majority population at early times. I_0 denotes the number of initial outbreak, which is much smaller than the whole population. I_0 must be a positive number for an infection to spread as given in Equation 2. R_0 represents the population immune to the disease.

We also can find a character in Equation 2. If we set Equation 2 to less than 0 or equal to 0, we can obtain the immunization criterion as follow.

$$S_0 < \frac{\gamma}{\beta} \tag{4}$$

If equation 4 is satisfied, I can never increase. This is the basis for treatment of disease. In the context of OSN, where OSN users are analogous to infective population, the immunization criterion represents a condition under which the OSN users will sharply decline.

The SIRS model can be applied to OSN with proper analogy. When applying SIRS model to OSN, the susceptible is equivalent to the population which could potentially join the OSN. The infective is analogous to OSN users. The infection process in epidemiology can be depicted as contact between OSN users and potential OSN users by invitation in physical world or through email. Finally, the removal represent the people who resist OSN and reject to join OSN. A summary of all OSN analogues to epidemic model parameters is listed in Table 1.

3.2   Infective-Removal SIRS Model

While using the traditional SIRS model to capture demographic prediction, the assumption of recovery rate is only related to population of the infective with a recovery rate γ in SIRS model is doubtful suitable for modeling the abandonment of OSN[9][10]. Different from the assumption of a recovery rate for disease, OSN users do not join in an OSN and then abandon it after a scheduled amount of time. OSN users are leaving an OSN mainly due to the leave of a majority of their friends which are newcomers of the removal. Every OSN users that joins the network expects to stay unquestionably, but eventually loses interest as their friends lose interest because OSN is a network laying emphasis on interaction. That is to say the leave of one user's friends may make influence on abandonment of OSN. That notion is supported by work predicting customer churn in mobile networks which show users are more likely to leave their network if their social group have left [11]. Another aspect we need to take into account is birth rate and death rate. In traditional SIRS model, birth rate and death rate are ignored. In the context of OSN, compartments in all categories may suddenly leave due to unexpected reason, such as account frozen, which is equivalent to accidental death in epidemic model. In OSN, the rate of account frozen is bigger than the rate of death as a result of spam users in OSN. Similarly, the rate of potential OSN users is bigger than the rate of birth due to the robot could register OSN user automatically. So it needs to consider birth rate and death rate in OSN. So it is necessary to modify the traditional SIRS model to include above-mentioned aspects, which provide a better description of OSN demographic prediction. The general transfer procedure for fixed irSIRS model can be depicted as follows in Fig 1.

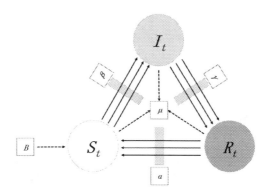

Fig. 1. General Transfer Diagram for the irSIRS Model with Birth and Death Rate

To include the removal's influence on the recovery procedure of the infective, the recovery rate must be in proportion to both I and R. So is it natural to let γIR denotes the recovery rate. So γI in Equation 2 need to be modified to γIR. To depict the robotic registration and sudden disappearance, birth rate B and death rate μ are introduced. Finally the infective-removal SIRS model with birth rate and death rate is demonstrated by the new ODEs as follows.

$$\frac{dS}{dt} = -\beta SI + BN - \mu S + \alpha R \tag{5}$$

$$\frac{dI}{dt} = \beta SI - \gamma IR - \mu I \tag{6}$$

$$\frac{dR}{dt} = \gamma IR - \mu R - \alpha R \tag{7}$$

In modified Equation 7, R_0 is no longer an arbitrary number in irSIRS model, because the magnitude of R is an important factor to recovery dynamics. Similar to the requirement of a small number of initial infective population in traditional SIRS model, the irSIRS model need a small quantity of initial removal population. If R_0 equals to 0, the recovery rate will be 0 permanently. None of the infective will be recovered since recovery requires contact with recovered individual. Mathematically, the Equation 7 will be 0 constantly and I will grow until all the S decline to 0, which means all the susceptible are infected. In context of OSN, R_0 could be the people who is indifferent to the OSN or resist joining the OSN altogether.

An immunization criterion can derived for irSIRS model in Equation 6 similar to SIRS model.

$$\beta S_0 < \gamma R_0 + \mu \tag{8}$$

If μ is not taken into consideration, the immunization criterion would be denoted as follows

$$\frac{S_0}{R_0} < \frac{\gamma}{\beta} \tag{9}$$

Similar to the implication of Equation 2 in traditional SIRS model, if immunization criterion is satisfied the I will never increase. It is also interesting to find a recovery criterion in Equation 7 if we let it equal to 0 or bigger than 0. Similar to immunization criterion, the recovery criterion is as follows

$$I_0 > \frac{\mu + \alpha}{\gamma} \tag{10}$$

Similarly, if μ is ignored, the recovery criterion would be as follows

$$I_0 > \frac{\alpha}{\gamma} \tag{11}$$

If the recovery criterion is satisfied, the removal category will increase constantly until $I(t)$ decline to the degree that recovery criterion is not satisfied any more.

3.3   Analysis of irSIRS Model

To analyse irSIRS model in context of OSN, it is important to adjust the factors, such as α, β, γ, B and μ and monitor the curve's change. That is to say it is need to discover what influence the parameters make on irSIRS model. We use the function *ode45* in MATLAB to make a numerical solution to ODEs of irSIRS model. *ode45* is a solver for nonstiff problem with medium accuracy, most of the time it could work well. To simulate a condition of a new OSN coming into being which is analogous to outbreak of an epidemic disease, it is reasonable to set S_0 with a large proportion of the population and set I_0 and R_0 with a small proportion of the population. Under that assumption, S_0, I_0, R_0 are assigned with 85, 10, 5. So the whole population $N = 100$. To simplify irSIRS model in this paper, we suppose birth rate is equal to death rate with 0.01. Fig 2(a) - Fig 2(d) is population evolution demonstration in OSN with irSIRS model by different α, β and γ.

In Fig 2(a), parameters are assigned with $\alpha = 1.7$, $\beta = 0.3$, $\gamma = 0.1$. In that condition, neither immunization criterion nor recovery criterion is satisfied. The number of the infective will grow fast to a peak then decline slowly. In Fig 2(b), parameters are assigned with $\alpha = 1.7$, $\beta = 0.005$, $\gamma = 0.1$. on that occasion, only immunization criterion is satisfied. So the disease cannot spread and the amount of the infective are gradually decrease to 0. In Fig 2(c), parameters are assigned with $\alpha = 1.7$, $\beta = 0.01$, $\gamma = 0.2$. Under the circumstances both immunization criterion and recovery criterion are satisfied, so it is similar to Fig 2(b). In the last condition, Fig 2(d), parameters are assigned with $\alpha = 1.7$, $\beta = 0.1$, $\gamma = 0.2$, only recovery criterion is satisfied. The infection is spreading and the infective are recovering simultaneously, so the distribution of the infective and the removal will float up or down and eventually approach to a constant.

Then we let α, β and γ be invariant and alter B and μ to observe the changes in irSIRS model. Fig. 2(e) - Fig. 2(h) show different curves with variant B and μ. In Fig 2(e), $B = 0.01$ and $\mu = 0.015$. It means the number of people suddenly leaving OSN is more than the number of people joining OSN, so the removal is declining, so does the population of N. On the contrary, in Fig 2(f), $B = 0.015$ and $\mu = 0.01$. Both the susceptible and the removal increase, so does N. In those two circumstance, B and μ are much smaller compared to α, β and γ, so it make little influence to irSIRS model. In Fig 2(g), $B = 0.15$, $\mu = 0.1$ and In Fig 2(h), $B = 0.1$, $\mu = 0.15$, it is obvious that B and μ make influence on irSIRS model when they are in the same order of magnitude.

4   Experiments

4.1   Methods and Dataset

To test the theory of epidemic model in OSN, an early work has been done by using publicly available Google search data as a proxy of users [12]. It is convenient to get those data from Google Search Engine, but the substitutes are not accurate as real data in OSN. In this paper, real data are available due

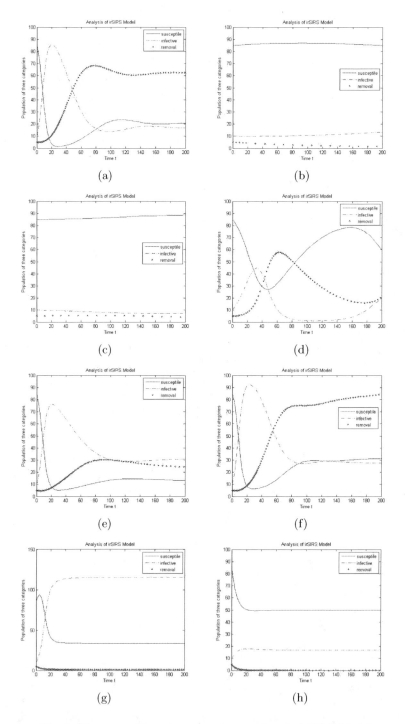

Fig. 2. Diagram with variant parameter values in irSIRS Model

to the API of OSN. It is naturally to suppose that a susceptible change to an infective when the user register an account or publish the first message in a OSN. So registration date or date of the first message are important parameters. It is acceptable to assume an individual has abandon a OSN if he or she doesn't update any information. In this paper, the period is set with number 30 days.

The Renren Network and Sina Weibo are two famous OSNs in China, they both have convenient APIs so the data is available. The Renren Network, formerly known as the Xiaonei Network is a Chinese social networking service that is dubbed as a Chinese copy of Facebook. It is popular amongst college students. Sina Weibo is a Chinese microblogging (weibo) website. Akin to a hybrid of Twitter and Facebook, it is one of the most popular sites in China, in use by well over 30 percents of Internet users, with a market penetration similar to what Twitter has established in the USA [13].

4.2   Data Curve Fitting

We use 138,952 users in the Renren Network and 347,278 users in Sina Weibo as a sample, those data are crawled through APIs randomly and we don't get rid of the spams. Then we normalize the number of users in different OSNs. The infective category is the most important part, so we only analyse the infective curve in this experiment.

The epidemic model is used to analyse the data by curve fitting the infective $I(t)$ population curve generated by irSIRS model to real data in OSNs. The curve fitting is carried out using MATLAB by the following approach to determine the best fit curve.

According to initial parameters S_0, I_0, R_0, α, β, γ, the ODEs of irSIRS model will generate corresponding $I(t)$. The sum of squared error (SSE) between real data points and generated $I(t)$ curve is calculated. The best curve fitting is defined as the $I(t)$ curve minimizes SSE. It is recommended to calculate the best fit curve by using *lsqcurvefit* function in MATLAB.

4.3   Result and Discussion

To validate irSIRS epidemic model in OSN demographic prediction, we make experiments on both Renren Network and Sina Weibo. In the case of Renren Network, we calculate the users from 2008. If a user has published a message in a month, we divide the user into the infective category. If a user who doesn't publish any messages in three months, the user will be classified into the removal category. If a user who doesn't update any messages in half a year, the user can be classified into the susceptible category.

With those prerequisite, we first apply the irSIRS model to Renren Network. As shown in Fig 3(a), blue curve represent real data in Renren Network and the green curve represent the prediction line with irSIRS model. The best fit curve parameters are listed in Table 2. The real data collected through API shows that from 2008 to 2010, the active users in Renren Network are increasing sharply. It may be a result of that Renren Network made a pre-IPO announcement in 2011.

After the peak of the curve, when the active users at the maximum point, the number of infective population declines slowly as time passing by. As discussed in sections before, recovery spreads infectiously, which means users begin to leave the OSN after their friends have left the OSN. So more and more users abandon the OSN then the number of active users in OSN will approach to a constant ratio. By the best fit curve, we can make a prediction that the infective population may fall to 40 percents by the year of 2015 and to 35 percents by the year of 2016. Due to the best curve fitting is sensitive to the recent data, so the prediction is more accurate if we get enough recent data.

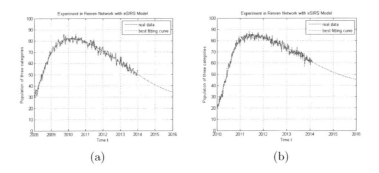

Fig. 3. Diagram with variant parameter values in irSIRS Model

Table 2. Best Curve Fitting Parameters in irSIRS Model

Fit	α	β	γ	S_0	I_0	R_0	SSE
Renren Network, irSIRS	0.0992	0.0286	0.0101	63.32	29.13	7.55	833.958
Sina Weibo, isSIRS	0.3739	0.0292	0.0112	74.56	20.13	5.31	1513.49

In Fig 3(b), we apply irSIRS model to Sina Weibo. As the biggest OSN in China established in 2009, the infective grow rapidly after 2010 as shown in real data. The number of infective reach to maximum value at around 2012. Then the proportion of the infective start to decrease tenderly as a result of infective removal theory. The registration of the spam also contribute to the decline of the infective as well, some spams update a status and may hibernate for a long time until they update the next information. We also can draw a best fit curve to predict the infective population in the future as in the diagram. By the year of 2016, the infective population will reduce to nearly 45 percents. But what we have to say is that irSIRS model is similar to the Markov chain, it is sensitive to the recent data. So we need more recent data to make eventual outcome of the infective population.

It is also important to understand the nature of the asymmetry of the prediction. Since the curves can be thought as equally probable future trajectories for the OSNs. It follows that the OSNs are more likely to deviate from the best fit curve on the side of slower decline. The reason for the asymmetry is the decline of the infective is entirely decided by the data after the peak. If the data after peak is ignored, the solution in which the OSNs continue with a constant infective population is possible with a finite SSE.

5   Conclusion and Future Work

In this paper we applied a modified epidemic model to describe the dynamics of demographic evolution of the online social network. Using the real data in Renren Network and Sina Weibo as a case study, we proposed a infective-removal SIRS model based on the traditional SIRS model to depict adoption and abandonment activity of the OSN. The best fit curve with irSIRS model suggests that both of the OSNs will undergo a slowly decline in the coming years. With the character of sensitivity to recent data in irSIRS model, we may be apply this model to prediction of dynamics of topics in future work.

References

1. Milgram, S.: The small world problem. Psychology Today 2(1), 60–67 (1967)
2. Kermack, W.O., McKendrick, A.G.: Contributions to the mathematical theory of epidermics (1927)
3. Kephart, J.O., White, S.R., Chess, D.M.: Computers and epidemiology. IEEE Spectrum 30(5), 20–26 (1993)
4. Pastor-Satorras, R., Vespignani, A.: Epidemic spreading in scale-free networks. Physical Review Letters 86(14), 3200 (2001)
5. Grabowski, A., Kosiński, R.A.: Epidemic spreading in a hierarchical social network. Physical Review E 70(3), 31908 (2004)
6. Krause, J., Croft, D.P., James, R.: Social network theory in the behavioural sciences: potential applications. Behavioral Ecology and Sociobiology 62(1), 15–27 (2007)
7. Brauer, F., Castillo-Chávez, C.: Basic ideas of mathematical epidemiology. Mathematical Models in Population Biology and Epidemiology, pp. 275–337. Springer, New York (2001)
8. Daley, D.J., Gani, J.: Epidemic Modelling: An Introduction (2005)
9. Wu, S., Das Sarma, A., Fabrikant, A., et al.: Arrival and departure dynamics in social networks. In: Proceedings of the Sixth ACM International Conference on Web Search and Data Mining, pp. 233–242. ACM (2013)
10. Cannarella, J., Spechler, J.A.: Epidemiological modeling of online social network dynamics. arXiv preprint arXiv:1401.4208 (2014)
11. Richter, Y., Yom-Tov, E., Slonim, N.: Predicting Customer Churn in Mobile Networks through Analysis of Social Groups. In: SDM, pp. 732–741 (2010)
12. Garcia, D., Mavrodiev, P., Schweitzer, F.: Social resilience in online communities: The autopsy of friendster. In: Proceedings of the First ACM Conference on Online Social Networks, pp. 39–50. ACM (2013)
13. Rapoza, K.: China's Weibos vs US's Twitter: And the Winner Is (2011)

Analysis on Chinese Microblog Sentiment Based on Syntax Parsing and Support Vector Machine

Ziyan Su[*], Bin Zhou, Aiping Li, and Yi Han

College of Computer, National University of Defense Technology, China
su.ziyan@163.com,
{bin.zhou.cn,apli1974}@gmail.com,
yihan@nudt.edu.cn

Abstract. Analysis on microblog sentiment is very important for microblog monitoring and guidance of public opinion. According to the problem that precision of Chinese microblog sentiment analysis was low, this paper proposed a model for analysis on Chinese microblog sentiment based on syntax parsing and the support vector machine algorithm. Firstly, this paper built a fundamental dictionary of emotion based on existing emotional words resources, and expanded the dictionary by computing similarity of words. Then, it computed emotional values of microblogs by syntax parsing and judged their emotional tendencies. Finally, it selected a certain percentage of positive and negative emotional microblogs as a training set. It utilized the support vector machine algorithm to classify microblogs that didn't belong to the training set and got all emotional tendencies of microblogs. Experimental results showed that the model proposed by this paper could achieve 77.3% precision. So the model is effective.

Keywords: fundamental emotional dictionary, word similarity calculation, syntax parsing, training set selection, support vector machine.

1 Introduction

Microblog is a social network platform for sharing information through broadcast of short attention mechanism and it is one of the most influential products in Web2.0 era. As the main media in the social networking site, microblog is short, pithy and fast. People tend to get information from microblog, such as news, opinions, comments and entertainment. They convey their ideas, trends and attitude through microblog [1]. Analysis on microblog emotion has extremely important practical significance for microblog monitoring and public opinion guidance. This paper divided microblog sentiment into positive emotion, negative emotion and neutral emotion and research the classification of microblog sentiment.

---

[*] Sponsored by National Key fundamental Research and Development Program No.2013CB329601.

W. Han et al. (Eds.): APWeb 2014 Workshops, LNCS 8710, pp. 104–114, 2014.

Analysis on English microblog sentiment mainly focuses on the information of the Twitter [2-5] and is generally divided into analysis of irrelevant theme and analysis of relevant theme. Chinese microblog need word segmentation processing and often produce new words. Therefore, methods of analysis on English microblog emotion can't be applied to Chinese microblog.

Analysis on Chinese microblog sentiment is still in its infancy, but it shows a rapid development tendency in the past two years. It includes the research of recognition of opinion bearing sentences and sentiment analysis. Microblog sentiment analysis mostly includes the method of manual annotation for microblog emotional tendency. As a result, the workload is huge and time-consuming. Lin Jianghao et al established a classification method of microblog sentiment based on Naive Bayes [6] and it achieved good classification results. However, the sentiment classifier needs a large training set to improve the classification performance and the training set still need manual annotation. Therefore, this paper proposed a model for analysis on Chinese microblog sentiment based on new words dictionary and syntax parsing. This model judged microblog emotional tendency through automatic calculation of microblog emotional value and had high accuracy. So it realized the purpose of reducing man-power and saving time.

This paper studied the resource construction of sentiment analysis, emotional tendency of words, discovery and sentiment analysis of Chinese microblog new words and Chinese microblog sentiment analysis based on syntax parsing. Finally, it developed and realized a model for analysis on Chinese microblog sentiment based on new words dictionary and syntax parsing.

This paper mainly includes the following three aspects:

1. Construction and expansion of Chinese microblog emotional dictionary

This paper built a fundamental dictionary of emotion based on existing emotional words resources and judged sentiment polarity and intensity of new words by semantic similarity calculation method based on "HowNet". Then it expanded the fundamental dictionary of emotion.

2. Analysis on emotional tendency of Chinese microblog based on syntax parsing

This paper pretreated Chinese microblog by word segmentation and removal of forwarding tags. Then it calculated emotional value of Chinese microblog by cumulative calculation of polarity based on syntax parsing. Finally, it judged emotional tendency of Chinese microblog through its emotional value.

3. Secondary classification based on support vector machine

This paper selected a certain percentage of positive and negative emotional microblogs judged by syntax parsing as a training set. Then, it utilized the support vector machine algorithm to classify microblogs that didn't belong to the training set and got all emotional tendencies of microblogs.

2   Construction and Expansion of Emotional Dictionary

As short text of 140 words, microblog mainly embodies its emotional tendency through emotional words. So the construction and expansion of emotional dictionary is the foundation of microblog sentiment analysis.

2.1   Fundamental Dictionary of Emotion

This paper built a fundamental dictionary of emotion which included commendatory terms and derogatory terms based on "Chinese/English Vocabulary for Sentiment Analysis (Beta version)" of "HowNet" [7], "National Taiwan University Sentiment Dictionary (NTUSD)" [8], "The Students Appraise Dictionary" [9] and "The Chinese Appraise Dictionary" [10].

This paper added positive emotional words of "HowNet", "National Taiwan University Sentiment Dictionary (NTUSD)", "The Students Appraise Dictionary" and "The Chinese Appraise Dictionary" to commendatory dictionary. In the same way, it added negative emotional words of "HowNet", "National Taiwan University Sentiment Dictionary (NTUSD)", "The Students Appraise Dictionary" and "The Chinese Appraise Dictionary" to derogatory dictionary. Finally, the fundamental dictionary of emotion it built included 5889 commendatory terms and 6255 derogatory terms.

2.2   Expansion of Emotional Dictionary

Calculation of Word Similarity
Word similarity is the degree that two words in different contexts can be used interchangeably without changing the syntactic and semantic structure of the text. Each word of "HowNet" can be expressed as a few concepts, and each concept can be described by some sememes [11].

For $Word_1$ and $Word_2$, assuming their concepts are $C_{11}, C_{12}, ..., C_{1n}$ and $C_{21}, C_{22}, ..., C_{2n}$, then their similarity was the maximum value of these concepts' similarity:

$$Sim(Word_1, Word_2) = \max_{i=1...n, j=1...m} Sim(C_{1i}, C_{2j}) \tag{1}$$

In "HowNet", the similarity of concepts C_1 and C_2 is constituted by first basic sememe, other basic sememes, relational sememes and relational symbols. They are $Sim_1(C_1, C_2)$, $Sim_2(C_1, C_2)$, $Sim_3(C_1, C_2)$ and $Sim_4(C_1, C_2)$. So the similarity of C_1 and C_2 was:

$$Sim(C_1, C_2) = \sum_{i=1}^{4} \beta_i \prod_{j=1}^{i} Sim_i(C_1, C_2) \tag{2}$$

Among them, $\beta_i (1 \leq i \leq 4)$ is adjustable parameter and meet: $\beta_1 + \beta_2 + \beta_3 + \beta_4 = 1$ and $\beta_1 \geq \beta_2 \geq \beta_3 \geq \beta_4$. Parameter β_i reflects the contribution of Sim_i for concept similarity.

For sememes P_1 and P_2, their similarity was calculated by their semantic distance:

$$Sim(P_1, P_2) = \frac{\alpha}{D(P_1, P_2) + \alpha} \tag{3}$$

$D(P_1, P_2)$ was the path length of P_1 and P_2 in sememe hierarchy system and α is adjustable parameter.

Calculation of Sentiment Polarity and Intensity for Unknown Emotional Words
The calculation process for the emotional value of unknown emotional word was shown in Fig. 1:

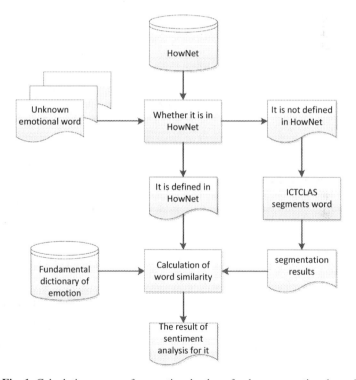

Fig. 1. Calculation process for emotional value of unknown emotional word

This paper calculated word similarity of unknown emotional words and words in the fundamental dictionary of emotion to judge its sentiment polarity and intensity.

It marked commendatory terms with *POS* and marked derogatory terms with *NEG*. Then, it judged unknown emotional word like *Word* as follows:

1. Firstly, this paper judged whether *Word* was in "HowNet". If *Word* was in "HowNet", it went to step 2. Otherwise, it went to step 3.

2. This paper calculated $Sim(Word, pos)$ which was the similarity of *Word* and each commendatory term *pos* in *POS*. It sorted them and stored them in array *WP*. Then it calculated $Sim(Word, neg)$ which was the similarity of *Word* and each derogatory term *neg* in *NEG*. It sorted them and put them in array *WE*. It set α_1 as a threshold. We concluded that two words are similar when their similarity was greater than α_1 which was 0.6 in this paper. Then, it counted N_1 which was the number of elements that were larger than α_1 and N_2 which was the number of elements that were larger than α_1. It set $N = \max(N_1, N_2)$. So the emotional value of was:

$$Sentiment(Word) = \frac{1}{N}\left(\sum_{i=0}^{N-1}WP[i] - \sum_{j=0}^{N-1}WE[j]\right) \quad (4)$$

This paper got the emotional value of *Word* according to the formula (4). Then it judged the emotional tendency and intensity of *Word* as follows:

$$Sentiment(Word)\begin{cases} >0: Word\ is\ positive,\ its\ intensity\ is\ |Sentiment(Word)| \\ <0: Word\ is\ negative,\ its\ intensity\ is\ |Sentiment(Word)| \\ =0: Word\ is\ neutral \end{cases} \quad (5)$$

3. This paper split *Word* into $Word_1$ and $Word_2$ so that $Word_1$ and $Word_2$ were in "HowNet". It calculated $Sentiment(Word_1)$ which was the emotional value of $Word_1$ and $Sentiment(Word_2)$ which was the emotional value of $Word_2$. Then, the emotional value of *Word* was:

$$Sentiment(Word) = \beta_1 Sentiment(Word_1) + \beta_2 Sentiment(Word_2) \quad (6)$$

β_1 and β_2 were adjustable parameters and they met: $\beta_1 + \beta_2 = 1$.

It calculated the emotional value of *Word* according to formula (6) and got the emotional tendency and intensity of *Word* according to formula (5). If $Word_1$ or $Word_2$ was named Entity, the emotional value of *Word* was 0. If $Word_i$ ($i = 1, 2$) was adjective and $Word_{3-i}$ was not adjective, it set: $\beta_i = 1$. Otherwise, it set: $\beta_1 = 0.5, \beta_2 = 0.5$.

3   Analysis on Microblog Sentiment Based on Syntax Parsing

3.1   Extraction of Emotional Modifier

Dexi Zhu in "Syntax Handouts" [12] defined adverbs as function words and considered adverbs included scope adverbs, degree adverbs, time adverbs, negative adverbs and reduplicative adverbs. It was generally believed that negative adverbs and degree adverbs could change the emotional polarity and intensity of emotional words which they modified. Therefore, negative adverbs and degree adverbs were emotional modifier.

Negative Adverbs

Min Liu in "Study of Chinese Negative Adverbs Origin and Diachronic Evolution" [13] pointed out that there were 43 negative adverbs in Chinese. Based on this, this paper added adverbs which contained negative sememes in "HowNet" to negative adverbs and built a word set which had 73 negative adverbs.

Degree Adverbs

Huang Lin etc. in "On the Characteristic, Range and Classification of Adverbs of Degree" [14] summarized 85 degree adverbs and classified them from low, medium, high to maximum. Therefore, we built a word set which had 85 degree adverbs.

3.2   Pretreatment of Chinese Microblog

This paper pretreated Chinese microblog as the following steps:

1. It removed meaningless forwarding tags in Chinese microblog.

2. It added degree adverbs and negative adverbs to ICTCLAS user-defined dictionary and then segmented Chinese microblog with labels.

3. It extracted time words, location words, orientation words, pronouns, numerals, quantifiers, prepositions, auxiliaries, interjections and modal particles which were unemotional from segmentation results to build the list of microblog stop words. In addition, 38 words of opinion in "Chinese/English Vocabulary for Sentiment Analysis (Beta version)" of "HowNet" and those unemotional words which had high frequency in segmentation results were added to the list of microblog stop words.

4. It extracted nouns, verbs and adjectives from segmentation results and calculated the emotional value of these words.

3.3   Analysis on Collocation pattern of Emotional Phrase

This paper analyzed phrase patterns of emotional words modified by negative adverbs and degree adverbs and microblogs which contained adversative conjunctions.

Negative adverb + Emotional word

The emotional word would change its emotional polarity and attenuate its emotional intensity when it was modified by negative adverb.

This paper took the weight of negative adverb for: $\gamma_{neg} = -0.8$.

Degree adverb + Emotional word

The emotional word would change its emotional intensity when it was modified by degree adverb.

This paper separately took the weights of maximum, high, medium and low degree adverbs for: $\gamma_{max} = 2.0, \gamma_{high} = 1.75, \gamma_{med} = 1.5, \gamma_{low} = 0.5$.

Negative adverb + Negative adverb + Emotional word

The emotional word would keep its emotional polarity and attenuate its emotional intensity when it was modified by two negative adverbs.

Therefore, this paper took the weight of double negative adverbs for: $\gamma_{neg} \times \gamma_{neg} = 0.64$.

Degree adverb + Negative adverb + Emotional word

Firstly, the emotional word would change its emotional polarity and attenuate its emotional intensity when it was modified by negative adverb. Then, the collocation pattern would change its emotional intensity when it was modified by degree.

Therefore, the weight of the modifier was the product of the weight of degree adverb and the weight of negative adverb.

Negative adverb + Degree adverb + Emotional word

Yanxia Shang in "A semantic and pragmatic analysis of concurrence of degree adv. and bu" [15] pointed out that "not + degree adverb + adjective" was attenuation of higher magnitude rather than negative.

Therefore, this paper took the weight of the modifier for: $\left|\gamma_{neg} \times \gamma_{low}\right| = 0.4$.

Clause 1 + adversative conjunction + clause 2

If two clauses are connected by adversative conjunction, the emotional value of the sentence was the same as clause2.

3.4   Analysis on Emotional Tendency of Chinese Microblog

This paper calculated the emotional tendency and intensity of Chinese microblog based on segmentation results. Firstly, it removed stop words according to stop word list. Then, it extracted n emotional words $Word_1, Word_2, ..., Word_n$ from each microblog. Finally, the emotional value of microblog $Tweet$ was:

$$Sentiment(Tweet) = \sum_{i=1}^{n} \gamma_i \times Sentiment(Word_i) \qquad (7)$$

Parameter γ_i was the weight of emotional word $Word_i$. If there was no modifier before $Word_i$, it set: $\gamma_i = 0$. If there was modifier before $Word_i$, it judged the collocation pattern of the modifier according to the 3.3 section and then decided the value of parameter γ_i. If $Tweet$ contained adversative conjunction C, the weight of $Word_i$ before C was 0.

The value of $Tweet$ could be calculated through formula (7). Then, the emotional tendency and intensity of $Tweet$ could be judged as follows:

$$Sentiment\left(Tweet\right)\begin{cases} > 0 : Tweet\ is\ positive\ ,\ its\ intensity\ is\left|Sentiment\left(Tweet\right)\right| \\ < 0 : Tweet\ is\ negative\ ,\ its\ intensity\ is\left|Sentiment\left(Tweet\right)\right| \\ = 0 : Tweet\ is\ neutral \end{cases} \quad (8)$$

4   Secondary Classification Based on Support Vector Machine

4.1   Support Vector Machine

Support vector machine is a machine learning method based on VC dimension theory of statistical learning theory and structure risk minimum principle. Support vector machine can find an optimal hyperplane which can ensure classification accuracy and maximize blank space on either side of the hyperplane. The algorithm can obtain a good statistical law even though the statistical sample size is less.

The basic idea of support vector machine is finding an optimal classification hyperplane of two classes of samples. For texts, the optimal classification hyperplane is the hyperplane which can separate two kinds of texts accurately and make the classification interval maximal. The former ensures the empirical risk minimum and the latter minimizes the confidence interval so that the real risk is minimal.

4.2   Secondary Classification of Microblogs

This paper calculated the accuracy rate of positive emotional microblogs and negative emotional microblogs and marked them with P_1 and P_2. It marked the number of positive emotional microblogs and negative emotional microblogs judged by the model with N_{pos} and N_{neg}. It set: $N_{train} = \min\left(N_{pos} \times P_1, N_{neg} \times P_2\right)$.

This paper sorted positive emotional microblogs according to their emotional values from 3.4 section and stored them in array $POSBLOG$. Then, it sorted negative emotional microblogs according to their emotional values from 3.4 section and stored them in array $NEGBLOG$. It stored neutral microblogs judged by 3.4 section in array $NEUBLOG$.

It selected maximum N_{train} emotional values of microblogs in $POSBLOG$ and minimum N_{train} emotional values of microblogs in $NEGBLOG$ to build the training set of support vector machine. Then, it utilized the support vector machine algorithm to classify microblogs that didn't belong to the training set. We selected document frequency as characteristic evaluation function of the support vector machine.

Finally, it got all emotional tendencies of microblogs through sentiment analysis model based on syntax parsing and support vector machine. When microblogs belong to the training set, it judged them through their emotional values as we presented in the 3.4 section. Otherwise, it judged them by support vector machine.

5   Experimental Results and Analysis

5.1   Experimental Data Set

The microblog data set of this paper came from task 3 of COAE2013 and included 51865 microblogs. COAE2013 was the fifth Chinese opinion analysis evaluation which was aimed at promoting the research and application of Chinese tendency analysis. Its objective was establishing and perfecting fundamental data set and evaluation of Chinese tendency analysis.

5.2   Experimental Results and Their Analysis

This paper analyzed sentiment of 51865 microblogs and calculated emotional value of each microblog. As COAE2013 annotated emotional tendency of some microblogs which contained keywords of "mengniu", it analyzed emotional tendency of 5052 microblogs which contained "mengniu" and calculated accuracy rate *precision*, recall rate *recall* and F value *F* of analysis results. The microblogs which were not annotated were marked manually.

The calculation formula of accuracy rate *precision* was:

$$precision = \frac{the\ correct\ number\ of\ the\ class\ judged\ by\ this\ paper}{the\ number\ of\ the\ class\ judged\ by\ this\ paper} \quad (9)$$

The calculation formula of recall rate *recall* was:

$$recall = \frac{the\ correct\ number\ of\ the\ class\ judged\ by\ this\ paper}{the\ number\ of\ the\ class} \quad (10)$$

The calculation formula of F value *F* was:

$$F = \frac{2 \times precision \times recall}{precision + recall} \quad (11)$$

Table 1. Experimental results of different methods

Analysis method	Evaluation indicators	Positive microblogs	Negative microblogs	Neutral microblogs
Syntax parsing	precision	0.570	0.940	0.600
	recall	0.950	0.521	0.750
	F value	0.713	0.670	0.667
Syntax parsing +Support vector machine	precision	0.690	0.844	0.600
	recall	0.788	0.763	0.750
	F value	0.736	0.801	0.667

This paper designed two experiments to compare effects of sentiment analysis based on syntax parsing and sentiment analysis based on syntax parsing and support vector machine. The experimental results were shown in Table 1.

It could be seen from table 5-1 that the sentiment analysis based on syntax parsing and support vector machine had a better overall effect than the sentiment analysis based on syntax parsing. Among them, the effect of positive microblog analysis was slightly improved. And the effect of negative microblog analysis was significantly improved. At the same time, the overall accuracy rate of microblog sentiment analysis increased from 69.3% to 77.3%. Therefore, the analysis method based on syntax parsing and support vector machine was more effective.

The precision of positive microblog analysis was low because of not considering ironies. And the precision of neutral microblog analysis was low because of not discussing rhetorical questions and general questions. In addition, this paper didn't add emotional dictionary of network to fundamental dictionary of emotion. Therefore, the effect of sentiment analysis was decreased.

Table 2. The optimal results of COAE2013 and the experimental results of model based on syntax parsing and support vector machine

Analysis method	Evaluation indicators	Positive microblogs	Negative microblogs	Macro average
Optimal results of COAE2013	precision	0.350	0.440	0.395
	recall	0.526	0.536	0.531
	F value	0.351	0.397	0.374
Syntax parsing +Support vector machine	precision	0.690	0.844	0.767
	recall	0.788	0.763	0.776
	F value	0.736	0.801	0.769

This paper also compared the optimal results of COAE2013 and the experimental results of model based on syntax parsing and support vector machine. They were shown in Table 2. Among them, macro average was the average of positive microblogs and negative microblogs.

It could be seen from Table 2 that the sentiment analysis based on syntax parsing and support vector machine had better effects of all indexes than the optimal results of COAE2013. Therefore, the sentiment analysis model based on syntax parsing and support vector machine is effective.

6   Conclusion

This paper proposed a model for analysis on Chinese microblog sentiment based on syntax parsing and the support vector machine. This model realized the function of automatic sentiment annotation for Chinese microblog and achieved 77.3% precision. Therefore, the model is effective.

The next step is to add rhetorical questions, general questions and exclamatory sentences to this model to improve the classification performance.

References

1. Tan, S., Liu, K., Wang, S., et al.: Overview of Chinese Opinion Analysis Evaluation 2013. In: Proceeding of the Fifth Chinese Opinion Analysis Evaluation (2013)
2. Kumar, A., Sebastian, T.M.: Sentiment Analysis on Twitter. International Journal of Computer Science Issues (IJCSI) 9(4) (2012)
3. Lima, A., De Castro, L.: Automatic sentiment analysis of Twitter messages. In: Proceeding of the 2012 4th International Conference on Computational Aspects of Social Networks, pp. 52–57 (2012)
4. Logunov, A., Panchenko, V.: Characteristics and predictability of Twitter sentiment series. In: 19th International COngress on Modelling and Simulation, pp. 1617–1623 (2011)
5. Wang, X., Wei, F., Liu, X., et al.: Topic sentiment analysis in twitter: a graph-based hashtag sentiment classification approach. In: Proceedings of the 20th ACM International Conference on Information and Knowledge Management, pp. 1031–1040. ACM (2011)
6. Lin, J., Yang, A., Zhou, Y., et al.: Classification of Microblog Sentiment Based on Naive Bayesian. Computer Engineering & Science 9, 035 (2012)
7. Dong, Z., Dong, Q.: HowNet (2013), http://www.keenage.com [EB/OL]
8. National Taiwan University Sentiment Dictionary (NTUSD), Simplified Chinese sentiment polarity dictionary (2013),
   http://www.datatang.com/datares/go.aspx?dataid=601972
9. Zhang, W., Liu, J., Guo, X.: The Students Appraise Dictionary. Encyclopedia of China Publishing House (2004)
10. Li, J.: Experimental Study on Sentiment Classification of Chinese Reviews. Tsinghua University, Beijing (2008)
11. Liu, Q., Li, S.: Word similarity computing based on How-net. Computational Linguistics and Chinese Language Processing 7(2), 59–76 (2002)
12. Zhu, D.: Syntax Handouts. The Commercial Press (1982)
13. Liu, M.: Study of Chinese Negative Adverbs Origin and Diachronic Evolution, Hunan Normal University (2010)
14. Lin, H., Guo, S.: On the Characteristic, Range and Classification of Adverbs of Degree. Journal of Shanxi University 26(2), 71–74 (2003)
15. Shang, Y.: A semantic and pragmatic analysis of concurrence of degree adv. and bu. Journal of Zhoukou Normal University 23(3), 101–104 (2006)

A Parallel and Scalable Framework for Non-overlapping Community Detection Algorithms

Songchang Jin[1,2], Yuchao Zhang[3], Yuanping Nie[1], Xiang Zhu[1], Hong Yin[1],
Aiping Li[1], and Shuqiang Yang[1]

[1] College of Computer, National University of Defense Technology, Changsha 410073, China
[2] Department of Computer Science, University of Illinois at Chicago, Chicago, IL 60607, USA
[3] Beijing Institution of System Engineering, Beijing 10010, China
{jinsongchang87,yuanpingnnie,zhuxiang19881117,
apli1974}@gmail.com
dragonzyc@163.com, yinhonggfkd@aliyun.com
sqyang9999@126.com

Abstract. Community detection has been wildly studied during the past years by varies of researchers, and a plenty of excellent algorithms and approaches have been proposed. But networks are becoming larger and higher complicated in nowadays. How to excavate the hidden community structures in the expanding networks quickly with existing excellent methods has become a challenge. In this paper, we designed a parallel community discovery framework based on Map-reduce and implemented parallel version of some excellent existing standalone community detection methods. Results of empirical tests show that the framework is able to significantly speed up the mining process without compromising the accuracy excessively.

Keywords: parallel, Map-reduce, community detection.

1   Introduction

In recent years, researchers have been carrying out plenty of productive studies focusing on the statistical properties of networked systems, such as social networks, technological networks and so on, and a few properties have been found that seem to be common to many networks: The small-world property [1], the scale-free feature [2] and community structure pattern [3]. Uncovering the community structures in networks is not only very important for studying organization structure and analyzing functional features, but also can help people to dig out valuable hidden information and can be used in lots of applications such as product recommendation.

Researchers in many disciplines, starting from their respective fields, have been performing in-depth and meticulous studies, and have proposed plenty of excellent algorithms to discover the hidden community structures. However, algorithms are all marked with time brands, and each algorithm is proposed to solve the problem of that period. Early algorithms are generally used to handle small real networks such as Zachary's karate club network and Dolphin social network, et al. Representative

W. Han et al. (Eds.): APWeb 2014 Workshops, LNCS 8710, pp. 115–126, 2014.

algorithms of that era are GN algorithm [1] and hierarchical clustering [4]. Although GN has a running time of $O(m^2 n)$, where m and n represent the number of links and nodes, respectively, it is acceptable because of the size of the small network to be tackled.

Nowadays, with the rapid advancements in science and technology, more and more systems with network representation are showing more and more complex structure and larger scale. Actually, many of today's complex networks consist of millions or billions of nodes. A recent report published by Business Insider [5] in 2014 says that, the largest social media application – Facebook has and holds more than 1.16 billion monthly active users. Nevertheless, traditional community detection methods have at least two deficiencies: 1) Bad scalability. For any community detection algorithm, the amount of resource consumption will increase as network size expands. But most of the traditional algorithms are designed on single-core CPU, and due to production technology restrictions, single-core CPU has encountered performance ceiling in both performance/price ratio and performance/power ratio. To make matters worse, the microprocessor industry has shifted from maximizing single-core performance to integrating multiple cores on a single processor, which restricts the scalability of tra-ditional methods. 2) Low efficiency. Current mainstream commercial servers are equipped with multi-core processors, but traditional methods can only use one of them, so the servers cannot be effectively put to good use. These restrictions will result in the existing excellent algorithms into trouble when confronted with larger scale networks. How to quickly and accurately identify the community structures in the expanding and complicating networks, has become a serious problem.

In this paper, we focus on designing a parallelization framework for existing excel-lent algorithms but not developing new methods. With the help of multilevel k-way partitioning algorithm and Map-reduce parallel programming framework [6], we pro-pose a parallel framework for community detection in complex networks.

The rest of this paper is organized as follows. Section 2 briefly reviews some concepts and related works. Section 3 provides detailed description of the framework. In Section 4, we conduct a couple of experiments and analyze the results. Finally, we summarize our work and give out future research directions in Section 5.

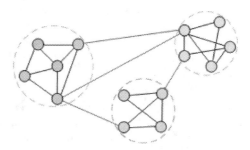

Fig. 1. A small network with three community structures surrounded by dashed circles

2   Related Works

In our framework, there are three key technologies: Multilevel graph partitioning, Map-reduce and non-overlapping community detection. They are playing different roles in the framework. We use the multilevel graph partitioning method to ensure the balance of data segmentation and the accuracy of the results. Map-reduce provides a natural framework for parallel computing, which is able to speed up the data processing and guarantee the scalability. Non-overlapping community detection, just as its name implies, is in charge of uncovering the hidden community structures in which there are no overlaps. In this section, we will briefly review the relevant studies.

2.1   Multilevel k-Way Graph Partitioning

A network can be mathematically represented as a graph $G = (V, E)$, where V contains all the nodes and E is the edge-set. Graph partition [7] problem is defined as partitioning a graph into smaller components with specific properties. For instance, k-way partition strategy divides the node-set into k smaller components or sets. For the purpose of non-overlapping community detection, a good partitioning method always tends to walk on the path between communities and avoid going across communities to destroy the structures. Take Fig.1 as example, excellent partitioning methods will prefer to cut off the red-color edges between the communities (inter-community links) rather than cut off the black-color edges within the communities (intra-community links).

Uniform graph partition aims to divid a graph into approximately equal sized components and there are few connections between the components. However, graph partition problems belong to the category of NP-hard problems. Solutions to these problems are generally using heuristics and approximation algorithms, such as spectral partitioning, geometric partitioning and multilevel graph partitioning method. Among the three types of methods, multilevel graph partitioning method is able to produces high quality partitions within a short period of time [8].

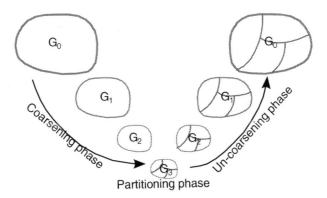

Fig. 2. Schematic diagram of multilevel k-way partitioning method

Based on the multilevel paradigm, Karypis and Kumar presented a *k*-way graph partitioning method, whose complexity is linear on the number of nodes in the graph [9]. More than that, it is able to tackle power-law networks. Briefly, graph partitioning process with this method is similar to the process of collapsing and expanding the directory tree, it consists of three phases which is shown in Fig.2: *Coarsening phase, Partitioning phase* and *Un-coarsening phase*. The process works as follows:

Coarsening phase: The original graph G_0 is transformed into several smaller graphs by merging adjoining nodes together, such as $G_1, G_2, ..., G_t$, and $|V_0| > |V_1| > ... > |V_t|$. The smallest graph G_t generated at the end of this phase only contains hundreds of nodes, and each node is marked with weight representing how many nodes are there.

Partitioning phase: Take G_t as a weighted network with only hundreds of nodes, and some simple clustering methods, such as Kernighan-Lin [10] and Fiduccia- Mattheyses [11], are able to divide it into *k* approximately equal partitions. In order to ensure the balance among the partitions and the integrity of the community structures, generally, the size of the partitions is allowed to fluctuate within a certain range. The rang is controlled by a unbalance coefficient - ξ, and for any final partition, it should meet the condition: $1 - \xi \le \frac{|P_i|}{|V_0|/k} \le 1 + \xi, 1 \le i \le k$, where $|P_i|$ means the number of nodes in partition P_i and $|V_0|$ represents the total number of nodes in the original graph.

Un-coarsening phase: Mapping the partitions in the smaller graph onto the larger graph to get the partitions in the larger graph, it reverses the coarsening phase. At last, we will get the final partitions on G_0.

2.2   Map-Reduce Programming Interface

Map-reduce adopts a flexible computation model with a simple interface consisting of "map" and "reduce" functions which can be customized by application developers. Hadoop is a java based open source implementation of the Map-reduce. The architecture of Map-reduce programming interface provided by Hadoop is shown in Fig.3.

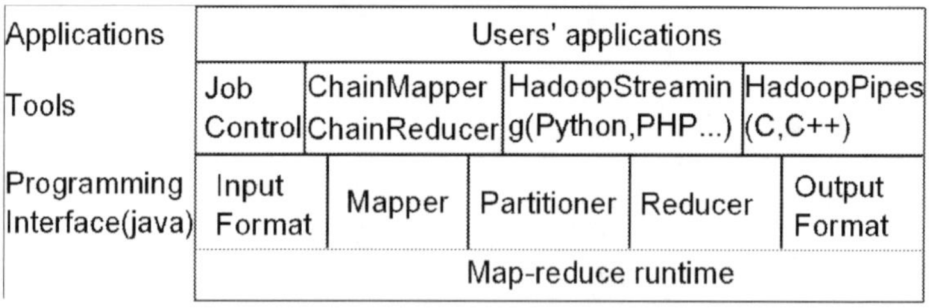

Applications	Users' applications			
Tools	Job Control	ChainMapper ChainReducer	HadoopStreaming(Python,PHP...)	HadoopPipes (C,C++)
Programming Interface(java)	Input Format	Mapper	Partitioner	Reducer	Output Format
	Map-reduce runtime			

Fig. 3. Architecture of Map-reduce programming interface

The entire programming model located between Map-reduce runtime and users' applications can be divided into two layers. The first layer is basic Java-based API and consists of five fundamental programmable components. The second layer is a tool layer offering at least four programming toolkit, it is designed mainly for the convenience of users to write complex Map-reduce applications and to increase platform compatibility for other programming language, such as PHP, C++, C and Python.

JobControl provides a method for users to write complex applications consisting of several dependent jobs and these jobs often constitute a directed graph.

ChainMapper/ChainReducer makes it convenient for users to write chain jobs, namely there are several Mappers during the *map* phase or *reduce* phase in the form below: [Mapper + Reducer Mapper*].

Hadoop Streaming allows users to specify an executable file or script as a Mapper or Reducer when using non-Java language.

Hadoop Pipes is a package specially designed for C/C++ programmers to write Map-reduce applications.

2.3 Non-overlapping Community Detection Algorithms

During the past decade, hundreds of community detection themed papers have been published and several surveys have been conducted in [12-15]. Technologies and requirements are complementary and mutually reinforcing, and both have time brand mark. Earlier methods are carried out to deal with the smaller networks with simpler structures and have reference value for subsequent methods. Generally, early methods are with high computational complexity and low accuracy. So, in this section, we only talk about some excellent non-overlapping community detection methods carried out in recent years and related to our work.

OSLOM (Order Statistics Local Optimization Method) [16] is a clustering algorithm designed and implemented by Lancichinetti et al in 2011. It is based on the local optimization of a fitness function expressing the statistical significance of clusters with respect to random fluctuations, which is estimated with tools of Extreme and Order Statistics. Statistical significance is defined as the probability of finding the cluster in a random null model [17], i. e. in a class of graphs without community structure. In OSLOM, they use the configuration model [18] as null model.

In 2007, Etienne Lefebvre developed a greedy optimization method to detect community structure. Then the method was improved and tested by Vincent Blondel, Jean-Loup Guillaume and Renaud Lambiotte. The method is now called Louvain method [19]. It attempts to optimize the "modularity", Q, of a partition of the network and is performed in two steps. In the first step, firstly, each node is considered as a community, then for a node, it looks in its neighbors to optimize modularity Q by removing its neighbor out from its community and by placing its neighbor in the community. In the second step, it aggregates nodes belonging to the same community and builds a new network whose nodes are the communities. These steps will be repeated iteratively until a maximum of modularity is gained and a hierarchy of communities is produced. The exact computational complexity of the method is not known, but the method seems to run in time $O(n\log n)$, and most of the computational effort is used in the first step.

InfoH (Infohiermap) [20] is proposed by Rosvall and Bergstrom in 2011 based on their previous work – Infomap [21] in 2008. InfoH uses multilevel coding strategy to encode every level. The lowest level is the nodes, next is community level consisting of nodes, and higher level is super-community level consists of communities. Every level is encoded by information coding strategy. It takes the minimal description length of information flow in the network as the target and transforms the problem of community detection into finding optimally coding schema problem. The process can be carried out with greedy search strategy and simulated annealing strategy.

3   Parallel and Scalable Non-overlapping Community Detection Framework

In this section, we will talk about the parallel and scalable framework for existing excellent non-overlapping community detection methods, which is shown in Fig.4.

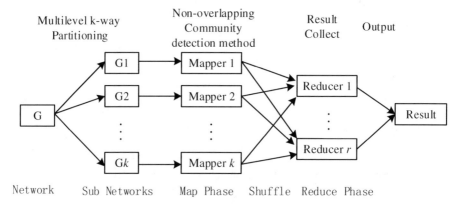

Fig. 4. Parallel and scalable non-overlapping community detection framework

In the framework, multilevel k-way partitioning method is used to divide network G into sub networks and all sub networks are stored in the HDFS. Then each mapper, a recreation of any excellent non-overlapping community detection method with Hadoop streaming, will take a sub network as an input split, read the key/value from the split into memory and carry out calculation on the data. Then the results (the community structures on all sub networks) from the mappers will be sent to reducers as input data, the reducers combine all the results together and write them into the HDFS. As we have introduced the multilevel k-way partitioning method in Sect.2, here we just talk about the data processing in Hadoop streaming, which is shown in Fig.5.

The components surrounded by the red dashed line are provided by Map-reduce and Hadoop Streaming. "Key/Value parse" is responsible for reading data from data split and translating it into key/value format and push it to the "mapper wrapper". The "mapper wrapper" is used to push data into the input stream "STDIN", "converter" reads the key/value data, converts it into the corresponding format required by the non-overlapping community detection method in "mapper1" and constructs a sub network in the memory, which is used as input for "mapper1". Here, we use "mapper1" and "reducer1" to differentiate them from the traditional "mapper" and "reducer. "Mapper1" is the community detection method which can be an executable file or script in any programming language, it is responsible for discovering the community structures in the sub network and outputting the result in the key/value format. The results from all the mappers will be pushed to the "reducer1" via the "shuffle". "Reducer1" allows users to customize what to output and in what format, it combines the results and write them into the output stream "STDOUT", and then the results flow towards the "Output Collector" and finally written into the HDFS.

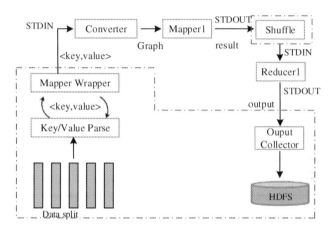

Fig. 5. Detail information of data processing with Hadoop Streaming

In the framework, we have implemented several kinds of "converter", such as Pajek, LFR and GML format converter, and some reducer1 models for users to customize. In the mapper1, the only thing users should do is to convert the output of the community detection method into the key/value format. So, theoretically, with our framework, users are able to parallelize any non-overlapping community detection method without understanding how it is designed and implemented but just should know how to read data from the STDIN and write data to the STDOUT with the corresponding programming language.

4   Experiments and Analysis

4.1   Experimental Environment and Data Sets

All the experiments in this paper are running on a Hadoop-1.1.1 cluster of Hunan Antivision Software Ltd. The cluster consists of 20 PowerEdge R320 servers (Intel Xeon CPU E5-1410 @2.80GHz, memory 8GB) with 64-bit NeoKylin Linux OS, servers are connected by a Cisco 3750G-48TS-S switch. We will test the accuracy and the speedup ratio of the framework with the methods mentioned in Sect. 2.3.

Table 1. Properties of four LFR generated networks used in the experiments

Data set	n	avg(d)	max(d)	γ	β	u	Community size
D1	200,000	35	100	2	1	0.3, 0.45, 0.60, 0.75	[30, 120]
D2	200,000	35	100	2	2	0.3, 0.45, 0.60, 0.75	[30, 120]
D3	200,000	35	100	3	1	0.3, 0.45, 0.60, 0.75	[30, 120]
D4	200,000	35	100	3	2	0.3, 0.45, 0.60, 0.75	[30, 120]

We use LFR [22] benchmark to generate several networks as data sets. With LFR, users are able to control some network properties: network size (n), degree distribution (γ, max(d)), community structure (β, u). γ is an exponent for the degree distribution ranging between [2, 3], max(d) means the maximal degree and β is an exponent for the community size distribution ranging between [1, 2]. u means the percentage of edges of a node connected with nodes in different communities, ranging between [0, 1]. NMI [23] is used to evaluate the accuracy of non-overlapping community detection methods, it is shown as below:

$$NMI(X;Y) = \frac{I(X;Y)}{\sqrt{H(X)H(Y)}} \tag{1}$$

Where $I(X;Y)$ means the mutual information between the random variable X and Y, and $H(X)$ and $H(Y)$ represent the entropy of X and Y, respectively. NMI calculation needs a lot of memory and CPU time, so we just set n=200,000 in our date sets.

4.2   Accuracy Test of the Framework

In this section, we first test the edge cut off ratio by the multilevel k-way partitioning method and then test accuracy of 3 excellent non-overlapping community detection methods parallelized by us with our framework.

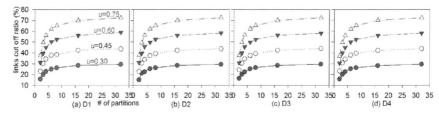

Fig. 6. Links cut off ratio during the partitioning process

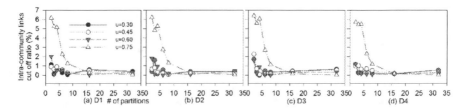

Fig. 7. Intra-community links cut off ratio during the partitioning process

From Fig.6, we find that the number of links cut off by multilevel k-way partition-ing method is positively related to the number of partitions, and the percentage of links cut off will get close to the parameter u with the increase of partition number. For a certain partition number, the larger the u is, the higher percentage of links will be cut off.

From Fig.7, we find that only few of intra-community links are cut off the parti-tioning method, which means that only few of the community structures are destroyed during the partitioning process, and when increase the partition number, the intra-community links cut off ratio will decrease. And only very large u is able to result in higher loss of intra-community links, such as $u=0.75$, but in this case, it means that for an arbitrary node, most of its neighbors are not in the same community with is, which is obviously meaningless for community detection. And besides, by contracting all the results in Fig.6 and Fig.7, we can infer that the performance of multilevel k-way parti-tioning method is fairly stable.

Next is the accuracy test of the framework. In the test, we compare the parallelized version of methods corresponding to the cases # >1, where # means partition number, with the standalone version (the case # =1). All the results are shown in Fig.8-Fig.10.

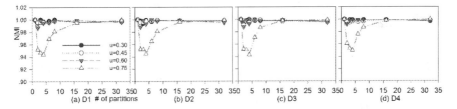

Fig. 8. NMI value of parallelized InfoH on different data sets

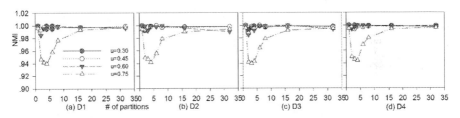

Fig. 9. NMI value of parallelized Louvain on different data sets

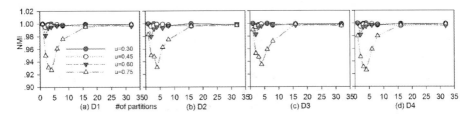

Fig. 10. NMI value of parallelized OSLOM on different data sets

From the results, we know that when the partition number is very small, such as #
=2, 3, 4, the accuracy will be not as good as when the number is larger. And when the
is small, the accuracy will decrease as u increase. But the accuracy loss is still very
low, no more than 2% when $u < 0.75$ and no more than 8% when u grows up to 0.75.
When # grows larger, the accuracy will be very close to that of standalone version.
Comprehensive analysis of Fig.7-Fig.10, we can get the following conclusions: intra-
community links cut off ratio plays a decisive influence on the effectiveness of parallel community detection and multilevel k-way partitioning method is excellent for
partitioning community structured networks.

4.3   Scalability Test of the Framework

In this section, we will test the scalability and speedup ratio of the framework with
InfoH method and Louvain method. The results are shown in Fig.11 and Fig.12.

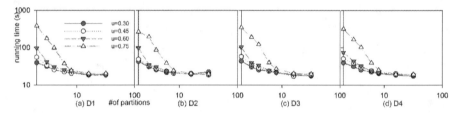

Fig. 11. Running time of parallelized InfoH on different data sets

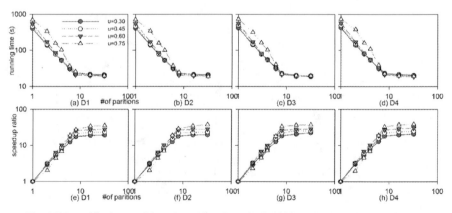

Fig. 12. Running time and speedup ratio of parallelized Louvain on different data sets

From the results, we can find that the running time of the parallelized method decreases linearly and the speedup ratio increases linearly with increasing partition
number before the partition number reaches such an "inflection point". Inflection
point comes out with the "long tail effect" for the following reason: 1) The total running time consists of at least two parts, the community detection running time and the
Map-reduce time including initialization time, shuffle time and output time. 2) Initialization time of Map-reduce is about a constant and shuffle time and output time are

concerned with the data size. 3) When # is small the community detection process will take much higher percentage of the total running time and the Map-reduce time can be negligible. But with the increase of #, the percentage of community detection running time will decrease which is contrary to the situation of Map-reduce time. And we can speculate that the inflection point will move to the right as the network size increases.

From all the experiments conducted in this section, we see that our framework is able to significantly accelerate the community mining process, and will not cause excessively performance degradation.

5 Conclusion and Future Work

Community detection has become a research hotspot in social network analysis and complex network research. But with the advent of the era of big data, network size expands rapidly and network structures become more complicated. Traditional excellent methods are designed for single-core CPU, and performance of single-core CPU is not likely to get massive upgrade in the next period of time, so all of them will face the scalability issues. In this paper, we design and implement a framework for non-overlapping community detection method. With the framework, we can easily achieve parallelization of existing excellent non-overlapping community detection algorithms without the knowledge of core logic of the algorithms. Experiments show that the framework is able to approximated linearly accelerate the community discovery process and without too high accuracy loss.

During the experiments, we find that Hadoop adopts a sorting-based data aggregation strategy during the shuffle phase which is non-essential for our work but takes up a lot of time. However, the strategy cannot be customized by users. But we'd like to do something to improve it, next.

Acknowledgements. We'd like to express our sincere gratitude to Mr. Dong Xicheng from Hulu.com and Dr. Lancichinetti A. from Amaral Lab, Northwestern University for providing technical support. Besides, this work was supported in part by the National High-Tech Research and Development Program of China (No.2012AA012600), Key Technologies Research and Development Program of China (No.2012BAH38B04, 2012BAH38B06), and National Natural Science Foundation of China (No. 60933005, 61202362).

References

1. Girvan, M., Newman, M.E.: Community structure in social and biological networks. Proceedings of the National Academy of Sciences 99(12), 7821–7826 (2002)
2. Watts, D.J., Strogatz, S.H.: Collective dynamics of 'small-world' networks. Nature 393(6684), 440–442 (1998)
3. Boccaletti, S., Latora, V., Moreno, Y., Chavez, M., Hwang, D.U.: Complex networks: Structure and dynamics. Physics Reports 424(4), 175–308 (2006)

4. Johnson, S.C.: Hierarchical clustering schemes. Psychometrika 32(3), 241–254 (1967)
5. Cooper, S.: The largest social networks in the world include some big surprises [Business Insider] (January 8, 2014), http://www.businessinsider.com/the-largest-social-networks-in-the-world-2013-12 (retrieved from)
6. Dean, J., Ghemawat, S.: MapReduce: simplified data processing on large clusters. Communications of the ACM 51(1), 107–113 (2008)
7. Feder, T., Hell, P., Klein, S., Motwani, R.: Complexity of graph partition problems. In: Proceedings of the Thirty-First Annual ACM Symposium on Theory of Computing, pp. 464–472. ACM (1999)
8. Karypis, G., Kumar, V.: A Coarse-Grain Parallel Formulation of Multilevel k-way Graph Partitioning Algorithm. In: PPSC (1997)
9. Karypis, G., Kumar, V.: Multilevel k-way hypergraph partitioning. VLSI Design 11(3), 285–300 (2000)
10. Kernighan, B.W., Lin, S.: An efficient heuristic procedure for partitioning graphs. Bell System Technical Journal 49(2), 291–307 (1970)
11. Fiduccia, C.M., Mattheyses, R.M.: A linear-time heuristic for improving network partitions. In: 19th Conference on Design Automation, pp. 175–181. IEEE (1982)
12. Fortunato, S.: Community detection in graphs. Physics Reports 486(3), 75–174 (2010)
13. Luo, Z.G., Ding, F., Jiang, X.Z., Shi, J.L.: New progress on community detection in complex networks. Journal of National University of Defense Technology 33(1), 47–52 (2011)
14. Porter, M.A., Onnela, J.P., Mucha, P.J.: Communities in networks. Notices of the AMS 56(9), 1082–1097 (2009)
15. Yang, B., Liu, D., Liu, J.: Discovering Communities from Social Networks: Methodologies and Applications. In: Handbook for Social Network Technologies and Applications, pp. 331–346. Springer US (2010)
16. Lancichinetti, A., Radicchi, F., Ramasco, J.J., Fortunato, S.: Finding statistically significant communities in networks. PloS One 6(4), e18961 (2011)
17. Newman, M.E., Girvan, M.: Finding and evaluating community structure in networks. Physical Review E 69(2), 026113 (2004)
18. Molloy, M., Reed, B.: A critical point for random graphs with a given degree sequence. Random Structures & Algorithms 6(2-3), 161–180 (1995)
19. Blondel, V.D., Guillaume, J.L., Lambiotte, R., Lefebvre, E.: Fast unfolding of communities in large networks. Journal of Statistical Mechanics: Theory and Experiment 2008(10), P10008 (2008)
20. Rosvall, M., Bergstrom, C.T.: Multilevel compression of random walks on networks reveals hierarchical organization in large integrated systems. PloS One 6(4), e18209 (2011)
21. Rosvall, M., Bergstrom, C.T.: Maps of random walks on complex networks reveal community structure. Proceedings of the National Academy of Sciences 105(4), 1118–1123 (2008)
22. Lancichinetti, A., Fortunato, S., Radicchi, F.: Benchmark graphs for testing community detection algorithms. Physical Review E 78(4), 046110 (2008)
23. Strehl, A., Ghosh, J.: Cluster ensembles—a knowledge reuse framework for combining multiple partitions. The Journal of Machine Learning Research 3, 583–617 (2003)

Microblog Sentiment Analysis Model Based on Emoticons

Shaojie Pei, Lumin Zhang, and Aiping Li

School of Computer, National University of Defense Technology, Changsha, China
`chenyue24psj@gmail.com`

Abstract. Millions of users share their opinions in microblog every day, which makes sentiment analysis in microblog an important and practical issue in social networks. In this paper, the problem of public sentiment analysis, and the construction of emoticon networks model are solved using emoticons. Based on large-scale corpus, FP-growth algorithm combined with the semantic similarity is proposed to aggregate similar emoticons. The construction of emoticon networks model is based on Pointwise Mutual Information. And a microblog orientation analysis framework for both emoticon messages and non-emoticon messages is presented. Experimental results have shown that our approach works effectively for microblog sentiment analysis.

Keywords: Sentiment analysis, Emoticon networks, Opinion mining, Microblog, Large-scale multimedia.

1 Introduction

With the rapid development of social network, microblog has become a popular and valuable social media to analyze user's opinions. Sentiment analysis, also known as opinion mining, aims to analyze people's attitude, viewpoint and feelings to entities, such as products, services, organizations, events, topics, and their properties [1]. Due to its important application, many researches have been conducted on this area. Those previous work will be briefly review in section 2.

Microblog contains more information than just a bag of words, such as emoticons, videos, hashtags, etc. Particularly, emoticons, which have strong relation with users subjectivity and sentiments, are becoming more and more popular for users to directly express their feelings, emotions and moods. In this paper, we tackle the problem of microblog sentiment analysis by leveraging emoticons in SINA microblog which contains richer emoticons than in Twitter. Our main contributions are as follows.

1) Based on large-scale microblog corpus, an Emoticon Networks model is presented. FP-growth algorithm combined with the semantic similarity is applied to aggregate similar emoticons, and calculate the interdependency between two types of emotions based on Pointwise Mutual Information(PMI). So in the emoticon networks model, each vertex represents a type of emoticon, and the edge between two nodes is the PMI values.

W. Han et al. (Eds.): APWeb 2014 Workshops, LNCS 8710, pp. 127–135, 2014.

2) More efficient methods for sentiment analysis is developed systematically. Naive Bayes Classifier is used to build the relation between microblog space and emoticon networks, and the relevant messages frequency algorithm is used to map emoticon networks and sentiment space. For messages those have no emoticons, a composite function is used to analyze sentiments. For emoticon messages, the text and emoticons is processed respectively, and a smooth function is used to get the final sentiment of the messages.

3) We conduct an final empirical study using a real dataset of 9,837,316 messages from SINA microblog. The result clearly shows that emoticons could help to perform sentiment analysis and our methods are effective for both emoticon messages and non-emoticon messages.

The rest of this paper is organized as follows. Section 2 describes related work briefly. Section 3 details our emoticon networks model and methods. Empirical study results are shown in Section 4, followed by a short discussion.

2   Related Work

There have been a large number of research papers in sentiment analysis and the main research approaches are usually based on two aspects, supervised learning and unsupervised learning.

Supervised learning methods, such as Naive Bayes or Support Vector Machines, were firstly applied to sentiment analysis in [2], They use supervised learning to classify movie reviews into two classes, positive and negative, and showed that using unigrams as features in classification performed quite well with either Naive Bayes or SVM. In [3], an exploring research is conducted on sentiment analysis to Chinese traveler reviews by Support Vector Machine algorithm.

As for unsupervised learning, [4] proposed a unsupervised classification method for document by building emotional vocabulary list. [5] is a typical method of this technique. Given a review, the algorithm computes the average sentiment orientation (SO) of all phrases in the review based on point-wise mutual information measures, and classified the review as positive if the average SO is positive and negative otherwise. In addition, lexicon-based method, which uses a dictionary of sentiment words and phrases with their associated orientations and strength, and incorporates intensification and negation to compute a sentiment score for each document, is also widely used.

Some researches have been made to explore the effects of emotional signals on sentiment analysis. An emoticon-based sentiment analysis system called MoodLens for chinese tweets is erected [6], they employed the emoticon related tweets to train a naive Bayes classifier for sentiment classification. In [7], they presented an automatic method of collect negative and positive sentiments based on the emoticons in Twitter. A comprehensive data analysis study of the role of emoticons in sentence level sentiment classification is proposed in [8]. [9] proposed to study the problem of unsupervised sentiment analysis with emotional signals. They investigated that the signals can help sentiment analysis by providing a unified way to model two main categories of emotional signals, and further incorporated the signals into an unsupervised learning framework for sentiment analysis. Aoki's study [10] is most relevant with ours methods, which proposes an automatic method to generate emotional vector of emoticons using blog articles. The two methods, however, have some fundamental differences. Aoki's methods focused on detecting the associate relation between emoticons and emotional words in blog while we target at microblog sentiment analysis. Besides, emotion Networks model is proposed, which is quite different from the vector in their work.

3   Model and Methods

3.1   Problem Definition

$D = \{d_1, d_2, d_3, \cdots\}$ is the microblog stream where $d_i = \{w_1, w_2, \cdots\}$ is a microblog message, and w_j is the feature item after segmentation and removing stop words.

$M = < V, E >$ is Emoticon Networks graph comprising a set V of emoticons together with a set E of edges. For $\forall v_i \in V$ represents a type of emoticon, and contains many emoticons $< q_i^1, q_i^2, \cdots >$. For $\forall e_{ij} = (v_i, v_j) \in E$ means there is association relationship between v_i and v_j. Let $\tau(e_{ij})$ represents the weight of the edge e_{ij}.

Given a message d, let

$$\delta_d^{v_k} = \begin{cases} 1 \text{ if } d \text{ contains emoticon } v_k \\ 0 \text{ else} \end{cases}$$

So the message corpus containing emoticon v_k can be defined as

$$D_{v_k} = \{d | d \in D \wedge \delta_d^{v_k} = 1\}$$

$S = < s_1, \cdots, s_k, \cdots, s_m >$ is the sentiment vector where s_k represents a sentiment. our purpose is to give a message d, detect its sentiment s_k, such that $s_k = \arg\max_{s_i \in S} P(s_i | d)$ by leveraging emoticons. In order to fully implement our approach, we need to complete the following aspects, including how to build the emoticon networks M, how to model sentiment space S, and how to construct the functions from message space D to M and S.

3.2   Method Description

The framework of our approach is illustrated in figure 1. First, a Emotion Networks model is constructed based on large-scale corpus. Then the following mapping is used to perform sentiment analysis: $D \Rightarrow M \Rightarrow S$.

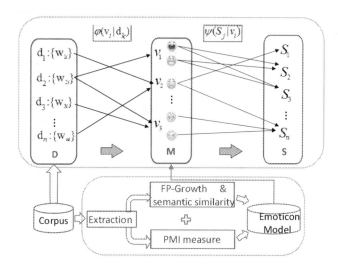

Fig. 1. The framework of our model

Emoticon Networks Model Construction. The key in this module is to determine the space of emoticons and the weight of edges in M. Each emoticon from large corpus is extracted as a vertex of M. Because there are a lot of similarities emoticons, we need to do a cluster for similar emoticons, and they are together as a vertex of M.

Here, we regard two emoticons q_1 and q_2 as similar emticons if they meets the following two characteristics: 1) they often appear together, which means that their co-occurrence reaches a certain frequent patterns; 2) they have strong correlation in semantic. Based on the large-scale corpus, we use FP-growth algorithm combining with Semantic similarity to cluster similar emoticons.

FP-growth algorithm is an efficient method to detect frequent patterns. It firstly counts occurrence of items in the dataset, and stores them to 'header table' and builds the FP-Tree structure. Please see [11] for a thorough review. According to lexical semantic similarity calculation method proposed reference[12], The semantic similarity of two emoticons q_1 and q_2 is defined as

$$SIM(q_1, q_2) = \sum_{i=1}^{4} \beta_i \prod_{j=1}^{i} Sim_j(q_1, q_2) \tag{1}$$

where $\beta_i (1 \leq i \leq 4)$ s an adjustable parameter, and $\beta_1 + \beta_2 + \beta_3 + \beta_4 = 1$, $\beta_1 \geq \beta_2 \geq \beta_3 \geq \beta_4$. $Sim_1(q_1, q_2) = \frac{\alpha}{d+\alpha}$, here, α is a variable parameter , d is path length of q_1 and q_2 in system of primitive ; $Sim_2(q_1, q_2)$ represents the similarity of the other primitives for q_1 and q_2; $Sim_3(q_1, q_2)$ represents the similarity of the correlative primitives for q_1 and q_2 ; $Sim_4(q_1, q_2)$ represents the similarity of the symbol primitives for q_1 and q_2.

As for the weight of edge $\tau(e_{ij})$, Mutual Information is used to measure the correlation of vertex v_i and v_j:

$$\begin{aligned}
\tau(e_{ij}) &= PMI(v_i, v_j) \\
&= \frac{1}{|v_i| * |v_j|} \sum_{k=1}^{|v_i|} \sum_{m=1}^{|v_j|} PMI(q_i^k, q_j^m) \\
&= \frac{1}{|v_i| * |v_j|} \sum_{k=1}^{|v_i|} \sum_{m=1}^{|v_j|} \log_2 \frac{p(q_i^k \& q_j^m)}{p(q_i^k)p(q_j^m)}
\end{aligned} \tag{2}$$

where $p(q_i^k)$ and $p(q_j^m)$ are the probabilities of emoticons q_i^k and q_j^m, respectively; and $p(q_i^k \& q_j^m)$ is the probability that q_i^k and q_j^m co-occur. $p(q_i^k)p(q_j^m)$ is a measure of the degree of statistical dependence between the words.

From Aug. 2013 to Feb. 2014, More than 60 million messages containing emoticons have been collected as our corpus from SINA microblog.And the algorithm of constructing emoticon networks model is shown in algorithm 1.

Algorithm 1. Emoticon Networks Model Construction

1: **Input**: message coupus D, minimum support threshold min_sup.
2: **Output**: Emoticon networks model M.
3: For $\forall d \in D$, extract the emoticons in d.
4: Calculate the frequent item set F, sort F according to item support descending to generate a list of frequent items L.
5: Rescan the corpus to construct FP-Tree, choose frequent 2-items of emoticons according to min_sup
6: Calculate the $SIM(q_i, q_j)$ in frequent 2-items using equation 1, and
7: **if** $SIM(q_i, q_j) > 0.5$ **then** aggregate q_i, q_j to the same category as one vertex
8: **end if**
9: **for** each $e_{ij} \in E$ **do** Calculate $\tau(e_{ij})$
10: **end for**

Sentiment Model. Here, we use sentiment vector model proposed in early work [13] combining the affective lexicon ontology proposed in [14] as our sentiment space. It contains 27466 affective words within seven categories, that is *joy, love, anger, sadness, fear, dislike* and *surprise*.

In addition, many new Internet words have appeared and used frequently, so methods in early work are used to collect new Internet words which can express user's sentiment, so that make rich our sentiment model. Depending to the idea which is n-gram segmentation, repeated mode detection and multi-platform cross validation[13], Finally, those words are classified into the existing seven categories, and got 23276 sentiment words. So the sentiment space S has seven elements, $S = \{s_1, s_2, \cdots, s_7\}$.

Sentiment Analysis. As shown in figure 1, function $\varphi : D \Rightarrow M$ is used to build the relation between microblog space D and emoticon networks model M, and $\psi : M \Rightarrow S$ to represent the mapping from M to sentiment space S. As one emoticon may hold different sentiments in different topics, sentiment analysis for each certain topic is presented here.

As for $\varphi : D \Rightarrow M$, given a message $d =< w_1, w_2, \cdots w_n >$, Naive Bayes Classifier is applied to determine the correlated emoticons v_i as

$$\varphi(v_j|d) = \arg \max_{v_j \in V} P(v_j, d)$$
$$= \arg \max_{v_j \in V} \{P(v_j) \prod_{i=1}^{n} P(w_i, v_j)\} \tag{3}$$

The prior probability $P(v_j)$ is obtained according to the prior estimation from the training corpus as

$$P(v_j) = \frac{message(v_j)}{\sum_{v_j \in V} message(v_j)} \tag{4}$$

where $message(v_j)$ represents the microbolgs belong to category v_j . The calculation method of the posterior probability $P(w_i, v_j)$ is shown in formula (5). here, $weight(w_i, v_j)$ represents frequency statistics of word w_i in category v_j.

$$P(w_i, v_j) = \frac{weight(w_i, v_j) + 1/N}{\sum_{i=1}^{n} weight(w_i, v_j) + 1} \quad (N = \sum_{j=1}^{|V|} \sum_{i=1}^{n} weight(w_i, v_j)) \tag{5}$$

As for $\psi : M \Rightarrow S$, the relevant messages frequency algorithm is applied to calculate the probability of $\psi(s_i|v_j)$ as

$$\psi(s_i|v_j) = P(s_i|v_j) = \frac{P(v_j, s_i)}{P(v_j)} = \frac{Num(v_j, s_i)}{Num(v_j)}$$

where $Num(v_j, s_i)$ represents the number of messages of v_j and s_i appear together, $Num(v_j)$ is the number of messages which contains v_j.

Different strategies of sentiment analysis is used for non-emoticon messages and emoticon messages.

+ **Non-emoticon messages**

Given a messages without emoticons, a composite function of $\Gamma = \varphi \circ \psi$ to applied to classify sentiments, and the equation is as follows:

$$\Phi(s_i|d) = \arg \max_{s_i \in S} \varphi \circ \psi = \arg \max_{s_i \in S} \sum_{j=1}^{k} \varphi(v_j|d) * \psi(s_i|v_j) \tag{6}$$

where k is the number of vertices in E.

+ **Emoticon messages**

When a message contains emoticons, it is divided into two parts: the text m_t and the emoticons m_e. The sentiment probability of text m_t is calculated in the formula 6. v_i is the vertex containing m_e. We choose all vertex set v_k meeting requirements $\tau(e_{ik}) > \theta$ where θ is a PMI threshold. So the sentiment probability of emoticons can be calculated by $\sum_k \psi(s_j|v_k)$. Then a parameter β is used to adjust the weights of text and emoticons as follows:

$$P(s_j|d) = \beta * \Gamma(s_j|m_t) + (1 - \beta) * \sum_k \psi(s_j|v_k) \tag{7}$$

4 Results and Discussions

4.1 Dataset

Our experimental platform based on SINA microblog. All messages are collected using SINA API. The corpus $setA$ contains 60,132,451 emoticons messages from October, 2013 to February, 2014 are to be used to construct emoticon networks. Then we have collected 9,837,316 messages from September, 2013 ,and those messages were divided the corpus $setB$ and the corpus $setC$. The corpus $setB$ contains 85% of emoticon messages, to be used to build the relation between microblog space D and emoticon networks model M and estimate the parameters of τ, The corpus $setC$ as test data contains the rest 15% emoticon messages and all non-emoticon messages.

Table 1. A small part of M

v_i	v_j	$\tau(e_{ij})$	v_i	v_j	$\tau(e_{ij})$
		3.74			2.11
		2.92			1.93
		3.78			2.71

4.2 Results and Discussions

We construct emoticon networks model based on the 60 million emoticon message and FP-growth algorithm combined with the semantic similarity. Set the minimum support threshold min_sup is 1.5%, we conducted algorithm 1, and finally 237 types of emoticons are obtained. And each pair of vertex is marked with $\tau(e_{ij})$. Table 1 illustrates a small part of the graph.

Based on the emoticon networks model, we analysed emoticon messages from September 2013. Figure 2 shows the frequency distribution of emoticons, and displays the proportion of emoticon messages by this month. Microblog users

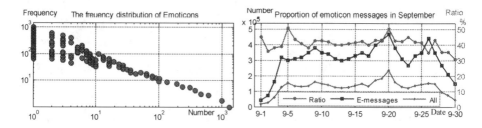

Fig. 2. The Emoticons analysis of September

usually use several frequent emoticons to express their feeling. However,the majority of emoticons are seldom applied, only part of emoticons was used. Emotions of statistic from the corpus, which the percentage of the message contained emoticons is approximate 43.2%, which indicates emoticons are very popular in microblog messages.

In this paper, Precision, Recall, F-value is applied to evaluate the result of our methods. The corpus $setB$ were selected as training data to determine the parameters of φ and τ. The corpus $setC$ were selected as testing data. As manually examining all the results is prohibitively expensive, we randomly chose 15000 messages, and manually judge whether the classification result was right or not. Table 2 illustrate the finally results compared with Naive Bayes model and Support Vector Machine model which directly map from microblog space D to sentiment space S.

Table 2. Precision of different models

Method	Precision(%)	Recall(%)	F-value(%)
Naive Bayes	65.92	69.37	67.60
SVM	73.43	75.84	74.62
no-emoticon messages	84.56	80.76	82.61
emoticon messages	89.81	83.39	86.48

As can be seen from Table 2, the precision of non-emoticon messages set is 84.56%, Recall and F-value reachs at 80.76%, 82.61% respectively. while that of emoticon messages is much higher, Precison, Recall, F-value run up to 89.81%, 83.39%, 86.48%. All these two value is higher than Naive Bayes model and SVM model.

5   Conclusions

In this paper, an emoticon network model for the microblog sentiment analysis is proposed for the microblog sentiment analysis. Based on large-scale corpus, FP-growth algorithm method combined with the semantic similarity is used to aggregate similar emoticons. With Pointwise Mutual Information, the emoticon networks model is presented naturally. The microblog orientation analysis framework proposed for both emoticon messages and non-emoticon messages improved the analysis accuracy. Experimental evaluations show that our approach could perform perfectly for microblog sentiment analysis.

Acknowledgments. This research is supported by National Basic Research Program (2013CB329601, 2013CB329602); National Key Technology R&D Program (No.2012BAH38B04, 2012BAH38B06).

References

1. Bing, L.: Sentiment analysis and opinion mining. Synthesis Lectures on Human Language Technologies 5(1), 1–167 (2012)
2. Pang, B., Lillian, L., Shivakumar, V.: Thumbs up?: sentiment classification using machine learning techniques. In: Proceedings of the ACL 2002 Conference on Empirical Methods in Natural Language Processing, pp. 79–86 (2002)
3. Wenying, Z., Qiang, Y.: Sentiment classification of Chinese traveler reviews by support vector machine algorithm. In: Third International Symposium on Intelligent Information Technology Application, pp. 335–338 (2009)
4. Li, T., Xiao, X., Xue, Q.: An unsupervised approach for sentiment classification. In: 2012 IEEE Symposium on Robotics and Applications (ISRA), pp. 638–640 (2012)
5. Turney, P.D.: Thumbs up or thumbs down?: semantic orientation applied to unsupervised classification of reviews. In: Proceedings of the 40th Annual Meeting on Association for Computational Linguistics, pp. 417–424 (2002)
6. Zhao, J., Dong, L., Wu, J., Xu, K.: Moodlens: an emoticon-based sentiment analysis system for chinese tweets. In: Proceedings of the 18th ACM SIGKDD International Conference on Knowledge Discovery and Data Mining, pp. 1528–1531 (2012)
7. Go, A., Huang, L., Bhayani, R.: Twitter entiment analysis. Entropy 17 (2009)
8. Min, M., Lee, T., Hsu, R.: Role of Emoticons in Sentence-Level Sentiment Classification. In: Sun, M., Zhang, M., Lin, D., Wang, H. (eds.) CCL and NLP-NABD 2013. LNCS, vol. 8202, pp. 203–213. Springer, Heidelberg (2013)
9. Hu, X., Tang, J., Gao, H., Liu, H.: Unsupervised sentiment analysis with emotional signals. In: Proceedings of the 22nd International Conference on World Wide Web. International World Wide Web Conferences Steering Committee, pp. 607–618 (2013)
10. Aoki, S., Uchida, O.: An automatic method to generate the emotional vectors of emoticons using blog articles
11. Han, J., Pei, J., Yin, Y.: Mining frequent patterns without candidate generation. In: ACM SIGMOD Record, pp. 1–12 (2000)
12. Qun, L., Sujian, L.: Word Similarity Computing Based on How-net. International Journal of Computational Linguistics & Chinese Language Processing 7(2), 59–76 (2002)
13. Zhang, L., Jia, Y., Zhou, B., Han, Y.: Microblogging sentiment analysis using emotional vector. In: Proceedings of the Second International Conference on Cloud and Green Computing (CGC), pp. 430–433. IEEE (2012)
14. Xu, L., Lin, H., Pan, Y., Ren, H., Chen, J.: Constructing the affective lexicon ontology. Journal of The China Society For Scientific and Technical Information 27(2), 180–185 (2008)

Detect and Analyze Flu Outlier Events via Social Network

Quanquan Fu[1], Changjun Hu[1], Wenwen Xu[1], Xiao He[1], and Tieshan Zhang[2]

[1] School of Computer & Communication Engineering,
University of Science & Technology Beijing, Beijing, China
[2] Information Office, China-Japan Friendship Hospital, Beijing, China
`fuquanquan06@sina.com, huchangjun@ies.ustb.edu.cn,`
`{xuwenwenustb,tie_shan}@163.com, hexiao83@gmail.com`

Abstract. The popularity of social networks provides a new way for constant surveillance of unusual events related to a certain disease. Some researchers have begun to use twitter to estimate the situation of public health, as well as predict disease trends. However, previous studies usually focused on the infection data but not the data judged as non-infection, which was usually filtered directly in their studies. We believe that the non-infection data is also essential for monitoring disease activity, because of their inherently subtle connections. Firstly, we construct a time series outlier model that can detect flu outlier events of different region in China with high precision and good recall by mining all the flu related data. Secondly, those outlier events are used to find out hot topics by SN-TDT and use the twice iteration classification method which is designed to analyze users' status who published a flu-related weibo. These results could provide science reference for deploying sickness prevention resources, and make recommendation about which place pose a high risk of getting infected.

Keywords: weibo, outlier events, time series, twice iterate classification.

1 Introduction

Sina Weibo is a social network platform for broadcasting real-time brief information. Having wide user groups range makes it a new way to monitor disease activity. By the end of 2013, Sina Weibo has 281 million registered users, and nearly up to 33% in all Chinese internet users. Among them more than 69.7% people use weibo mobile client, which make it has good real-time features than the traditional way reported by disease prevention and control institutions. Superior to other platforms, such as the search engines, each item weibo contains 140 words brief content and metadata with a semi-structure, just like time and location information, by which we are able to get richer information than the number of infections. Recent work has demonstrated that micro-blogging data can be used to track levels of disease activity and public concern (Alessio Signorini 2010) [1], predict flu trends (Harshavardhan Achrekar 2011) [2], fine-grained predict the health of specific people (Adam Sadilek 2012) [3], and detect health conditions (Victor M. Prieto 2014) [4].

W. Han et al. (Eds.): APWeb 2014 Workshops, LNCS 8710, pp. 136–147, 2014.
© Springer International Publishing Switzerland 2014

So far, most of the research is focused on classification for the infection of weibo, ignoring the research on the non-infection. However, more than 80% are non-infection ones in flu related weibos, which contains a wealth of user information. In previous studies, non-infection weibos were filtered directly, only a very small part of the infection weibos were analyzed, which made it difficult to reflect the integrity of the flu event. A different kind of problem one can pose, however, is to combine with infection the non-infection weibos to analyze those complete flu activities.

This paper discusses the detection of abnormal flu related events and analyses the relationship between infection and non-infection weibos. Two conditions should be satisfied in detection of abnormal weibos events. Frist, the outlier detection algorithm must meet the real-time requirement, that is to say, outliers can be found immediately once they happens. Second, outlier detection algorithm should be able to find out both of the patterns outlier and single point outlier.

We combine both reachable neighbor outlier factor and time sequence outlier detection methods to analyze the quantity changes of flu related weibos in each province every day. Outlier patterns and point are detected in the time sequence. Since the attention differences on every event, there would be large changes in weibo quantity, which require that the detection algorithm should have good adaptability to make continuous dynamic adjustment. Then we analyze a certain number of weibos within the time scope of the outlier. The SN-TDT model is adopted based on the short text characteristic of social network. The cluster analysis is made to find out the topic of the outlier event and mining users' focus. In this way, flu related weibos are divided into four stages, which include caring about the news, taking precautions, anxious about illness and finally infection. For this purpose, we designed twice iterative classifier to process and separate weibos.

Analysis and detection of outlier events have practical significance. The research results can provide reference for government grasping public opinions, and correctly guide the dissemination of information, avoid panic due to information asymmetry. According to user status, when the attentions of prevention measures become high in some place, government can increase the deployment of medical resources and promote the correct disease prevention knowledge. What's more, advice could provide to citizen that where is the high risk area.

This paper is organized as follows. Section 2 introduces related works and section 3 describes the study method which mainly consists of three parts. The first part describes outlier detect model. The sliding window-reachable neighbor detection method is introduced in this part. The second part introduces the SN-TDT algorithm. The third part describes the steps how to classify flu related weibos into four categories of public concern states. Section 4 is the experiments and results. Finally, the conclusion and directions for future work are given in section 5.

2   Related Works

A number of studies have been conducted using different forms of social networks to monitoring and prediction hot social events. Like Takeshi Sakaki et al. (2010) [5] investigate the real-time interaction of earthquakes in Twitter. They consider each Twitter user as a sensor. When the user feels the earthquake occurrence, he may make a tweet to broadcast the fact. False-positive ratio is used to calculate the probability of earthquake occurrence when there are n items of weibo classified as positive. The alarm will be set when the probability exceed the threshold. The author records each user's longitude and latitude marked in weibos. Then Kalman filtering and particle filtering are applied to find the center and the trajectory of the event location.

The abnormal event detection is rarely used in flu detection. The outlier detection technique based on relative density and the outlier detection technique based on time sequence are two common outlier detection technologies. The outlier detection technique based on density believe that normal data is in high density area, while the abnormal point in the low density area. But if the data is in the variable density region, the technique based on relative density will greatly affected. In order to solve this problem, Markus M. Breunig et al. [6] defined LOF (Local Outlier Factor) as the ratio of average density of K nearest neighbor and the density of the data itself, Using LOF as the outlier degree. Zakia Ferdousi et al. [7] use Peer Group Analysis (PGA), which is an unsupervised technique for fraud detection. PGA characterize the expected pattern of behavior around the target sequence in terms of the behavior of similar objects, and then to detect any difference in evolution between the expected pattern and the target.

Classification systems have also been used to filter the original data to obtain better results. Adam Sadilek et al. [3] predicted disease transmission from geo-tagged Micro-Blog data. They worked on SVM classifier to classify health-related text messages and detecting illness-related messages such as flu, sick, headache, stomach etc.

3   Methods

3.1   Detect Flu Related Weibos' Outlier

In order to detect abnormal events, we collect flu weibo data every day continuously. The number of flu weibos is analyzed in the stage of outlier detection. We think that the total number of flu weibo should change smoothly and continuously if there not any outlier happened. The sudden appearance of the point deviation and mode change can be judged as a flu outlier. We collected flu related weibo number every day from March 11th, 2013 to May 31th, as shown in Figure 1.

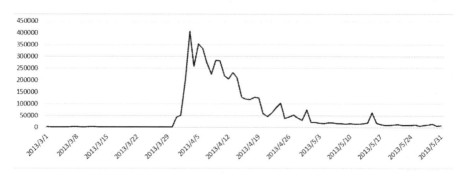

Fig. 1. The quantity of flu related weibos

We can see that the rapid growth of the flu weibo in April in figure 1. It leads to the number of weibo in March cannot be measures by the same magnitude of that in April. This requires outlier detection algorithm should aim at dealing with local data, and have the ability to adapt to new data dynamically. The sooner we discover the outliers, the early corresponding measures could be implemented. All of that make the algorithm should be in real time, the result could be found immediately once the outlier happens.

Take the daily amount of flu weibo data of every provinces and the record time as an ordered set $X = \langle x_1=(t_1,v_1), x_2=(t_2,v_2),\dots x_n=(t_n,v_n) \rangle$, the recording time is strictly increasing ($i < j \Leftrightarrow t_i < t_j$). The number of flu weibo series drawn a series of discrete points in a daily interval, and a straight line can be got by connecting the two adjacent points, which makes a time series. Sub pattern can be divided by determining the edge points using the slope in analytic geometry. Method for judging whether a day is an edge point is:

Calculate the slope k_1 by the number of weibo x_i in day i and the number of weibo x_{i-1} in day i-1, and slope k_2 by the number of weibo x_{i-1} in day i-1 and the number of weibo x_{i-2} in day i-2. If $|k_1 - k_2| > maxSlope$, it can be judged as the edge point of a sub model p. Use its length, intercept and mean value as characteristics of this sub model. In this way, the pattern density of time series can be converted into the density of data objects in data set D. When pattern length is set to 1, pattern outlier becomes single point outlier.

The density of p is defined as following:

Definition 1: *k-distance* of an object p

For any positive integer k, the k-distance of object p, denoted as k-dist(p), is defined as the distance d(p, o) between p and an object o $\in$ D such that:

 (1) For at least k objects o' $\in$ D\{p}, it holds that d(p, o') $\leq$ d(p, o) and

 (2) For at most k-1 objects o' $\in$ D\{p}, it holds that d(p, o') < d(p, o).

 Definition 2: k-distance neighborhood of an object p:

 $Nk(p) = \{q \in$ D\{p} | d(p, q) $\leq$ k-dist(p)\}

 Definition 3: k- reachable neighbor of p:

$\forall p \in$ D, if q $\in$ D\{p}, there is p $\in$ Nk(q), we call Nk(q) is one of p's k- reachable neighbor about k-dist(p), marked as RNK(p). It is can be seen that a smaller |RNK(p)| stands for a smaller chance the object p in other objects' reachable neighbor. Otherwise, object p is in an intensive position.

Definition 4: the density of an object p:

$$RNrk\ (p) = \frac{1}{|N_k(p)|} \sum_{o\ \in N_p(p)} \frac{|RN_k(p)|}{|RN_o(p)|}$$

RNrk (p) is the average ratio of the k- reachable neighbor of object p and others which reflect the local density of object p.

Definition 5 the k-reachable neighbor outlier factor of an object p:

 $RNOFk\ (p) = \max\ \{1\text{-}RNrk(p),\ 0\};$

The bigger RNOF, the more likely p is an outlier mode.

Outlier detection of flu activity requires an online, real-time-alarming algorithm. We introduce the sliding fixed size window model. When the number of object is greater than the window length, each insertion of a new object means the deletion of an old object from the window tail.

Flu outlier event monitoring definition and algorithm are given as following in table 1:

Time series = $\langle v_1, v_2 \dots v_m \rangle$, edge point $= \{X_{i_1}, X_{i_2} \dots X_{i_n}\}$. Sub pattern p = L $(X_{i_j}, X_{i_{j+1}})$. Use the sub model to represent the time series = $\langle L(X_{i_1}, X_{i_2}),$ $L(X_{i_2}, X_{i_3}), \dots L(X_{i_{n-1}}, X_{i_n}) \rangle$

Table 1. flu outlier events detection algorithm

---

while((v_i = Time series data) is valid){

if(v_i is the edge point){

new pattern $p = L(X_{i_j}, X_{i_{j+1}})$

if(time window is full)

Delete the old object from the tail of the queue, insert the new object in the header.

else

Insert a new object

for(each object p in window W){

calculate the k-dist and k-distance neighborhood of p

make every p and k- reachable neighbor of p in $N_k(p)$ increased by 1.

}

for(each object p in window W)

for(each object in $N_k(p)$){

calculate the outlier factor $RNOF_k$ of p

arrange $RNOF_k$ in ascending order

if(p $\in$ ranking of the k top   && slope of p > 0)

judge p as outlier

}

}

the next point in the time sequence

}

3.2   Short Text Topic Detection and Tracking

After the outliers in flu time series has been detected, we extract hot topics to make further study about the content of those events. For that purpose, we capture most the weibos in the days which are judged as a sub pattern outlier.

When detecting the outliers, we only give alarms if the slope of an event is greater than 0. The reason lies in that people's attention on an event usually has the properties of mutual exclusion. When someone pays attention to something in the near period of time, his attention on the other thing will drop. So when the quantity of flu weibos drops sharply, there may be due to other events, affecting the users' attention on flu. Because we can't collect all the information in Weibo platform, neither can we judge what causes the drop, so we do not discuss when the slope of an event is smaller than 0.

The topic detection and track algorithm based on the characteristics of social network, short for SN-TDT, proposed by Liubao Yu is adopted to extract top N topic flu events.

SN-TDT algorithm can be described as follows:

Step 1. Read the new text, if there is no new text, jump to step 7, else to step 2.

Step 2. Mark the feature words of this text.

Step 3. Judging the correlation degree between text and topics, if there are not any existing topic associated with the text, in other words, the correlation degree between the topic and text is lower than a given threshold, jump to step 4, otherwise to step 5.

Step 4. Create a new topic, add the text to the newly created topic, go to step 6.

Step 5. Select the topic most associated with the text, and add the text into the topic.

Step 6. Adjust each attribute of the topic, and jump to step 1.

Step 7. Sort all the topics and the top N are the hot topics.

3.3   Capture Users' Different States

Most of flu related weibos are non-infection. If people are not sick, it is what their mood would be when they send these weibos that is focused in our next research.

Compared with other ways such as search engine, weibo has a wealth of contextual information. Natural language processing is used to analyze users' different states.

Most of non-infection weibos can be summarized as the following three types, news, preventive measures and anxiety. In this paper, raw weibo data searched by keyword "flu" are classified into four categories. Class A for related news, class B for precautions, class C for anxiety and class D stands for influenza infection activities. At class A flu begins to draw public attention, gradually upgraded to fear and panic at class C. The other weibos generally consists of the word like "潮流感" which takes a very small proportion of the flu weibos, so it can be ignored when designing the classifier. ICTCLAS is used in weibos word segmentation. The most of news and precautions weibos begin with "[" and "]", so we can simply filter out other punctuations during word segmentation.

Select the Feature Terms. In order to obtain better classified results, three different feature evaluation functions are used for scoring feature terms w to find the most representative characteristic of text category.

Information gain. A key measure of information gain is how much information feature terms can bring to the classification system [8]. The more information feature term brings, the more important it is. For a given term w, the difference between the prior entropy of text and posterior entropy is the information gain it brought to the system. The information gain of term "w" is:

$$IG(W) = H(C) - H(C \mid W)$$

$$= -\sum_{j=1}^{m} P(C_j) \log_2 P(C_j) + P(w) \sum_{j=1}^{m} p(C_j \mid w) \log_2 P(C_j \mid w)$$

$$+ P(\overline{w}) \sum_{j=1}^{m} p(C_j \mid \overline{w}) \log_2 p(C_j \mid \overline{w}) \tag{1}$$

C_j ranges from 1 to 4.

Chi-Square Statistic. Chi-square statistic (χ^2) quantify the importance of the feature terms by estimating the correlation between terms and classes [9]. The terms should be selected with the highest correlation.

$$\chi^2(c, w) = \frac{n(P(c,w)P(\bar{c},\bar{w}) - P(c,\bar{w})P(\bar{c},w))^2}{P(c)P(w)P(\bar{c})P(\bar{w})} \tag{2}$$

Taking the maximum value of class A to class D as the chi-square statistic result of the feature item "w".

Document frequency. Document frequency is the easiest method to select the feature item, which is defined as the frequency of feature item "w" appearing in the set D [10]. Set the minimum and maximum threshold, and calculate the document frequency of each feature term. If the term's frequency is greater than the maximum threshold or less than the minimum threshold, this feature item would be deleted, otherwise be retained.

With the clear distinction, the byte stream length is defined as one of the features to describe the weibos. The news category are usually released by Sina verified accounts called "big V", such as government agencies, the media, scholars and celebrities. And then those weibos will be forwarded by ordinary users in a large quantity. Because the influenza infection weibos usually released to express the illness of themselves or their close relatives, only the original weibo can be marked as class D when judging flu infections. And whether a weibo is released by Sina verify account or original account is taken as a feature item.

Calculate the Feature Weights. TFIDF is the most widely used weight calculation algorithm in text processing field [11]:

$$T(w) = f_i(w) \times \log \left(\frac{n}{n_k} + l \right) \tag{3}$$

In this expression, $f_i(w)$ is the number of times that term w occurs in a weibo, n_k stands for the number of weibos containing the term. Taking the impact of a weibo's length, we normal the weight to [0, 1]. And "l" takes the empirical value 0.01.

$$T(w) = \frac{f_i(w) \times \log \left(\frac{n}{n_k} + 0.01 \right)}{\sqrt{\sum_1^n (f_i^2(w) + \log^2 (\frac{n}{n_k} + 0.01))}} \tag{4}$$

Twice Iterative Classification. The open source data mining platform WEKA [12] is used to train microblogging classifier which is used for target classification. WEKA provides classifier based on different algorithms such as Navie Bayes algorithm, KNN algorithm, neural networks algorithm, SVM algorithm [13] and their improved algorithms. A variety of classification algorithms are investigated. When weibos are divided into four categories directly, the results are not satisfactory. To make sure weibos can be correctly classified we propose a twice iterative classification method.

Temporarily mark all class B and class C instances in dataset D1 as class A. Select the feature terms. Calculate the feature weights and trained classifier to classify D1. The data belonging to class A at first classification is marked as D2. Recovery the label of class A, class B, class C in D2 and label misclassified class D as class C. Reselect feature terms of D2. Calculate weights and twice iteration classify D2. The process is shown in Figure 2.

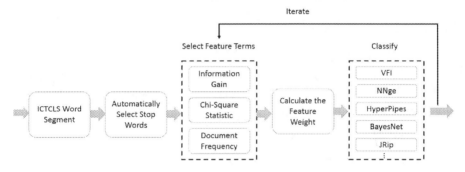

Fig. 2. Twice iterative classification process

4   Experiment and Result

In this section, we evaluate our approach by a number of experiments. The experimental data acquires 34 provinces and cities nationwide daily flu related weibos from March 1, 2013 to May 31, 2013. The sliding window-reachable neighbor algorithm proposed above is tested for anomaly detection with the flu weibos data in Beijing, Shanghai and Guangdong, and when the parameter k=3, the window length = 5 can get a better result. Take outlier threshold $\lambda = 0.5$, the range of outlier factor [0, 0.5]. Because couldn't be determine what happens when weibo data drops, the sub model will be not shown on the diagram if its slope is less than 0.

We can see that using sliding window – reachable neighbor algorithm can basically detect outlier events in Beijing, Shanghai and Guangdong, but at the beginning of the time sequence, the algorithm performance was not very good. In the time series of Beijing form March 3rd to March 5th, flu weibo data increases rapidly, but the detection algorithm fails to find it out. When the data window is full, the algorithm shows a high rate of accuracy and good recall with basically no errors and omissions. The algorithm is designed for outlier patterns in sliding window, therefore the outlier factors are equal within the same sub pattern, and that is to say, we care about both the growth process of outliers in sub models and the peak value.

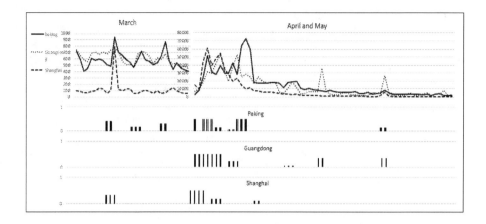

Fig. 3. Experiment of weibo outlier events detection

About 180000 (110M) flu outlier weibos are detected, and Fig. 4 shows the mined hot topics by the SN-TDT algorithm, which includes provincial and topic content.

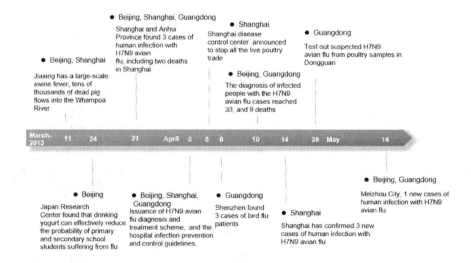

Fig. 4. SN-TDT hot topic mining results

We can see the reason why flu weibos appears mass exceptions in April is that Shanghai and Anhui confirmed 3 cases of human infection with the H7N9 avian flu virus in March 31st, which is the first cases of H7N9 human infection in world, caused widespread concern. All the outlier events from April to May is about H7N9.

For monitoring flu user status, we use twice iterative classification method to classify flu related weibos of Shanghai in the March 31st. We randomly select 500 item of news, prevention, anxious and infection weibos to label as training set manually from March to May. And randomly select 2000 from flu weibos of

Shanghai labeled as a test set, including 921 news, 364 prevention, 662 anxious, and only 53 infection.

Feature words selected by three different feature evaluation functions are shown in Table 2.

Table 2. Feature items extracted by different methods

Methods	Feature Items
Information Gain	[,], disease, infection, avian flu, H7N9,diagnosis, reports, prevention, methods, attention, express, death, epidemic, use, type, not, institutions, city,
χ^2 test	disease, avian influenza,H7N9, diagnosis, prevention, death, infection, insist, epidemic, notification, report
Document Frequency	[,], H7N9, avian flu, report, attention , infection, virus, body , people, vaccines, anti, effective, drink, hospitals,

Multiple classification algorithm are tested to find a suitable flu weibos classification system. When using WEKA classification directly on the test set, the classification accuracy can be up to 71.4%. System accuracy increases to 79.64% with twice iterative classification. Table 3 compares the results of the various classification systems.

Table 3. TP rate, FP rate, precision, recall, F-Measure, ROC Area of each system

System	TP Rate	FP Rate	Precision	Recall	F-Measure
IG-VFI-VFI	0.7686	0.0418	0.7704	0.7686	0.7695
χ^2-VFI-VFI	0.8055	0.0341	0.7964	0.8145	0.8055
DF-VFI-VFI	0.6993	0.0396	0.7623	0.6993	0.7308
IG-NNge-HyperPipes	0.7137	0.0473	0.7452	0.7137	0.7295
χ^2-NNge-HyperPipes	0.7875	0.0352	0.7596	0.7515	0.7556
DF-NNge-HyperPipes	0.7326	0.055	0.738	0.7326	0.7353
IG-BayesNet-JPip	0.7686	0.0418	0.7695	0.7686	0.7691
χ^2-BayesNet-JRip	0.7281	0.055	0.7263	0.7263	0.7272
DF-BayesNet-JRip	0.729	0.055	0.7335	0.729	0.7313

Comparison of three characteristic evaluation functions, chi-square statistic classification accuracy is better than information gain and document frequency. Chi-square selects local terms. However, information gain and document frequency can only select the global terms. The system χ^2-VFI-VFI have the best performance.

In order to observe the classification results more intuitive, the map of national influenza attention in March 31st is drawn, and the flu weibos classification results in

Shanghai are shown in Figure 5. As can be seen from the diagram, although there is a high degree of attention on Shanghai, but the actual number of infections is not so much. People pay more attention to the news. The number of Sina weibos classified as infections in March 31st accounted for 3%, about 420 cases.

Even after filtration, the number of influenza infections is still larger than NHFPC measured (99 case in March). There are three possible reasons: first of all, weibos cover a wider range than NHFPC data collection agency. Everyone can publish weibos wherever they could access to the internet by PC or the mobile phones. Secondly, the system cannot guarantee completely correct classification. A certain number of non-infection flu weibos are divided into class D. Thirdly, the authenticity of each flu weibo also cannot be promised.

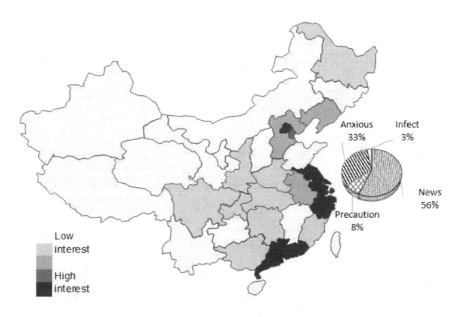

Fig. 5. The map of national flu attention in March 31st distributed in China

5   Conclusions and Future Work

This work focuses on both injection and non-injection flu weibos. By analyzing the number of flu weibos, we can detect outlier events of any province timely and effectively. Not only the point of unexpected events can be detected, outlier patterns caused by high continued attention can also be detected by using the sliding window - density detection method. Some of the detected event has a nationwide influence, taking March 31, 2013 as an example, the first H7N9 diagnosis lead to a surge in the number of flu weibos, but some only caused attentions in the local area. For instance March 24, 2013 the news "Japan research center found that drinking yogurt can effectively reduce the probability suffering from flu" attract users in Beijing attention.

According to the characteristics of social networks, the designed SN-TDT and twice iteration classification method can extract the hot topic and the analyze users' states accurately. In the future, we will study the flu weibos on temporal and spatial, and capture how the disease spread geographically.

References

1. Signorini, A., Segre, A.M., Polgreen, P.M.: The Use of Twitter to Track Levels of Disease Activity and Public Concern in the U.S. during the Influenza A H1N1 Pandemic. PLoS One 6(5), e19467 (1946), doi:10.1371/journal.pone.0019467
2. Achrekar, H., Gandhe, A., et al.: Predicting Flu Trends using Twitter Data. In: The First International Workshop on Cyber-Physical Networking Systems, pp. 702–707 (2011)
3. Sadilek, A., Kautz, H.: Vincent Silenzio: Predicting Disease Transmission from Geo-Tagged Micro-Blog Data. In: Proceedings of the Twenty-Sixth AAAI Conference on Artificial Intelligence, pp. 136–142 (2012)
4. Prieto, V.M., Matos, S., Alvarez, M., et al.: Twitter: A Good Place to Detect Health Conditions. PLoS One 9(1), e86191 (2014), doi:10.1371/journal.pone.0086191
5. Sakaki, T., Okazaki, M., Matsuo, Y.: Earthquake Shakes Twitter Users: Real-time Event Detection by Social Sensors. In: 19th International Conference on World Wide Web, pp. 851–860 (2010)
6. Breunig, M.M., Kriegel, H.-P., et al.: LOF: Identifying Density-Based Local Outliers. In: Proceedings of the 2000 ACM SIGMOD International Conference on Management of Data (2000)
7. Ferdousi, Z., Maeda, A.: Unsupervised Outlier Detection in Time Series Data. In: The 22nd International Conference on Data Engineering Workshops, 0-7695-2571-7 (2006)
8. Lee, C., Lee, G.G.: Information gain and divergence-based feature selection for machine learning-based text categorization. Information Processing & Management 42(1), 155–165 (2006)
9. Chen, Y.T., Chen, M.C.: Using chi-square statistics to measure similarities for text categorization. Expert Systems with Applications 38(4), 3085–3090 (2010)
10. Azam, N., Yao, J.: Comparison of term frequency and document frequency based feature selection metrics in text categorization. Expert Systems with Applications 39(5), 4760–4768 (2012)
11. Joachims, T.: A Probabilistic Analysis of the Rocchio Algorithm with TFIDF for Text Categorization, Carnegie-Mellon Univ. Pittsburgh Pa Dept of Computer Science. No. CMU-CS 96-118 (1996)
12. Hall, M., Frank, E., Holmes, G., Pfahringer, B., Reutemann, P., Witten, I.H.: The WEKA data mining software: an update. ACM SIGKDD Explorations Newsletter 11(1), 10–18 (2009)
13. Joachims, T.: Learning to classify text using support vector machines: Methods, theory and algorithms, p. 205. Kluwer Academic Publishers (2005)

An Advanced Spam Detection Technique Based on Self-adaptive Piecewise Hash Algorithm

Junxing Zhu and Aiping Li

School of Computer, National University of Defense Technology, Changsha, 410073, China
`daishanshen5036@sina.com`, `13017395458@163.com`

Abstract. Nowadays, email spam problems continue growing drastically and many spam detection algorithms have been developed at the same time. However, there are several shortcomings shared by most of these algorithms. In order to solve these shortcomings, we present an advanced spam detection technique(ASDT). It is based on the extremum characteristic theory, Rabin fingerprint algorithm, modified Bayesian method and optimization theory. Then we designed several experiments to evaluate ASDT's performance, including accuracy, speed and robustness, by comparing them with SFSPH, SFSPH-S, the famous DSC algorithm and the Email Remove-duplicate Algorithm Based on SHA-1(ERABS). Our extensive experiments demonstrated that ASDT has the best accuracy, speed and robustness on spam filtering.

Keywords: Spam Detection, Rabin Fingerprint, Bayesian method.

1   Introduction

Nowadays, Internet spam has been observed in many domains such as email, instant messaging, web pages, internet telephony, social network information, etc [1], [2]. Since almost all of the current spam filtering techniques can perform well when dealing with clumsy spam, which has duplicate content with suspicious keywords or is sent from an identical notorious server, the next stage of spam detection research should focus on coping with cunning spam which evolves continuously [3].

Generally speaking, email spam filtering techniques can be divided into semantic dependent techniques and semantic-free techniques. The semantic dependent techniques filter spam by analyzing the semantic information of the emails, and its typical cases are the keywords detection method [4] and the Latent Semantic Analysis (LSA) method [5]. Semantic-free techniques can't deal with semantics of the suspicious spam. The DSC technique, Hash digest matching technique, blacklisting technique and senders' features analysis technique [6] can be representatives of this group.

Semantic dependent techniques were among the first spam detection methods to be applied [6]. They rely on indicative keywords, or unusual distribution of punctuation marks and capital letters to identify spam from regular emails [7]. However, as the translation of email is based on MIME [8], sometimes we should waste time to retranslate the MIME contents into their original format before spam detection.

W. Han et al. (Eds.): APWeb 2014 Workshops, LNCS 8710, pp. 148–157, 2014.

Semantic-free techniques can deal with MIME content without retranslation. Among them, Hash digest matching techniques generate a corresponding fixed length hash value for each email, and filter spam by comparing it with the hash values in spam information database[9]. However, if even a single bit of the email content is changed, the generated hash value can be radically different. It means fixed-length hash methods can do nothing about cunning spam that evolves continuously. Nicholas Harbour developed a piecewise hash in 2002 [10], it divided the email contents into several fixed-size segments, and generates hash values for them, thus it can deal with evolved spam. But when there are several insertions or deletions in the original spam, many fixed-size segments behind these revisions can also be changed, that will reduce the accuracy of spam detection.

Document Syntactic Clustering (DSC) [11] view a document as a stream of tokens. It can only deal with the contents that can be segmented by its content structure. DSC-SS is the super method of DSC, it reduces consumption of the operation time, but it is powerless while dealing with short contents. I-Match overcomes the short-content reliability problem of DSC-SS and is very efficient to implement, but its robustness is very poor [12], [13]. A. Kołcz and his teammates present a randomization-based technique to increase its signature stability, but its effectiveness as a countermeasure to word substitutions is still smaller [14].

In this paper, we formed an advanced spam detection technique(ASDT) basing on piecewise hash algorithm to deal with the shortcomings of these methods we listed before. The rest of this paper is organized as follows. The next section describes the structure free self-adaptive piecewise hashing algorithm. Section 3 gives the spam identification algorithm, Section 4 presents and analyzes the experiments, and Section 5 concludes.

2   Structure Free Self-adaptive Piecewise Hashing Algorithm

Sun Ji-zhong and his workmates describes a kind of content splitting method that can split the email content into segments of variable length just by relying on the information of each content character in 2010 [15]. Here, we modified their method in order to split the email content into segments of variable length without relying on semantic or specific structure information and at the same time to make it quicker and better adapted to spam detection works.

Supposing that an email E's content is expressed as a stream of characters $<C_1, C_2, \cdots, C_n>$, C_k is the k-th character, function $Asc(C_k)$ will return the ASC-II binary value of C_k. So we can define the left consistent sequence of C_k as $seq^-(C_k)$, here:

$$\begin{cases} seq^-(C_k) = <C_{k-i}, C_{k-i+1}, \cdots, C_k> \\ Asc(C_{k-j}) \leq Asc(C_{k-j+1}), 0 < j \leq i \\ Asc(C_{k-i-1}) > Asc(C_{k-i}) \end{cases} \tag{1}$$

Then we can get the right consistent sequence of C_k as $seq^+(C_k)$, here :

$$\begin{cases} seq^+(C_k) =< C_k, C_{k+1}, \cdots, C_{k+l} > \\ Asc(C_{k+j}) \geq Asc(C_{k+j+1}), 0 \leq j < l \\ Asc(C_{k+l}) < Asc(C_{k+l+1}) \end{cases} \quad (2)$$

Then we define the characters counting function as following:

$$\begin{cases} Num(C_k) = \min\{Num^+(C_k), Num^-(C_k)\} \\ Num^+(C_k) = l \\ Num^-(C_k) = i \end{cases} \quad (3)$$

Now we should define *Maxwinl* as the upper bound of $Num(C_k)$. Thus we can cut an email content into short strings with variable length.

The process of setting boundaries for short string segments can be shown in figure. (see Fig.1)

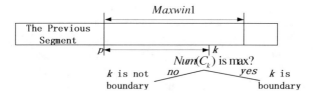

Fig. 1. The process of boundary generating

Although string segments in $< S_1, S_2, \cdots, S_m >$ are variable length, their lengths are often too short for spam detection. So next, we should devise a method to combine these segments and get a collection of bigger segments. Supposing that S_k is the k-th string segment of $< S_1, S_2, \cdots, S_m >$, function $Asc(S_k)$ will return the ASC-II binary value of the first character in S_k. So we can define the left consistent sequence of S_k as $seq^-(S_k)$, here :

$$\begin{cases} seq^-(S_k) =< S_{k-i}, S_{k-i+1}, \cdots, S_k > \\ Asc(S_{k-j}) \leq Asc(S_{k-j+1}), 0 < j \leq i \\ Asc(S_{k-i-1}) > Asc(S_{k-i}) \end{cases} \quad (4)$$

And we can get the right consistent sequence of S_k as $seq^+(S_k)$, here :

$$\begin{cases} seq^+(S_k) =< S_k, S_{k+1}, \cdots, S_{k+l} > \\ Asc(S_{k+j}) \geq Asc(S_{k+j+1}), 0 \leq j < l \\ Asc(S_{k+l}) < Asc(S_{k+l+1}) \end{cases} \quad (5)$$

Then we define the strings counting function as following:

$$\begin{cases} Num(S_k) = \min\{Num^+(S_k), Num^-(S_k)\} \\ Num^+(S_k) = l \\ Num^-(S_k) = i \end{cases} \tag{6}$$

Now we should define $Maxwin2$ as the upper bound of $Num(S_k)$. Thus we can combine these small segments into bigger segments with variable length.

After email content segmentation, we should choose a hash algorithm to encode these segments. However, Popular hash functions, such as MD5 and SHA-1, have been designed for cryptographic purposes, not for hash-based lookup or string matching, whose main concern is the throughput [16]. So we should first choose a suitable hash algorithm. It is lucky that Michael O. Rabin has devised a fingerprint generating algorithm (Rabin hash algorithm), which can satisfy our needs [17].

Supposing $f(B)$ is the fingerprint of a binary string B, and B_i is the binary string of BS_i, we can get the vector of piecewise hash values of email E's content as $R(E) =< f(B_1), f(B_2), \cdots, f(B_m) >$.

3   Spam Identification Algorithm

Bayesian method can predict the occurrence possibility of an event according to the distribution of its occurrence times [18]. By using Bayesian method, we can effectively filter the spam from normal emails.

Assuming that Sd_1 is the collection of spam information, Sd_2 is the collection of regular email information. $P(Sd_i | E)$ is the possibility of an email E belonging to the type of information in Sd_i, E is the email under identification, e_k is a segment from $< e_1, e_2, \cdots, e_n >$, and $< e_1, e_2, \cdots, e_n >$ is the piecewise hash segments vector we get via the methods in section 2 from E. Then we can see:

$$\begin{cases} P(Sd_i) = \dfrac{the\ amount\ of\ emails\ used\ to\ build\ Sd_i}{the\ total\ amount\ of\ emails\ used\ to\ build\ Sd_1\ and\ Sd_2} \\[2mm] P(Sd_i | E) = \dfrac{P(E | Sd_i) P(Sd_i)}{\sum\limits_{j=1}^{2} P(E | Sd_j) P(Sd_j)} \\[2mm] P(E | Sd_i) = \prod\limits_{k=1}^{n} P(e_k | Sd_i) = \prod\limits_{k=1}^{n} \dfrac{Count(e_k, Sd_i)}{\sum\limits_{j=1}^{2} Count(e_k, Sd_j)} \\[2mm] Count(e_k, Sd_i) = 1 + the\ frequency\ of\ e_k\ in\ Sd_i \end{cases} \tag{7}$$

If the amount of emails used to build Sd_1 equals to the amount of emails used to build Sd_2 ,then:

$$P(Sd_i \mid E) = \frac{\prod_{k=1}^{n} Count(e_k, Sd_i)}{\sum_{j=1}^{2} \prod_{k=1}^{n} Count(e_k, Sd_j)} \tag{8}$$

According to Kosmopoulos A's work in 2008 [19], conventional Bayesian method would identify E as spam if only $P(Sd_1 \mid E) > P(Sd_2 \mid E)$, but this identification is not precise enough, and its false positive rate as well as false negative rate are a little high. To overcome this problem, we modify the identification progress of conventional Bayesian method by setting an evaluation function, which is as following:

$$B(E) = \frac{P(Sd_1 \mid E)}{P(Sd_2 \mid E) + P(Sd_1 \mid E)} = \left(1 + \prod_{k=1}^{n} \frac{Count(e_k, Sd_2)}{Count(e_k, Sd_1)}\right)^{-1} \tag{9}$$

If the $B(E)$ is bigger than a given threshold R_0, then E would be identified as spam. Thus the value of R_0 can determine the result of identification, and should be carefully set. So it is important to get an optimal value for R_0 through a proper way. Shao Jian-feng and his teammates show us a reasonable method to get the optimal threshold basing on linear programming as follows [20]:

$$\max \sum_i \text{sgn}(R_i^{(1)} - R_0) + \sum_j \text{sgn}(R_0 - R_j^{(2)})$$

$$s.t. \begin{cases} \text{sgn}(x) = \begin{cases} 1 & x>0 \\ 0 & x=0 \\ -1 & x<0 \end{cases} \\ R_i^{(1)} \in \{B(E_i) \mid E_i \in Sd_1\} \\ R_j^{(2)} \in \{B(E_i) \mid E_i \in Sd_2\} \\ a < R_0 < b \end{cases} \tag{10}$$

Here (a,b) is the value range of R_0. But sometimes, the theoretical optimal threshold can have more than one value. It is because to their method, the optimal threshold can be any number in a specific region, in which all the numbers can satisfy the optimizing function. However, in this specific region, there must be a most reasonable threshold. In order to get it, we modify their algorithm as follows:

$$\max \sum_i \text{sgn}(R_i^{(1)} - R_0)\sqrt{\left|R_i^{(1)} - R_0\right|} + \sum_j \text{sgn}(R_0 - R_j^{(2)})\sqrt{\left|R_0 - R_j^{(2)}\right|} \tag{11}$$

4   Performance Evaluation

This section presents experimental evaluations of our method, including their accuracy, speed and robustness evaluations. Here we make comparison of our advanced spam detection technique(ASDT) with DSC algorithm[12] and the Email Remove-duplicate Algorithm Based on SHA-1(ERABS)[14]. We will describe our methodology before evaluating experimental result. All the spam emails we use are collected from 2006 TREC Public Spam Corpora[1], and all the regular emails are collected by ourselves from Gmail.

4.1   Experiment Parameters

Firstly, we should set parameter-values for these 3 algorithms. Here we set $Maxwin1 = 12; Maxwin2 = 12$ for ASDT. Then we set $W = 6; Mod = 4$ for DSC algorithms, and suppose the word's max length of this algorithm is 12, which means if there comes a string with more than 12 chars and can't be divided into words by semantics or syntax theories, the string would be sliced into words with 12 chars automatically.

Secondly, we should use the method we have described in Section 3 to get optimal threshold for each algorithm. Before doing this, we use 1000 spam emails to construct our spam information database Sd for each algorithm (for ASDT, the spam database is Sd_1), and use 1000 regular emails to construct our regular information database Sd_2 for ASDT. Then choose 100 emails from the 1000 spam emails with some modifications(such as insertion, deletion, replacement, etc.) on the contents together with 100 regular emails to be the emails used to get optimal threshold.

Finally, we get these optimal thresholds as: 0.5923for ASDT, 0.1629 for DSC and 0.5070 for ERABS.

4.2   Accuracy Evaluation and Comparison

We use several criteria to evaluate an algorithm's accuracy, namely precision and recall. Firstly, let's review the relationship between true positive, true negative, false positive, and false positive, shown in Table 1.

Table 1. The relationship review

Actual Label	Predicted Label	
	Positive	Negative
Positive	True-Positive (*TP*)	False-Negative (*FN*)
Negative	False-Positive (*FP*)	True-Negative (*TN*)

---

[1] http://plg.uwaterloo.ca/~gvcormac/treccorpus06/

The definitions of precision(P), recall(R) are based on above terms, and are given by the following formulas:

$$P = \frac{TP}{TP + FP}; R = \frac{TP}{TP + FN} \tag{12}$$

Although P and R can show the accuracy degree of a spam identification algorithm, however, when we are comparing the accuracy of all these algorithms, we have to evaluate P and R separately. So at last, in order to combine the accuracy information of P and R, we set an accuracy weight Q and use it for algorithms' accuracy comparison. Here:

$$Q = \frac{2P \times R}{P + R} \tag{13}$$

Here, we choose 500 normal emails and 500 modified spam emails as experimental samples. By using all the data we've set in section 4.1 to get the values of P, R and Q for these 3 algorithms, we can known each algorithm's accuracy. The experiment results are shown Table 2.

Table 2. The experiment result

Algo-	Accuracy Indexes		
rithms	P	R	Q
ASDT	99.40%	95.58%	97.45%
DSC	53.80%	87.91%	66.75%
ERABS	99.40%	98.42%	98.91%

From Table 2, we can see the accuracy weight of ASDT is significantly higher than DSC and only a little lower than ERABS. It means that ASDT's accuracy is good enough for spam identification.

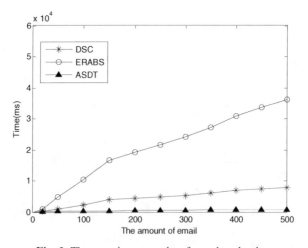

Fig. 2. The experiment results of speed evaluation

4.3   Speed Evaluation and Comparison

Running speed is also an important factor on evaluating the performance of a spam detection algorithm. Here we construct our experimental environment basing on java language and Eclipse SDK, and compare these 3 algorithms' running speed on it. All the parameter-values we use are the same as what we've defined in Section 4.1, and these 500 emails we chose for speed evaluation are from the spam set and the regular email set that we've collected before. The results can be seen in Fig.2.

As it shows in Fig.2, although ERABS has been proved to be the most accurate algorithm, it costs more running time than any other algorithms, and ASDT is the most time saving method and at the same time it can achieve high accuracy.

4.4   Robustness Evaluation and Comparison

Here, a spam filtering algorithm's robustness is the capability of resisting the effects of modifications in original spam from disturbing the outcomes of spam identification. While the modifications of an original spam from our spam database haven't successfully disturbed the outcome of spam identification, the outcome of spam identification function will be the max number 1.0. When the modifications have slightly disturbed the identification, although the outcome of spam identification function is larger than the optimal threshold, it is less than 1.0. When the modifications have severely disturbed the identification, the outcome of spam identification function is smaller than the optimal threshold.

we judge the robustness of a spam detection algorithm by a variable *Rob*, whose value varies in the range of -1 to 1, and the higher it is, the better the algorithm can resist the effects of modifications in original spam, thus it has a better robustness. The formula used to calculate it is as follows:

$$
Rob_i = \begin{cases} \dfrac{B(A_i) - R_0}{1 - R_0} & B(A_i) - R_0 > 0 \\[2mm] \dfrac{B(A_i) - R_0}{R_0} & B(A_i) - R_0 \leq 0 \end{cases} \tag{14}
$$

Rob_i represents the value of *Rob* get from the i-th email A_i. R_0 is the optimal threshold of the current algorithm, bayesian evaluation function $M(A_i, Sd)$ represent the membership of A_i on the spam database Sd and for ASDT, $M(A_i, Sd) = B(A_i)$. Then we choose n different emails $\{A_1, A_2, \cdots, A_n\}$, and get the *Rob* value set $\{Rob_1, Rob_2, \cdots, Rob_n\}$ for each of the four algorithms. By analyzing and comparing their *Rob* values' sets, we can get our judgment on the five algorithms' robustness.

Since the modification models for original spam emails to pass through spam filters including insertion, deletion, replacement and interchange, and different modification model would have different effect, we should compare the robustness of these 3 spam identification algorithms to each of the spam modification models. Here we select spam emails with 1kb and 2 kb length text content, and do different times of a certain modification on it to get our experiment email set, then we use these 3 algorithms to detect them, and get their *Rob* values (see Fig.3).

Form Fig.3, we can see ASDT has the best robustness to these four modification models it is dealing with. That may because the modified Bayesian method in Section 3 focuses on comparing the feature of an unidentified email E with the common feature of a type of emails, so effects of modifications to E would be suffered by all the emails in a certain type, that would make the outcome of ASDT better than the other 2 methods, whose robustness to modifications on E is determined by a certain email in spam database that helps to compute the biggest membership value of E. In addition, Our modification on the traditional Bayesian method also helps to enforce the robustness of ASDT.

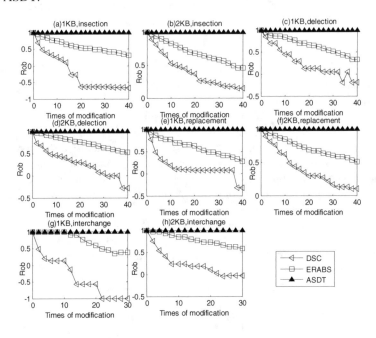

Fig. 3. The experiment results of robustness evaluation

5   Conclusion

In this paper we proposed our ASDT methods, it's significantly faster than DSC and ERABS method, and at the same time keeps satisfying accuracy and robustness. However, there are still many works remain to be completed:

Firstly, although ASDT can deal with non-text media contents theoretically, their performances are still need to be tested. So it is important to devise a series of experiments to do these tests. Secondly, since ASDT's potential values are proved in our experiments, thus their effects on other related fields such as microblog, SMS Center, etc. also needs to analyze.

Our further works will concentrate on these works, and try to solve the problems of them.

Acknowledgments.   The authors acknowledge the financial support by State Key Development Program of Basic Research of China (No. 2013CB329601), National Key Technology R&D Program (No. 2012BAH38B-04), "863" program(No. 2010AA -012505, 2011AA010702). The author is grateful to the anonymous referee for a careful checking of the details and for helpful comments that improved this paper.

References

1. Gyongyi, Z., Garcia-Molina, H.: Web spam taxonomy. In: First International Workshop on Adversarial Information Retrieval on the Web (AIRWeb 2005) (2005)
2. Hayati, P., et al.: Definition of spam 2.0: New spamming boom. In: 2010 4th IEEE International Conference on Digital Ecosystems and Technologies (DEST). IEEE (2010)
3. Moniza, P., Asha, P.: An assortment of spam detection system. In: 2012 International Conference on Computing, Electronics and Electrical Technologies (ICCEET). IEEE (2012)
4. Sebastiani, F.: Machine learning in automated text categorization. ACM Computing Surveys (CSUR) 34(1), 1–47 (2002)
5. Whitworth, B., Whitworth, E.: Spam and the social-technical gap. Computer 37(10), 38–45 (2004)
6. Xu, Q., et al.: Sms spam detection using noncontent features". IEEE Intelligent Systems 27(6), 44–51 (2012)
7. Hidalgo, G., María, J., et al.: Content based SMS spam filtering. In: Proceedings of the 2006 ACM Symposium on Document Engineering. ACM (2006)
8. Resnick, P.: RFC 2822: Internet message format. IETF (Standards Track) Request for Comments 2822 (2001)
9. Kornblum, J.: Identifying almost identical files using context triggered piecewise hashing. Digital Investigation 3, 91–97 (2006)
10. Breitinger, F., Baier, H.: Performance issues about context-triggered piecewise hashing. In: Gladyshev, P., Rogers, M.K. (eds.) ICDF2C 2011. LNICST, vol. 88, pp. 141–155. Springer, Heidelberg (2012)
11. Broder, A.Z., et al.: Syntactic clustering of the web. Computer Networks and ISDN Systems 29(8), 1157–1166 (1997)
12. Kołcz, A., Chowdhury, A., Alspector, J.: Improved robustness of signature-based near-replica detection via lexicon randomization. In: Proceedings of the Tenth ACM SIGKDD International Conference on Knowledge Discovery and Data Mining. ACM (2004)
13. Zhang, M., Li, B.C., Chen, L.: Email Remove-duplicate Algorithm Based on SHA-1. Computer Engineering 11, 098 (2008)
14. Kołcz, A.: Lexicon randomization for near-duplicate detection with I-Match. The Journal of Supercomputing 45(3), 255–276 (2008)
15. Sun, J.Z., Ma, Y.Q., Li, Y.H.: Data Chunking Algorithm Based on Byte-fingerprint Extremum Characteristics. Computer Engineering 8, 26 (2010)
16. Zhong, Z., Li, K.: Speed Up Statistical Spam Filter by Approximation. IEEE Transactions on Computers 60(1), 120–134 (2011)
17. Rabin, M.O.: Fingerprinting by random polynomials. Center for Research in Computing Techn. Aiken Computation Laboratory, Univ. (1981)
18. Luo, Q., Qin, Y.-P., Wang, C.-L.: Anti-spam technology review. Journal of Bohai University (Natural Science Edition) 4 (2008)
19. Kosmopoulos, A., Paliouras, G., Androutsopoulos, I.: Adaptive spam filtering using only naive bayes text classifiers. In: Proceedings of the Fifth Conference on Email and Anti-Spam (CEAS) (2008)
20. Shao, J., Yan, X., Shao, S.: SNR of DNA sequences mapped by general affine transformations of the indicator sequences. Journal of Mathematical Biology 67(2), 433–451 (2013)

DNFStore: A Distributed Netflow Storage System Supports Fast Retrieval

Fengguang Gong[1], Wenting Huang[2], Hao Luo[1], and Hailong Zhu[2]

[1] Beijing University of Posts and Telecommunications, Network Information Center,
Beijing, 100876, China
fggong@bupt.edu.cn
[2] National Computer Network Emergency Response Technical Team
Coordination Center of China
Beijing, 100029, China
hwt_chn@163.com

Abstract. Network anomaly detection or network optimization based on Netflow plays an important role in current high-speed network management. Storage and analysis of high-speed continuous Netflow are hot and difficult issues in network security research and industry communities. Existing solutions, although useful in above areas, have several drawbacks in well handling Netflow records generated by large-scale backbone networks. In this paper, we design and implement distributed Netflow storage system DNFStore. First, DNFStore dispatches high-speed Netflow records to efficient storage nodes through independent dispatchers to achieve parallel storage and uses interactive protocol to realize the goal of dynamically join and exit of nodes. Second, DNFStore merges Netflow records returned from storage nodes through query agent, and provides users with a fast unified query interface. We deploy DNFStore on a Gigabit network. The experimental results show that DNFStore can handle 20 million Netflow records per second and process multiple backbone networks concurrently. Besides, DNFStore can provide 40 times speed of query response than centralized storage under the same conditions.

Keywords: Distributed system, Dynamic expansion, Query agent, Load-balance algorithm.

1 Introduction

With the development of network, the network abnormal behaviors become more and more widespread, and it is more difficult to manage IP network. Storage and analysis of streaming network data can provide more secure and reliable network environment, better Qos, precisely charge based on traffic, network upgrades. Such work cannot be trusted when lacking of long-term continuous and large-scale traffic [1]. For considerations of network security and management, storage of long-term continuous and large-scale traffic is of great significance, while it is not easy. When recording

W. Han et al. (Eds.): APWeb 2014 Workshops, LNCS 8710, pp. 158–166, 2014.

actual packet content, something results in prohibitive repository and also severely compromise user privacy. To overcome these limitations, researchers proposed flow-based technologies, in which Netflow [2] is a typical approach.

Long-term and large-scale traffic brings great challenges to the storage and retrieval of Netflow, which can be described in the following aspects: 1) in the backbone networks, the number of concurrent connections usually reaches up to millions [3], so it requires efficient network dispatcher; 2) long-term Netflow leads to large disk consumption; 3) large-scale Neflow makes it difficult for interactive query. For example, Netflow records collected from a backbone network during one day consume about 3.78TB of disk space.

The contributions of our work can be summarized as follows:

1) We design efficient dispatcher and storage node, by which DNFStore can handle Netflow from multiple backbone networks concurrently.

2) The nodes of DNFStore can dynamically join and exit, which provide efficient system scalability.

3) DNFStore provides unified query interface to clients through transparent implementation.

The rest of this paper is organized as follows: we present related work in Section 2. The system architecture is described in Section 3. Experimental results and evaluation are presented in Section 4 and we conclude in Section 5.

2   Related Work

Netflow storage and analysis technologies can be summarized from the network and database perspective. In this section, we review some solutions relevant and highlight the differences to our approach.

1) Data Stream Management Systems (DSMS), such as Gigascope [4] and Telegraph [5] pay too much attention on real-time filtering, while ignoring storing the entire data on disk.

2) Flow-based Systems, such as SiLK [6], nfdump [7] and flowtools [8], though with high insertion rate of Netflow, are typically centralized solutions with limited scalability.

3) Flow-based, column-oriented systems, such as NET-Fli [9] and NetStore [10] adopt the ideas of BigTable [11]. Their indexing mechanisms not only increase the consumption of disk, but also decrease the insertion rate of data. In addition, low scalability restricts their use.

4) Flow-based distributed architectures, such as MIND [12] and DipStore [13] are based on P2P, deployed on the wide-area network for flow collection, storage and analysis. P2P routing of each query could tremendously increase the query time. Xu Fei et al. [14] propose a P2P group algorithm, which is a good illustration of this problem. Besides, the open security problem of P2P platform [15] also limits the use of this kind of system.

5) New type of distributed database, such as Hbase [16] supports high-speed and massive data storage, and has good scalability. Primary key is the only way to locate data in Hbase table, so its retrieval mechanism cannot meet the requirements of describing of complex network issues.

In conclusion, all of the above approaches have drawbacks. The scalability of storage system based on network flow record is limited; Column-oriented solutions reduce the storage performance, while increasing disk consumption; The query time of distributed system based on P2P restricts its use; Simple query interface of Hbase cannot meet the demand of complex analysis issues. In our solution, we present DNFStore whose storage nodes are as efficient as SiLK, and it provides dynamic expansion and efficient query interface.

3   Overview of DNFStore

In this section, we describe the architecture and implementation of DNFStore. DNFStore can easily obtain its operational status and statistical information. DNFStore provides good scalability, efficient storage and retrieval performance.

Dispatchers send Netflow records to independent storage nodes through Load-balancing algorithm. Storage unit is the foundation of DNFStore's storage performance. Query agent processes query request by communicating with the management agent.

3.1   Storage Unit

Fig. 1 depicts   the storage unit designed in this paper, which mainly contains six modules: 1) receive module: which is responsible for receiving Netflow; 2) preprocess module: which takes charge of verifying the correctness of Netflow; 3) cache module: which is in charge of cache results and providing high storage performance; 4) serialization module: which is responsible for writing cache results to disk with unstructured and time-based file system; 5) statistical module: which takes charge of collecting various Netflow information processed by storage unit; 6)query module: which is primarily responsible for retrieving Netflow.

In order to increase performance of storage unit, Netflow process can be executed concurrently (Fig. 1).

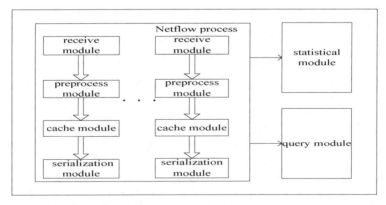

Fig. 1. Storage unit of DNFStore

3.2   Storage Engine

Storage engine (Fig. 2) contains the following elements: 1) dispatcher: dispatchers collect Netflow records from network and dispatch them to independent storage nodes with load balance algorithm. Independent dispatchers can process backbone networks concurrently, which provide the ability to handle wide-ranging Netflow records. 2) Storage node: storage node consists of storage unit and management agent, where management agent is responsible for detecting the status of storage unit and collecting statistical information. 3) Query agent: query agent receives the query request, sends it to multiple management agents, and merges query results. By shielding the implementation details of DNFStore, query agent provides a uniform query interface.

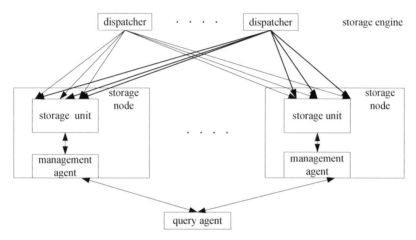

Fig. 2. Storage engine of DNFStore

DNFStore uses zookeeper [18] to manager global consistency information, such as statistical information, fault status, and debug information.

3.3   Netflow Load-Balancing Algorithm

Netflow processed in DNFStore is generated through the heartbeat mechanism which sampling on a certain amount of packets. Different heartbeat mechanism generates different number of Netflow records and loss rates. When using hash algorithm, no matter how much the frequency of the heartbeat is, the loss rate of Netflow is $\frac{1}{n}$, where n is the number of storage nodes. While taking use of Round-robin algorithm, the loss rate is $\frac{1}{n} \times \frac{1}{m}$, where m is the average number of Netflow records of each session. DNFStore supports both Round-robin and hash algorithms [17].

3.4   Query Interface

With Round-robin dispatch algorithm, the query agent polls all storage nodes, receives Netflow records, sorts the results and then returns them to client. However, as for hash algorithm, if the query condition corresponds to that of the hash algorithm, the query request can be forwarded to a specific storage node determined by hash algorithm. For example, if condition of hash and query are both source IP, a query task involves only one storage node. Otherwise, it will poll each storage node, which is in accordance with the Round-robin algorithm.

4   Evaluation

We deploy our DNFStore prototype system on 10 nodes connected via a Gigabit network, each of which is configured as follows: Inter® Xeon® CPU E7-4820; 4*8 core; 128GB memory; Linux. We evaluate DNFStore on four aspects: storage performance of storage unit, storage performance of storage engine, loss rate of Netflow and query time.

4.1   Storage Performance of Storage Unit

Fig.3 shows the storage unit's performance with increasing of the number of concurrent Netflow process threads. Write IO and Disk IO in Fig. 3 represent the storage performance of storage unit and the max storage performance of the storage node respectively.

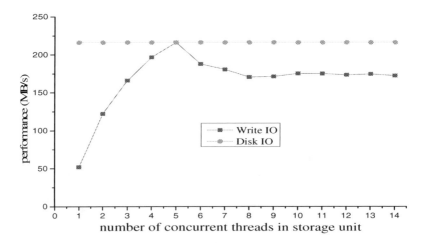

Fig. 3. Storage performance of storage unit

Fig. 3 indicates that with the increasing of the concurrent threads, the performance of storage unit increases as well. A storage unit can handle 4 million Netflow records per second when the number of concurrent Netflow process threads is 5. While the number of parallel threads is more than 5, the performance of storage unit decreases, which is due to the frequent scheduling of threads. We can see that Disk IO is a determining factor in storage unit performance metrics.

4.2   Storage Performance of Storage Engine

Fig. 4 reports the performance of storage engine. According to the conclusion of Fig. 3, the number of concurrent Netflow process threads of storage node in Fig. 4 is 5.

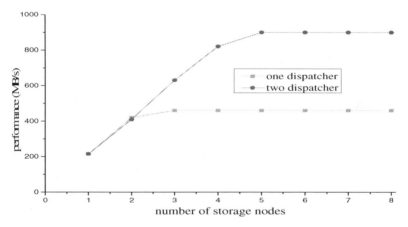

Fig. 4. Storage performance of DNFStore

When the number of storage nodes is less than 3, one dispatcher is sufficient. Otherwise, two dispatchers are needed. However, when the number of storage nodes is more than 5, due to the limitation of the performance of dispatchers, the storage performance of DNFStore tends to be stable. DNFStore can handle 20 million Netflow records per second when the number of monitors is 2 and the number of storage nodes is 5.

4.3   Loss Rate of Netflow

We illustrate the average number of packets in each session in Fig. 5. The sessions are based on 5TB data collected from backbone networks, and contain 8 million IPs.

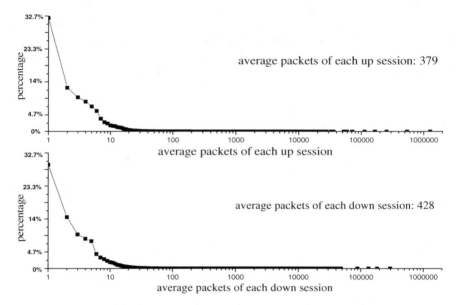

Fig. 5. Average number of packets in each session

From Fig. 5, we can conclude that the average number of packets in each up session and down session is 379 and 428, respectively.

According Fig. 5, Fig. 6 reports the comparisons of loss rate of hash and Round-robin algorithm.

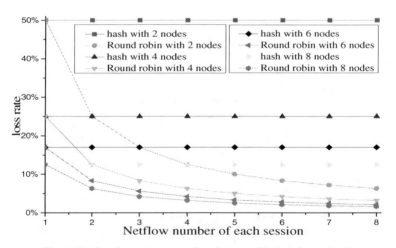

Fig. 6. Netflow loss rate comparison between Hash and round robin

As depicted in Fig. 6, when the number of Netflow records in a session is greater than 1, Round-robin can obtain much lower loss rate than that of hash algorithm.

4.4   Query Time

Fig. 7 shows the comparison of retrieval performance between centralized storage system and DNFStore, where the query condition is source IP. With the increase of the storage node number in DNFStore, the retrieval time decreases gradually. From Fig. 7, we can observe that the query time of centralized storage is always longer than that of DNFStore. When the Netflow number reaches 8 billion, the retrieval time of centralized storage increases sharply up to 371s, while the retrieval time of DNFStore with 8 storage nodes is 9s.

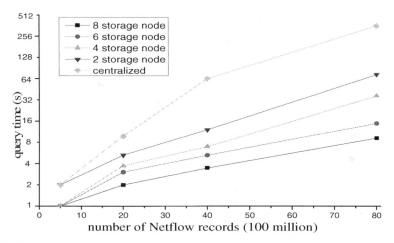

Fig. 7. Comparision of retrieval time between centralized storage system and DNFStore

Fig. 7 implies that DNFStore can reduce query time dramatically compared with the centralized storage.

5   Conclusions

Nowadays, Netflow has been widely used in network optimization and network anomaly detection. Netflow storage has attracted a lot of attentions from research and industry communities. In this paper, we design and implement DNFStore, a distributed Netflow storage system supporting fast retrieval. DNFStore can handle Netflow records generated by multiple backbone networks concurrently with efficient independent dispatchers and storage nodes. DNFStore can adjust its storage capacity according to the size of the network by joining and exiting nodes dynamically. In the case with the same storage size, DNFStore provides 40 times query response time than that of centralized storage. We test DNFStore on a Gigabit network. The experimental results show that DNFStore can handle approximately 20 million Netflow records per second through its efficient storage engine. Based on these results, we infer that DNFStore can deal with the backbone network traffic.

Acknownledgement. Supported by the National Key Technologies Research and Development Program of the Ministry of Science and Technology of China under Grant No.2012BAH37B02; the National Information Seurity 242 Project of China under Grant No.2013A012; the National Natural Science Foundation of China under Grand No.61272536.

References

1. Claffy, K.C.: Internet as emerging critical infrastructure: what needs to be measured. In: The OECD ICCP Workshop on the Future of the Internet (2006)
2. Cisco System. Netflow services and application. White Paper (1999)
3. Estan, C., Varghese, G.: New directions in traffic measurement and accounting. ACM (2002)
4. Cranor, C.D., Johnson, T., Spatscheck, O., Shkapenyuk, V.: Gigascope: A Stream Database for Network Application. In: Proc. of SIGMOD, pp. 647–651 (2003)
5. Chandasekaran, S., Cooper, O., Desphpande, A., Franklin, M.J., Hellerstein, J.M., Hong, W., Krishnamurthy, S., Madden, S.R., Reiss, F., Shah, M.A.: TelegraphCQ: continuous dataflow processing. In: Proc. of SIGMOD, pp. 668–668 (2003)
6. SiLK: Monitoring for Large-Scale Networks, http://tools.netsa.cert.org/silk/
7. Haag, P.: Nfdump, http://nfdump.sourceforge.net/
8. Chandasekaran, S., Cooper, O., Desphpande, A., Franklin, M.J., Hellerstein, J.M., Hong, W., Krishnamurthy, S., Madden, S.R., Reiss, F., Shah, M.A.: TelegraphCQ: continuous dataflow processing. In: Proc. of SIGMOD, pp. 668–668 (2003)
9. Fusco, F., Stoechklin, M.P., Vlachos, M.: NET-FLi: on-the-fly compression, archiving and indexing of streaming network traffic. Proceedings of the VLDB Endowment 3(1-2), 1382–1393 (2010)
10. Giura, P., Memon, N.: An efficient storage infrastructure for network forensics and monitoring. In: Recent Advances in Intrusion Detection, pp. 277–296. Springer, Herdelberg (2010)
11. Chang, F., Dean, J., Ghemawat, S., et al.: Bigtable: A distributed storage system for structured data. ACM Transactions on Computer Systems (TOCS) 26(2), 4 (2008)
12. Li, X., Bian, F., Zhang, H., et al.: MIND: A Distributed Mutil-Dimensional Indexing System for Network Diagnosis. In: INFOCOM (2006)
13. Giura, P., Memon, N.: An efficient storage infrastructure for network forensics and monitoring. In: Recent Advances in intrusion Detection, pp. 277–296. Springer, Herdelberg (2010)
14. Xu, F., Yang, G.W., Ju, D.P.: Design of distributed storage system on peer-peer structure. Journal of Software 15(2), 268–277 (2004)
15. Liu, Y., Li, Q., Li, Z.-J.: Survey of Network Security and Defense Mechanism. Computer Science (4), 15–19 (2013)
16. HBASE, http://www.hbase.apache.org
17. Luo, Y.-J., Li, X.-L., Sun, R.-X.: Summarization of the Load-balancing Algorithm. Sci-Tech Information Development & Economy (23), 140–142 (2008)
18. Zookeeper, http://zookeeper.apache.org/

A Novel Road Topology-aware Routing in VANETs

Hongli Zhang and Qiang Zhang

School of Computer Science and Technology, Harbin Institute of Technology, Harbin, China
zhanghongli@hit.edu.cn, zhangqiang@pact518.hit.edu.cn

Abstract. Path connectivity may change due to node mobility in Vehicular Ad hoc Networks (VANETs). When the current routing path is disconnected, finding another connected routing path quickly is important for excellent routing performance in VANETs. In this paper, we propose a novel road topology-aware routing (RTR) protocol for VANETs. RTR establishes two junction-disjoint paths in order to utilize connected routing paths sufficiently. RTR alternately transfers data packets through each established routing path and dynamically changes the routing path based on the connectivity of the current path. Simulation results have shown that the proposed RTR improves the performance compared to single path routing GPSR.

Keywords: topology-aware, junction-disjoint, routing, VANET.

1 Introduction

Routing protocol plays a key role for performance of communications between vehicles in the VANETs. So the efficient and robust routing protocol is very important for VANETs. However, traditional node-based routing protocols are unsuitable for VANETs because of node mobility. In recent years, geographic routing is studied specially to cope with node mobility in VANETs. The geographic routing can select the next hop node flexibly by using position information of nodes. And nodes do not need to maintain routing information to forward packets. Though geographic routing has better performance for VANETs, the performance still needs to be improved for some real-time and emergency applications. Therefore the design of high-performance routing protocol is a challenge.

Researchers have proposed several geographic routing protocols [1,2,3] for VANETs, and most of them are single path routing. And their researches focus on the easily disconnected path and aim to find a stable route in the case of high mobility. In [4], performance of node-disjoint multipath routing is studied, and their conclusion is that, through careful path selection, node-disjoint multipath routing gains an advantage over single path routing in terms of packet delivery ratio and end-to-end delay. Meanwhile, it is pointed out that path coupling affects performance of multipath routing. On the other hand, in order to enhance the stability of routing, some multipath routing protocols [5,6] have been proposed. And when the current route fails, they employ multiple paths to recover normal route. However, multipath routing has large communication overhead owing to the redundancy of data transmission. So how to

W. Han et al. (Eds.): APWeb 2014 Workshops, LNCS 8710, pp. 167–176, 2014.

enhance the stability of routing with low communication overhead in VANETs is a challenge. In this paper, we propose a novel road topology-aware routing (RTR) protocol in VANETs. RTR is not only a road-based routing, but also a path-based routing. Based on position information of road and vehicles, two junction-disjoint routing paths are established. Source node uses one of the established paths rather than all for one packet forwarding so that communication overhead is reduced. When both two junction-disjoint paths are connected, they are chosen alternately to transfer data packets. When one of two junction-disjoint paths is disconnected, the other path is chosen. RTR utilizes two junction-disjoint paths to transfer data packets, which avoids network congestion and link failure in a single routing path. Hence when the current routing path is disconnected, RTR is able to forward data packets through the other routing path.

2   Related Work

GPSR [7] is a geographic routing. Due to node mobility, greedy forwarding is used to improve routing performance. But it does not consider the topology of the roads in city, so a local maximum usually happens. Then perimeter mode is started, and the packet is routed around the perimeter of the region. In addition, GPSR does not detect the connectivity of routing path ahead, and so it often finds that the routing path is disconnected when a packet has been forwarded through several hops.

FROMR (Fast Restoration On-demand Multipath Routing) [5] extends the AODV [8] protocol to discover multiple paths between the source and the destination in every route discovery. While their objective is to primarily design a multipath routing framework for providing enhanced robustness to node failures. FROMR was proposed to build an alternate path when current route is broken. But it is a node-based routing protocol unsuitable for VANETs. And when a source node is sending data, only one routing path is used.

RMRV (Road-based QoS-aware Multipath Routing) [6] was proposed to find a stable route so as to improve routing performance. RMRV uses a space-time planar graph approach to predict the connectivity of each road section in a path, and then estimates path's future lifetime. Hence source node may dynamically choose a path with the longest lifetime. And only one routing path is used by each source node.

In [4], performance of node-disjoint multipath routing in VANETs was studied. The authors concluded that path coupling is an important factor for multipath routing performance. However, they did not present an algorithm for finding two minimum-interference paths between a source-destination pair. And for packet load, the source node splits the generated load into two parts, and each part distributed over one path has half of the whole load. So the packet load on each path has the same amount of packets. However when of the two routing paths is disconnected, if half of packets loads is distributed on the disconnected path, it is not reasonable.

3   Road Topology-Aware Routing

A.   Model and Problem Statement

Our system assumes that the amount of connected path between source node and destination node in VANETs is greater than or equal to 2. Each routing path may become disconnected from connected due to node mobility. Each vehicle is able to obtain road topology and its own position information as well as the destination node. In addition, periodical beacon messages are broadcasted by each vehicle to get neighboring information.

B.   Two Junction-Disjoint Paths Establishment

First, source node sends a route request message RREQ, and then the RREQ message is broadcasted to the entire network to collect multiple paths connected to destination node. When a node receives a RREQ message, if the node has not received the same RREQ message, it rebroadcasts the RREQ message. Otherwise, the node discards the RREQ message. Further, we record junction list in the RREQ. When a RREQ message enters into a new road section, the junction list in the RREQ message is updated. The junction list represents the path that the corresponding RREQ message has passed through. When some RREQ messages attached with different junction lists are received by destination node, the junction list in the RREQ message corresponding to the minimum delay is chosen as the first routing path. And then the destination node checks whether the routing path (junction list) in the RREQ message corresponding to the second minimum delay and the first routing path are disjoint i.e. the two junction lists have not the same junction. If yes, the second routing path is chosen by the destination node. If not, the destination node continues to check the path in next RREQ message. If the destination node does not find two junction-disjoint paths finally, then only the first routing path is used to forward packets. Otherwise, a route reply message RREP attached with the first routing path information is forwarded along the first routing path, and another route reply message RREP attached with the second routing path information is forwarded along the second routing path.

As shown in Figure 1, source node S can obtain two junction-disjoint paths i.e. the first routing path (S-J_a-D) and the second routing path (S-J_b-J_c-J_d-D) when it receives two RREP messages from destination node D.

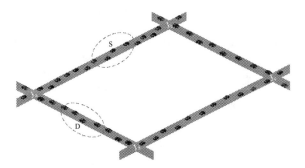

Fig. 1. Two junction-disjoint paths

C.   *Routing Algorithms in RTR*

In this section, we describe how a source node selects its next hop to forward data packets and how an intermediate node forwards data packets. Before the demonstration of algorithms, we give some notations and some preliminary introductions. Let P_i denotes one of the two junction-disjoint paths and P_{-i} denotes the other routing path. The data packet in RTR contains position information of junctions in order and the destination node's position, which means that the order of junctions in RREQ is the same with data packets. Source node alternately uses P_i and P_{-i} to transmit data packets. Source node and intermediate node utilize greedy forwarding to transfer data packets based on position information of junctions and destination node in each data packet.

For a source node, the corresponding routing strategy is shown as follows:

(1) Judge whether the destination node is the source node's neighbor. If yes, transmit data packets to the destination node directly. If no, go to step (2).

(2) We assume that the current routing path is P_i (i.e., it is the chance of P_i to be used for data transmission). If there is a neighbor nearer to the first junction than the source node, then the source node selects that neighbor as the next hop. If no other neighbor is nearer to the first junction than the source node, then the source node selects the neighbor nearest to the second junction in the data packet as the next hop. Meanwhile, source node needs to delete the first junction in the data packet and transfer. If no other neighbor is nearer to the second junction than the source node, go to step (3).

(3) Add P_{-i} into the data packet (i.e., replace P_i by P_{-i}). If there is a neighbor nearer to the first junction in P_{-i} than the source node, then the source node selects that neighbor as the next hop. If no other neighbor is nearer to the first junction than the source node, then the source node selects the neighbor nearest to the second junction in the data packet as the next hop. Meanwhile, source node needs to delete the first junction in the data packet, and then transfer it. If no other neighbor is nearer to the second junction than the source node, then the source node discards the data packet.

For an intermediate node, the corresponding routing is shown as follows:

(1) Judge whether the destination node is the intermediate node's neighbor. If yes, transmit data packets to the destination node directly. If no, go to step (2).

(2) Judge whether there is a neighbor nearer to the first junction in the data packet than the intermediate node. If yes, forward the data packet to the neighbor. If no, go to step (3).

(3) Judge whether there is a neighbor nearer to the second junction in the data packet than the intermediate node. If yes, the intermediate node needs to delete the first junction in the data packet, and then forward it to the neighbor. If no, the intermediate node discards the data packet.

We design Algorithm 1 to implement routing strategy for source node in RTR.

Algorithm 1.

INPUT: *Two junction-disjoint paths{P_i, P_{-i}}*
  Destination node id=D, Source node id=S
  Node set G={S, all neighbors of S}
  Current chance is P_i

1: **if** D is a neighbor of S
2:   S transfers the data packet to D;
3: **else**
4:   **if** X_1^1 in G nearest to the first junction of $P_i = S$
5:     **if** X_1^2 in G nearest to the second junction of $P_i = S$
6:       modify the data packet using P_{-i};
7:       **if** X_2^1 in G nearest to the first junction of $P_{-i} = S$
8:         **if** X_2^2 in G nearest to the first junction of $P_{-i} = S$
9:           discard the data packet;
10:           restart routing paths establishment;
11:         **else**
12:           S deletes the first junction of P_{-i};
13:           S transfers the data packet to X_2^2;
14:       **else**
15:           S transfers the data packet to X_2^1;
16:     **else**
17:       S deletes the first junction of P_i;
18:       S transfers the data packet to X_1^2;
19:   **else**
20:     S transfers the data packet to X_1^1;

For better understanding our routing algorithm, we demonstrate 5 cases shown in Figure 2. We assume that in all cases the current chance of routing path is P_i. In case 1, source node selects A_3 as the next hop based on the position of J_i^1. In case 2, source node selects A_2 as the next hop based on the position of J_i^2. In case 3 and 4, source node selects A_2 as the next hop and uses the routing path P_{-i}. In case 5, source node is not able to find a suitable next hop based on the current routing paths P_i and P_{-i}, so source node discards the data packet and restart routing paths discovery and establishment.

The main idea of routing strategy for intermediate node is similar to that of source node. Based on the position information of the first two junctions in the data packet, an intermediate node selects the next hop to achieve greedy forwarding. In addition, the different part of routing strategy between source node and intermediate is that routing approach for intermediate node is not able to change the current routing path based on position information. However, the routing strategy for source node is able to dynamically change routing path according to the current connectivity of the first hop. So, a source node can avoid routing failure greatly. For brevity, we do not demonstrate the specific cases on intermediate node routing.

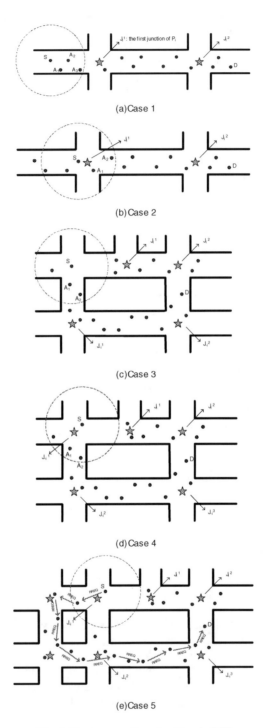

(a) Case 1

(b) Case 2

(c) Case 3

(d) Case 4

(e) Case 5

Fig. 2. Cases on routing algorithm in RTR

We design Algorithm 2 to implement routing strategy for intermediate node in RTR as follows.

Algorithm 2.

INPUT: *Two junction-disjoint paths$\{P_i, P_{-i}\}$*
　　　　Destination node id=D, Intermediate node id=F
　　　　Node set E={F, all neighbors of F}
　　　　Current chance is P_i
1: **if** D is a neighbor of F
2:　　F forwards the data packet to D;
3: **else**
4:　　**if** X_1^1 in E nearest to the first junction of $P_i = F$
5:　　　**if** X_1^2 in E nearest to the second junction of $P_i = F$
6:　　　　discard the data packet;
7:　　　**else**
8:　　　　F deletes the first junction of P_i;
9:　　　　F transfers the data packet to X_1^2;
10:　**else**
11:　　　F forwards the data packet to X_1^1;

4　Performance Evaluation

A.　Experimental Setting

We use NS-2 for simulation. The communication range of a node is 250 *m*. A total of 30 nodes are deployed in a region of size 1500 *m* x 1500 *m*. Two-ray ground model is used as the radio model. IEEE 802.11 DCF is selected as the medium access control (MAC) layer protocol. The MAC layer data rate is set to 2Mbps. We use VanetMobiSim [10] to generate movements of nodes. The node speed is in the range of [5, 20] *m/s*. And we consider the range of [5, 10] *m/s* as the scenario of low speed. The range of [10, 15] *m/s* and the range of [15, 20] *m/s* are respectively corresponding to medium speed and high speed. The simulation time is 200s. We randomly select three source-destination pairs in each scenario. The generated constant bit rate (CBR) is in the range [5, 20] *packets/s*. The data packet size is 512 bytes. And the beaconing frequency is 2.0 *s*. The queue length of each node is set to 50 packets.

B.　Experimental Results

First, we discuss the performance metric of routing control overhead. The communication overhead in RTR includes beacon messages, RREQ and RREP messages. The amount of beacon messages is identical in all routing protocols for VANETs. As beacon message is broadcasted periodically and frequently, the amount

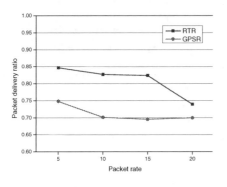

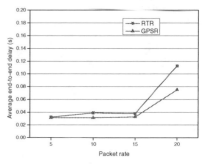

(a) Packet delivery ratio versus packet rate for low speed

(b) Average end-to-end delay versus packet rate for low speed

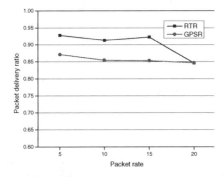

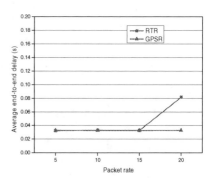

(c) Packet delivery ratio versus packet rate for medium speed

(d) Average end-to-end delay versus packet rate for medium speed

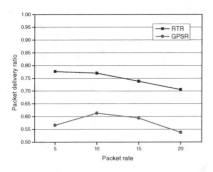

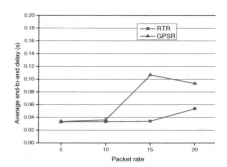

(e) Packet delivery ratio versus packet rate for high speed

(f) Average end-to-end delay versus packet rate for high speed

Fig. 3. Performance evaluation in terms of packet delivery and average end-to-end delay

of beacon messages is relatively large compared to that of other routing control messages. In addition, the number of RREQ messages is also equal to that of other routing protocols that employ route discovery mechanism, for example, CMGR [11]. And the number of RREP messages in the proposed RTR is about twice that of single path routing protocols. However, a RREP message is only forwarded through a path not like a RREQ message that should be rebroadcasted into the entire network. So the number of RREP messages is relatively small compared to that of beacon and RREQ message. Therefore the routing control overhead of the proposed RTR is similar to that of other routing protocols, and we do not provide the specific numerical value.

Second, we discuss the performance metrics of packet delivery ratio and average end-to-end delay. We compare the proposed RTR with single path routing that only use one path. As shown in Figure 3, the packet delivery ratio of RTR outperforms that of single path routing. When the node speed is high, the average end-to-end delay of RTR is less than that of single path routing. For the above results, it is mainly because the routing path composed of fast vehicles is not stable, and when a single routing path is disconnected the routing performance is affected badly. However, RTR is able to dynamically and alternately use two junction-disjoint routing paths to transfer data packets. For other cases including low and medium speed, their average end-to-end delays are about the same.

5   Conclusion

In this paper, we propose a novel road topology-aware routing (RTR) protocol for VANETs. The proposed RTR is able to establish two junction-disjoint paths to forward packets based on road topology. Source node uses one of the established paths rather than all for one packet forwarding. When each first hop node of the two junction-disjoint paths is connected to source node, source node alternately uses two paths to transfer data packets. RTR judges the connectivity of the first hop in the current routing path before transferring each data packet, if it is disconnected the other path is chosen by source node. Finally, the experimental results show that RTR enhances the packet delivery ratio greatly compared to GPSR. Meanwhile, RTR guarantees non-redundant data transmission.

Acknowledgment. This research was partially supported by the National Basic Research Program of China (973 Program) under grant No. 2011CB302605, the National High Technology Research and Development Program of China (863 Program) under grant No. 2011AA010705, the National Science Foundation of China under grants No. 61100188.

References

1. Naumov, V., Gross, T.R.: Connectivity-aware routing (CAR) in vehicular ad hoc networks. In: Proc. of IEEE International Conference on Computer Communications (Infocom), pp. 1919–1927 (2007)
2. Nzouonta, J., Rajgure, N., Wang, G., Borcea, C.: VANET routing on city roads using real-time vehicular traffic information. IEEE Transactions on Vehicular Technology 58(7), 3609–3626 (2009)

3. Yang, Q., et al.: ACAR: Adaptive connectivity aware routing for vehicular ad hoc networks in city scenarios. Mobile Networks and Applications 15(1), 36–60 (2010)
4. Huang, X., Fang, Y.: Performance study of node-disjoint multipath routing in vehicular ad hoc networks. IEEE Trans. Veh. Technol. 58(4), 1942–1950 (2009)
5. Wu, C.-S., Hu, S.-C., Hsu, C.-S.: Design of fast restoration multipath routing in VANETs. In: Proc. of International Computer Symposium (ICS), pp. 73–78 (2011)
6. Hsieh, Y.-L., Wang, K.: A Road-based QoS-aware Multipath Routing for Urban Vehicular Ad Hoc Networks. In: Proc. of the IEEE Global Communications Conference (GLOBECOM), pp. 207–212 (2012)
7. Karp, B., Kung, H.: GPSR: greedy perimeter stateless routing for wireless networks. In: Proc. 2000 ACM/IEEE International Conference on Mobile Computing and Networking (2000)
8. Perkins, C.E., Belding-Royer, E., Das, S.R.: Ad-Hoc On-demand Distance Vector (AODV) routing (July 2003), http://www.ietf.org/rfc/rfc3561.txt, RFC 3561
9. NS-2 Simulator, http://www.isi.edu/nsnam/ns/
10. Härri, J., Filali, F., Bonnet, C., Fiore, M.: VanetMobiSim: generating realistic mobility patterns for vanets. In: Proc. 3th Int. Wksp. Vehicular Ad Hoc Networks (VANET 2006), New York, NY, USA, pp. 96–97 (2006)
11. Shafiee, K., Leung, V.: Connectivity-aware minimum-delay geographic routing with vehicle tracking in VANETs. Ad Hoc Networks 9(2), 131–141 (2011)

Toward Identifying and Understanding User-Agent Strings in HTTP Traffic

Ye Xu[1,2], Gang Xiong[2,*], Yong Zhao[2], and Li Guo[2]

[1] University of Chinese Academy of Sciences
[2] Institute of Information Engineering, Chinese Academy of Science, 10093, China
{xuye,xionggang,zhaoyong,guoli}@iie.ac.cn

Abstract. Since HTTP is responsible for more than 25% of the total traffic volume in the Internet, we are curious about what happens in the http traffic and what it contains. We collected an anonymized HTTP head only data from a DSL line of a campus's border for continuous 11 days. In this paper, we focus on the user-agent (UA) strings. We first improve a well know open source UA-Sparser, identifying 34.4% (9.8%) more user-agent types (operator system) of overall transactions. In addition, our analysis of user-agent strings shows that Windows XP contributes to more than half of personal computer (PC) HTTP traffic. Mobile devices account for more than 16% of total transactions, and automatic programs share at least 20%. These information leads to following conclusions: PC devices in China are threatened to a huge risk as security support for Windows XP has been stopped recently; mobile devices occupy a big proportion in total transactions of a DSL lines, larger than other researchers [10] noticed 4 or 5 years ago; Android and IOS devices dominate the mobile transactions; automatic programs take up at least one fifth of all HTTP traffic and that means the researcher should no longer ignore the influence of them.

1 Introduction

HTTP has become one of the most popular protocols for many Internet services. Internet users with all kind of devices can take the advantage of HTTP to download files, watch video from You Ku, share messages on social networks, get news, send and read emails with Gmail or even study program on Codecademy online. As a consequence, many researchers have focused on HTTP traffic analysis [1-8]. Often, these studies would face the problem of identifying HTTP user types, the devices or browsers they use. This problem can be solved by User-Agent strings in http head filed of each request. The user-agent (UA) request header field contains information about the user agent originating the request [9]. We find that many researchers use an open source, UASparser[1], to parse UA strings [2, 12-13]. Many of them simply suggest that the result is correct and choose to ignore unknown UA strings. However, we find that the result of UASparser is far from satisfactory, with more than 6.5% unidentified UA strings which share 23% of total HTTP transactions.

---

[1] http://user-agent-string.info/download/UASparser

W. Han et al. (Eds.): APWeb 2014 Workshops, LNCS 8710, pp. 177–187, 2014.

So we developed a tool named R-UASP (Refined-UASparser), which is derived from UASparser. The R-UASP can identify 8% more distinctive user-agent strings than UASparser does. These UA strings contribute to identifying 9.8% more operator system types and 34.4% more UA types judged by total transaction times. In this paper, we usually use total request times (transactions) to represent the number of UA strings other than unique UA strings. Because the most requests are originated by only a few portion of UA strings. Figure 3 shows this phenomenon and we will discuss it in Section 3.

Moreover, we analyzed user-agent strings in our dataset. We observe that browser contributes more than 58% (33%) HTTP transactions (unique UA string), while mobile browser has become the second biggest http user in the Internet, accounting for 11% (36%) of the total requests (unique UA string). The proportion of unique UA strings and request times suggest that mobile devices and browsers are much more fragmental than PC ones. At least 20% of HTTP requests are produced by automatic programs, such as web crawler or update program. We also find that Windows clearly dominates, with more than 64% transactions. More than half computers are running on Windows XP, as security support for Windows XP has been stopped since April 8 2014. In addition, Android devices produce the most transactions in mobile devices, more than twice as apple devices produce. More than 80% of Android devices are running on Android 4.x and 50% of IOS devices are running on IOS 7, while only 2.7% computer choose Windows 8, indicating that mobile users are more willing to update OS of their devices or replace old devices. At last, we find that even though IE, Chrome, and Firefox occupy the top three positions in transactions, automatic programs, such as web crawlers, updates clients and programs that called library to request, should not be ignored. These programs take 6 positions in the top-20-request clients list.

The remainder of this paper is organized as follows. We briefly touch on related work in Section 2. We then describe our data collection and some of its characteristics in Section 3. We next consider various aspects of UA string distributions, such as popularity of operator system and UA type in Section 4. In Section 5, we briefly summarize our work in this paper.

2   Related Work

Callahan et al. [1] analyzed three and a half years of HTTP traffic observed on a small research institute, he use the longitudinal data to study various characteristics of traffic from both client and server behavior characteristics. Dhungana et al. [3], Maier et al. [10] and Falaki et al. [11] analyzed the traffic, especially HTTP traffic on mobile devices or smartphones. They revealed browsing contribution of the HTTP traffic, operator system distribution and application usage on mobile devices. Since their dataset was collected 4 or 5 years ago, the user situation in mobile devices has significantly changed, the most dramatic changes is the declination in Nokia and Symbian devices from the most popular mobile devices to less than 0.8% of all mobile devices, and Android along with Apple devices has dominate the market, accounting for more than 95% (64% and 32% respectively in our dataset) HTTP transactions of all mobile devices.

Fabian et al. [2] find three pitfalls which can render a HTTP analysis study flawed. In his work, he used UASparser as a tool to extract browser type and operator system information. Holley et al. [13] describe implementation of a classifier for User-Agent strings using Support Vector Machines. In his paper, he collected 53,829 UA strings from different web resources, while we collect 318,247 unique UA strings from DSL line of a campus's border. He used UASparser and uaParser[2] to annotate the UA strings they got. Even though he found a little improvement using the linear kernel, the classifier could not work very well if the annotation process had some pitfalls. Because the tools he use missed many popular UA strings

As mentioned above, the previous works have their time limitation since the user-agent types are changing dramatically year by year, especially in mobile browsing and application usage. As a consequence, it is essential to provide a modern view of the user-agent and web application usage in HTTP and compare it to the previous findings. Unlike previous studies, we do not entirely rely on open sources to parse UA strings. We first refined one popular open source to better identify UA strings before we start our analysis work. Thus, the refined tool helps us to understand HTTP traffic from UA perspective better and more precisely.

3   A General View of Dataset

In this section, we present our data sets before describing our experiment. We summary this section by introducing the general information of our dataset, such as size, packet and flow number, browser and OS usage.

3.1   Anonymized HTTP-Head-Only(HHO) Data

Table 1. Summary of HTTP

beginning of data time	18 April 2014
End of data time	28 April 2014
Carrier link speed	120Mbps
Original data size	6.180TB
HTTP volume	1.56TB
HHO data size	50.69GB
Number of http packets	98,165,427
Number of http flows	32,885,418

Our data was collected in a DSL line of a campus's border for 11 days. To preserve as many privacy as we can only HTTP-head information, length of payload and

---

[2]   Browserscope. Uaparser. Source Repository:
https://github.com/dinomite/uaParser

the two sides of IP address and ports were preserved for further observation. Moreover, we replaced all prefix of all IP address of clients to hide user information and preserve privacy. The bandwidth of DSL line is about 120Mbps and 90,000 Packets/s. Since the bandwidth is not wide, and the data we preserved is only a very small portion—http without payload—of it, we typically do not experience any packet loss. We generalize the characteristics, including start time, over time, original size, http volume and preserved dataset in Table 1.

3.2   More Characteristics of the Dataset

We first take a look at the composition of protocols used in the Original data in the DSL line. Figure 1 shows the different usage of protocols in the DSL line in which we collected dataset. We identified the protocols by an open source DPI software--nDPI[3]. We find that HTTP is the most significant protocol, accounting for more than 25% of bytes and 18% of packets. We also find that P2P Download (including BT, eDonkey and Xunlei etc.) share 14% of bytes but only 7% of the packets. This interesting answer indicates that P2P traffic has larger average packets than other normal protocols. On the contrary, Chat (including skype, QQ and MSN etc.) accounts for only 6.5% of bytes but more than 10% of packets. To our surprise, Remote control (including ssh, telnet, ftp,RDP etc.) accounts for 6%   of bytes. That might because the DSL line we used is most filled by university campus users, where ftp is a popular means of downloading and sharing files.

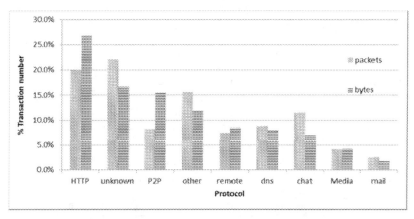

Fig. 1. Percentage of protocol usage in the HHO dataset

Figure 2 shows the number of different transaction types were requested in a day. In our dataset, the majority of observed transactions—more than 92%—are requests for data (GET requests). Almost 6% of the requests are contributed to POST requests and 1.5% of the requests are contributed to HEAD requests.

---

[3] http://www.ntop.org/products/ndpi/

We got 318,247 unique UA strings in HHO dataset, for each UA string. We also record the number of times they have appeared. Figure 3 shows the distribution of UA strings in transactions. The top 1% UA strings contribute to more than 84% of total transactions, while the top 2% contribute to 88.5% and top 3% contribute to 90.7%. The last 67.3% UA strings share only 1% of all transactions. The distribution evidently illustrates the fact that the most transactions are produced by the very minority of UA strings. And that's why we choose to use total transactions of UA strings other than unique UA strings as a scalable base in most of our experiments.

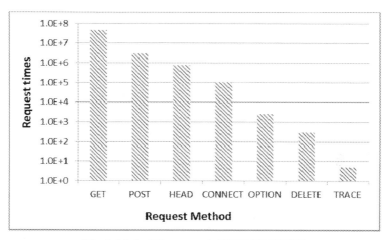

Fig. 2. Method Frequency in HTTP REQUEST

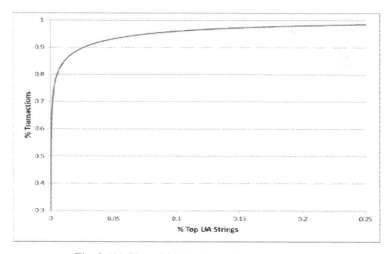

Fig. 3. UA String Distribution in Total Transactions

4   Analyze User Agent in HTTP Transactions

The "User-Agent (UA) request header field" contains information about the user agent originating the request. From UA, we can identify browser type, operate system type and device information, and thus, analyze clients' characteristics in the dataset. In this section we first introduce "Refined UASparser", a tool we developed based on an open source: UASparser—an engine used for parsing User-Agent string of almost all frequent used browser and operator systems. Then we focused on the features in UA distribution of our dataset.

4.1   Refined UASparser

We got 318,247 unique UA strings in HHO dataset, for each UA string we also record the number of times they have appeared. Then we tried to use UASparser to parse UA strings. UASparser can extract information like operator system, browser type, and UA type (browser, library etc.). However, we find that more than 6.5% of the user agents in our dataset cannot be recognized, and these user agents contribute to more than 23% of the requests packets. Since the answer is far from satisfactory, we decided to refine the UASparser by identifying the most significant unknown UA strings of UASparser. Table 3 shows the increment after we improved the tool. It is important to notice that the number is not the accumulation of unique UA strings but the number of times each UA string appeared in requests, as we care more about the proportion of overall transactions. The improvement is significant as 9.8% more transactions are assigned to their operator system types and 34.4% more to their browser types.

Table 2. Comparasion of UASparser and R-UASparser

Tools UA-infomation	UASparser (request times)	Refined-UASparser (request times)	Improvement (%)
Operator System	33,669,987	36,960,690	9.8%
Browser type	29,677,693	39,898,156	34.4%

Figure 4 shows the top 10 newly identified client types that contribute to the improvement.

OS Improvement. The most improvement in recognizing operator system is due to the traffic of windows updates—about 72% of the total improvement. The other improvement is in Windows library, Microsoft CryptoAPI and so on, Apple IOS and Symbian. Even Symbian devices is rare in these days, it still contributed to about 0.2% of total requests and 1% of requests from mobile devices, and UASparser ignored most Symbian UA strings. When it comes to browser type, we find even more improvement was brought by our Refined tool.

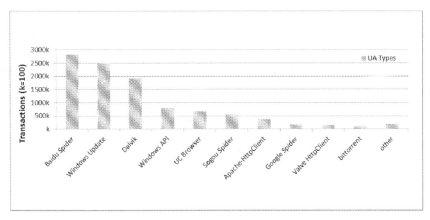

Fig. 4. Rank of Improved UA Types

UA Improvement. We find that UA types in Figure 4 are mostly 'none-browser', many of which are spiders, Windows update and Dalvik. Dalvik is a process virtual machine that runs on Android operation system, used to runs apps on Android devices. According to this result, we find UASparser is weak in finding none-browser UA strings like spider, update, and app UA strings. But in our dataset browsing requests contributes only 59% of the total requests—we will discuss it in next subsection (section 4.2). Thus, it evidenced the urgency for us to add none-browsing UA strings to UA library of the open source.

4.2   User-Agent Analysis

User-Agent Type
We classified UA strings to 8 types—browser: UA strings represent browser on the PC or laptops; mobile browser: browser on mobile devices; spider: crawler or robot programs of search engines; update: windows and other update traffic; mobile app: applications running on mobile devices; library: library API (Python-urllib, libwww-perl, etc.) that automatically request for data; P2P: P2P download traffic in http; unknown: unidentified UA strings.

Figure 5 shows the constitution of UA strings in HTTP Request. Sub-figure b presents the proportion of distinctive UA strings of each UA type, and sub-figure b presents the accumulative request number of proportion for each UA type. We can tell that browser still dominate, with more than 58% http requests, while mobile browser has become the second biggest http user in the Internet, accounting for 11% of the total requests. It infers a huge growth of mobile device usage since 2009 when compared with Gregor Maier's work [2].

It is interesting to notice that Spider (crawler or robot) accounts for more than 8% requests, showing that robot programs now have a dramatic influence to the http volume. More than 6% is made by windows update programs and 4% is sent by library programs. A library type is a client program that provoke library API (Python-urllib, libwww-perl, etc.) to automatically request for data. Putting former three

UA types together, we surprisingly find that almost at least 20% requests were sent by automatic programs. Since the number of Library (12 different Library) and Spider (8 different spider) are limited and some programs can blend itself in the normal requests by disguise its UA string as a popular browser. We believe that the proportion of program made requests is far more than 20%. It is important to identify these none-browser HTTP traffic when we analyze http traffic and user behavior, because they are not what many researchers care about and can lead to pitfalls and render a deliberately executed research flawed.

Secondly, we find that mobile devices contribute to the most unique UA strings while they only account for 11% total requests. That is due to the fragmentation of mobile devices, especially in Android devices. We find Android devices accounts for 80% distinctive UA strings of mobile browser. This finding conform to the fact that Android has a numerous number of different versions and is now running on devices produced by hundreds of different companies.

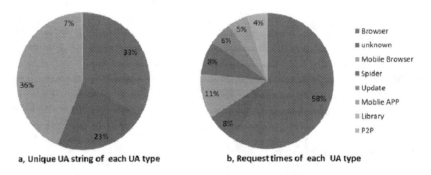

a, Unique UA string of each UA type   b, Request times of each UA type

Fig. 5. Unique UA strings and UA request times of each UA type

Operator System Popularity

Figure 6 shows the constitution of HTTP traffic of each operator system. And Table 4 further describes the contribution of different versions of Windows, Android and IOS system, which are the top 3 requests makers and contribute to 80% traffic in total. We have included all user-agent strings in the dataset and thus, the unknown part—take up 16% of traffic—is mainly library and spider that cannot be classified by operator system.

In terms, Windows clearly dominates in popularity of operator system, with more than 64% http requests traffic. We find that even though Microsoft has stopped the support to Windows XP since April 8 2014, it still account for 56% of all windows traffic. This means that Windows XP is still leading in operator system in Chinese computers. Since Windows takes up more than 90% PC market and PC traffic in our dataset. This finding suggests that more than half of Chinese computers are now in the threatening situation with no security support and unsolved vulnerabilities. Windows 7 contributes to 29.5% and the newest version Windows 8 only 2.7%, even less than Windows 2000 (4%).

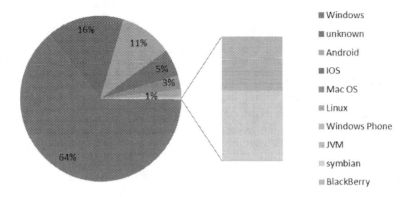

Fig. 6. Operator System Distribution

Mobile operator systems account for about 16% of total http transactions, in which Android devices take up 11% and IOS devices take up 5%. Other mobile operator systems—including Windows Phone, Symbian, BlackBerry and Palm OS—occupy only 1% transactions. The answer indicates that the former two mobile devices dominate the smartphone and tablet (or Pad) market.

Table 3. Distribution of Operator system version

Operator System	Percentage in Total requests	Operator System version	Percentage in its own type
Windows	64%	Windows XP	56.2%
		Windows 7	29.5%
		Unknown	5.5%
		Windows 2000	4.4
		Windows 8	2.7%
		Other	1.7%
Android	11%	Android 4.x	84.2%
		Android 2.x	8.3%
		Other	7.5%
ISO	5%	IOS 7	52.5%
		Unknown	23.3%
		IOS 6	10.8
		Other	13.6%

Table 3 also shows that Android 4.x account for 84.2% of all Android transactions and IOS 7 account for 52.5% of all IOS traffic. Comparing with Windows which old version XP dominates the proportion, we find that people are willing to use or update the latest system version on their mobile device and reluctant to update outdated operator system on their computers.

HTTP Client Popularity

Figure 7 presents the most popular http clients in our dataset.   IE, Chrome and Fire-fox Browser rank the first three positions, sharing more than 46% of overall and al-most 80% of browser transactions. The pink bar means transactions made by web crawlers (spiders).The first pink bar, Baidu spider, surprisingly takes the fourth posi-tion, means that 5% of total transactions are made by Baidu spider program.   The answer also illustrates that search engines, in order to crawl most web pages in the Internet and index them, creating a huge number of transactions. The second pink bar, Sogou spider, takes only one fifth of the number of Baidu spider. And Google spider shares one third of the number of Sogou spider, means that google spider is relatively inactive in China. That can be explained by the fact that Google Company quit China in 2010, and it market share in China is now less than 2%.

Blue bars represent clients as programs using library like ApacheHttp-Client or curl to automatically request data from servers. The appearance of these library users in Figure 7 manifests that the traffic created by program other than browser should not be ignored in a research of HTTP analysis.

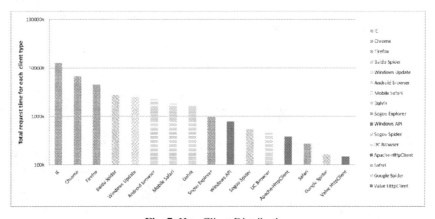

Fig. 7. Http Client Distribution

5   Conclusion

In this paper, we focus on the user-agent (UA) strings. We refined a popular open source UASparser, identifying 34.4% (9.8%) more user-agent types (operator system) of overall transactions. Moreover, we find that Windows dominate in HTTP clients, accounting for 64% of total transactions, and Windows XP takes up more than half of PC HTTP traffic, mobile devices share more than 16% of total HTTP traffic and au-tomatic programs at least 20%. Android devices share 65.5% of mobile transactions while IOS devices share 32.5%. These information leads to following consequences: PC devices in China obviously dominate the HTTP traffic, but more than half of them are threatened to a huge risk as security support for Windows XP has been ended recently; mobile devices occupy a big proportion in total transactions of a DSL lines; and Android and IOS devices dominate the mobile transactions, with more than 98%

of mobile transactions; automatic programs take up at least one fifth of all HTTP traffic and that means the researcher should no longer ignore the influence of them.

We also find that more than 80% of Android devices are running on Android 4.x and 50% of IOS devices are running on IOS 7, while only 2.7% computer choose Windows 8, indicating that mobile users are more willing to update OS of their devices or replace old devices.

Acknowledgements. This work is supported by the National High Technology Research and Development Program (863 Program) of China (No. 2011AA010703); the National Science and Technology Support Program (No. 2012BAH46B02); the Strategic Priority Research Program of the Chinese Academy of Sciences (No. XDA06030200) and National Natural Science Foundation of China (No. 61070184).

References

1. Callahan, T., Allman, M., Paxson, V.: A longitudinal view of HTTP traffic. In: Krishnamurthy, A., Plattner, B. (eds.) PAM 2010. LNCS, vol. 6032, pp. 222–231. Springer, Heidelberg (2010)
2. Schneider, F., Ager, B., Maier, G., Feldmann, A., Uhlig, S.: Pitfalls in HTTP traffic measurements and analysis. In: Taft, N., Ricciato, F. (eds.) PAM 2012. LNCS, vol. 7192, pp. 242–251. Springer, Heidelberg (2012)
3. Dhungana, S.: Mobile Web Usage: A Network Perspective (2013)
4. Ager, B., Schneider, F., Kim, J., et al.: Revisiting cacheability in times of user generated content. In: IEEE INFOCOM IEEE Conference on Computer Communications Workshops, vol. 2010, pp. 1–6 (2010)
5. Schneider, F., Agarwal, S., Alpcan, T., Feldmann, A.: The new web: Characterizing AJAX traffic. In: Claypool, M., Uhlig, S. (eds.) PAM 2008. LNCS, vol. 4979, pp. 31–40. Springer, Heidelberg (2008)
6. Schneider, F., Feldmann, A., Krishnamurthy, B., et al.: Understanding online social network usage from a network perspective. In: Proceedings of the 9th ACM SIGCOMM Conference on Internet Measurement Conference, pp. 35–48. ACM (2009)
7. Augustin, B., Mellouk, A.: On Traffic Patterns of HTTP Applications. In: 2011 IEEE Global Telecommunications Conference (GLOBECOM 2011), pp. 1–6. IEEE (2011)
8. Ihm, S., Pai, V.S.: Towards understanding modern web traffic. In: Proceedings of the 2011 ACM SIGCOMM Conference on Internet Measurement Conference, pp. 295–312. ACM (2011)
9. Fielding, R., Gettys, J., Mogul, J., et al.: Hypertext transfer protocol–HTTP/1.1, 1999. RFC2616 (2006)
10. Maier, G., Schneider, F., Feldmann, A.: A first look at mobile hand-held device traffic. In: Krishnamurthy, A., Plattner, B. (eds.) PAM 2010. LNCS, vol. 6032, pp. 161–170. Springer, Heidelberg (2010)
11. Falaki, H., Lymberopoulos, D., Mahajan, R., et al.: A first look at traffic on smartphones. In: Proceedings of the 10th ACM SIGCOMM Conference on Internet Measurement, pp. 281–287. ACM (2010)
12. Glitho, R.H., Hamadi, R., Huie, R.: Architectural framework for using java servlets in a SIP environment. In: Lorenz, P. (ed.) ICN 2001. LNCS, vol. 2094, pp. 707–716. Springer, Heidelberg (2001)
13. Holley, R., Rosenfeld, D.: MAUL: Machine Agent User Learning

Cloud-Oriented SAT Solver Based
on Obfuscating CNF Formula

Ying Qin, Shengyu Shen, Jingzhu Kong, and Huadong Dai

College of Computer, National University of Defense Technology, ChangSha, China
{yingqin,syshen}@nudt.edu.cn, {kongjinzhu,hddai}@vip.163.com

Abstract. Propositional satisfiability (SAT) has been widely used in hardware and software verification. Outsourcing complex SAT problem to Cloud can leverage the huge computation capability and flexibility of Cloud. But some confidential information encoded in CNF formula, such as circuit structure, may be leaked to unauthorized third party.

In this paper, we propose a novel Cloud-oriented SAT solving algorithm to preserve privacy. **First**, an obfuscated CNF formula is generated by embedding a Husk formula into the original formula with proper rules. **Second**, the obfuscated formula is solved by a state-of-the-art SAT solver deployed in Cloud. **Third**, a simple mapping algorithm is used to map the solution of the obfuscated formula back to that of the original CNF formula.

Theoretical analysis and experimental result show that our algorithms can significantly improve security of the SAT solver with linear complexity while keeping its solution space unchanged.

Keywords: SAT solver, CNF formula, Privacy, Obfuscating, Cloud computing.

1 Introduction

Propositional satisfiability [1] (SAT) has been widely used in hardware and software verification [2][3]. With the rapid increase of the hardware and software system size, the scale of SAT problem generated from verification also increases rapidly. On the other hand, Cloud computing can provide elastic computing resource, which make outsourcing hard SAT problem to public Cloud [19][20] very attractive. However, unauthorized access [11] to outsourced data prevent it from widely deployed.

In formal verification, circuit, code and properties will first be converted into CNF (Conjunctive Normal Form) formula by Tsentin coding[4] before SAT solving. After that, circuit structure and other confidential information are still existed in CNF formula. To prevent unauthorized user from accessing these confidential information, it is necessary to obfuscate CNF formula before outsourcing.

This paper presents a novel Cloud-Oriented SAT solver based on Obfuscating CNF formula. **First**, the original CNF formula S_1 is obfuscated into formula S by embedding another CNF formula S_2 with unique solution. And the embedding

W. Han et al. (Eds.): APWeb 2014 Workshops, LNCS 8710, pp. 188–199, 2014.

rules guarantee that S's graph structure are significantly different from that of S_1. **Second**, S is sent to a SAT solver in Cloud, which returns a solution R_O. **Third**, the solution R of S_1 is obtained from the solution of R_O by projection. its correctness can be guaranteed by the embedding rules in the first step.

The major contributions of this paper are: **First**, by obfuscating, confidential information in the original CNF formula, such as circuit structure, will be destroyed in the obfuscated CNF formula; **Second**, the obfuscated CNF formula can be solved by state-of-the-art SAT solver. **Third**, the theoretical analysis and experiments shows that obfuscating and solution recovery algorithms are linear complexity, reducing the impact on the overall performance of SAT solving.

The remainder of this paper is organized as follows. Background material is presented in Section 2. Section 3 gives the description of the problem, while the implementation of Cloud SAT solver based on obfuscating is presented in Section 4. Section 5 analyzes correctness, effectiveness and complexity; Section 6 describes the related work; Section 7 gives the experimental results, while Section 8 concludes this paper.

2 Preliminaries

2.1 SAT Solving

The Boolean value set is denoted as $B = \{T, F\}$. For a Boolean formula S over a variable set V, the propositional satisfiability problem (abbreviated as SAT) is to find a satisfying assignment $A : V \to B$, so that S evaluates to T. If such a satisfying assignment exists, then S is satisfiable; otherwise, it is unsatisfiable. A clause subset of an unsatisfiable formula is an unsatisfied core. A computer program that decides the existence of a satisfying assignment is SAT solver[10].

Normally, a SAT solver requires the formula to be conjunctive normal form (CNF), in which a formula is a conjunction of its clause set, and a clause is a disjunction of its literal set, while a literal is a variable or its negation.

Φ in Equation (1) is a CNF formula with four variables x_1, x_2, x_3, x_4, and four clauses $x_1 \vee \neg x_2$, $x_2 \vee x_3$, $x_2 \vee \neg x_4$, $\neg x_1 \vee \neg x_3 \vee x_4$. Literal x_1 is positive literal of variable x_1 in clause $x_1 \vee \neg x_2$, while $\neg x_2$ is a negative literal.

$$\Phi = (x_1 \vee \neg x_2) \wedge (x_2 \vee x_3) \wedge (x_2 \vee \neg x_4) \wedge (\neg x_1 \vee \neg x_3 \vee x_4) \tag{1}$$

The number of literals in clause C is denoted as $|C|$. The number of clauses in a CNF formula F is denoted as $|F|$. For example $|x_1 \vee \neg x_2| \equiv 2$, while $|\phi| \equiv 4$

2.2 Tseitin Encoding

In hardware verification, circuits and properties are converted into CNF formula by Tseitin encoding[4], and then CNF formula is solved by SAT solver. Circuits can all be expressed by a combination of gate AND2 and INV, so we only lists Tseitin encoding of gate AND2 and INV here.

For gate INV $z = \neg x$, its Tseitin encoding is $(x \vee z) \wedge (\neg x \vee \neg z)$. For gate AND2 $z = x_1 \wedge x_2$, its CNF formula is $(\neg x_1 \vee \neg x_2 \vee z) \wedge (x_1 \vee \neg z) \wedge (x_2 \vee \neg z)$. For a complex circuit C expressed by a combination of AND2 and INV, its Tseitin encoding $Tseitin(C)$ is a conjunctive of all these gates' Tseitin encoding.

For a circuit C with an INV $d = \neg a$ and an AND2 $e = d \wedge c$, its Tseitin encoding is shown in Equation (2).

$$Tseitin(C) = \left\{ \begin{array}{c} (a \vee d) \\ \wedge (\neg a \vee \neg d) \end{array} \right\} \wedge \left\{ \begin{array}{c} (\neg e \vee c) \\ \wedge (\neg e \vee d) \\ \wedge (e \vee \neg c \vee \neg d) \end{array} \right\} \tag{2}$$

3   Threat Model of Cloud-Based SAT Solving

Solving SAT in Cloud includes three steps: **first** generating and uploading CNF formula to Cloud, **then** solving CNF in Cloud, **finally** downloading solution. Thus, unauthorized access[11] to CNF formula in Cloud may result in leakage of confidential information.

On the other hand, verifying the result of SAT is simple. If CNF formula is satisfiable, the solution can be substituted into CNF to check if it is satisfiable; If CNF formula is unsatisfiable, an unsatisfiable core returned by the SAT solver can be verified by resolution. In both case, verifying the result are linear complexity.

Thus, in this paper we assume that Cloud servers are honest but curious. That is, the CNF will be solved correctly, but also may be analyzed to recover confidential information. Algorithms [6–9] have been proposed to recover circuit from CNF. Before discussing them, some concepts should be introduced first.

Definition 1 (CNF Signature). *CNF signature of gate g is its Tseitin encoding $Tseitin(g)$. Each clause in CNF signature is called characteristic clause. A characteristic clause containing all variables in CNF-signature is a **key clause**. Variables correspond to output of a gate is called **output variable**.*

For AND2 in Equation (2), $\neg e \vee c$ is a characteristic clause. Clause $e \vee \neg c \vee \neg d$ is a key clause. e is an output variable.

With such encoding rules, gates with the same CNF signature will be encoded into the same set of clauses. Potential attackers can exploit structural knowledge to recover the circuit structure. Some such algorithms[6–9] are based on concept of directed hyper-graph and bipartite graph described below.

Definition 2 (Hypergraph of CNF and Directed Hypergraph of CNF). *A Hypergraph $G = (V, E)$ of a CNF formula Σ is*

1. *each vertex of V corresponds to a clause of Σ;*
2. *each edge $(c_1, c_2) \in E$ corresponds to two clauses c_1 and c_2 containing the same variable or its negation;*
3. *each edge is labeled by the variable;*

A Directed Hypergraph is a Hypergraph with each endpoint of edge labeled by ↑ when clause contains non-negative variable, or † when clause contains negative variable.

Definition 3 (Bipartite graph of CNF). *A Bipartite graph $G = (V, E)$ of a CNF formula Σ is*

1. *each vertex of V corresponds to clause and variable of Σ;*
2. *each edge $(c, v) \in E$ corresponds to a clauses c that contains variable v;*
3. *each edge is labeled by ↑ when clause contains non-negative variable, or † when clause contains negative variable.*

With these definitions, Roy et al. [8] first converts the CNF to an Hypergraph G, and then matches the CNF signatures of all types of gates in G to recover gates by subgraph isomorphism, finally creates a maximal independent set (MIS) instance to represent the recovered circuit.

Fu et al.[9] presents another algorithm that first detects all possible gates with pattern matching, and then constructs a maximum acyclic combinational circuit by selecting a maximum subset of matched gate.

In Cloud computing paradigm, attacker may use these algorithms to recover the circuit structure from CNF formula. Therefore, it is essential to prevent information leakage before outsourcing CNF formula to Cloud.

4   Cloud-Based SAT Solver Preserving Privacy

In this paper, we present a Cloud-oriented SAT solver based on obfuscating CNF formula, which prevent information leakage by hiding the structure in CNF. The obfuscating algorithm is based on the following three facts:

1. First, changing CNF signature of gates in CNF formula will make circuit recovering based on pattern matching or subgraph isomorphism impossible.
2. Second, current SAT solver with conflict analysis [10] is very efficient. So we would like to use them directly, instead of developing a new one like [12].
3. Third, the solution of obfuscated CNF formula should be easily mapped back to original formula.

The proposed algorithm is based on the following definition.

Definition 4 (Husk Formula). *Husk formula is a CNF formula with a unique solution, and assignment of variables is non-uniform, that is, not all 0 or all 1.*

The overall framework of the Cloud SAT solver is shown intuitively in Figure 1. In Step 1, GENERATOR algorithm generates a husk formula with unique solution R_H; In Step 2, OBFUSCATOR algorithm obfuscates S_1 to obtain a new CNF formula S; In Step 3, S is solved in Cloud; In Step 4, MAPPER algorithm maps solution of S to that of S_1.

The GENERATOR, OBFUSCATOR and MAPPER algorithm will be described in Subsection 4.1,4.2 and 4.3 respectively.

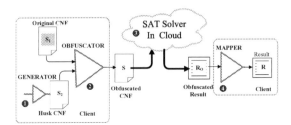

Fig. 1. The Cloud SAT solver based on obfuscating CNF

4.1 Generating Husk Formula

Husk formula is constructed based on prime factorization method: **First**, given a prime p represented by a binary vector $p = <p_1, p_2, \ldots, p_n>$, we assigning its square p^2 to the output of a multiplier M, with constraint $p \neq 1$. **Second**, We convert the multiplier M into CNF formula $Tseitin(M)$.

To satisfy $Tseitin(M)$, the two inputs of M must all be $p = <p_1, p_2, \ldots, p_n>$, which makes the p the unique solution of $Tseitin(M)$.

With these discussion, GENERATOR algorithm to generate Husk formula is shown in Algorithm 1.

4.2 Constructing Obfuscated Formula

The proposed OBUFSCATOR algorithm generates a new CNF formula S by embedding Husk formula S_2 into the original formula S_1 with proper rules. By adding new clauses and new literals, OBUFSCATOR algorithm changes the clause set and literal set in clauses of S_1, to prevent its structure from being recovered.

In order to keep solution space unchanged, when inserting variables of S_2 into clauses of S_1, the following rules must be followed, as depicted in Line 5 and 18 of OBFUSCATOR algorithm: Variables assigned F in R_H will be inserted as positive literals; While variables assigned T will be inserted as negative literals.

Algorithm 1. GENERATOR

Data: NULL
Result: Husk CNF S_2 and Husk result R_H
1 **begin**
2 | Generating a prime number p ;
3 | $\Phi = M(I_1 \neq 1, I_2 \neq 1, O = p^2)$;
4 | $S_2 = Tseitin(\Phi)$;
5 | $R_H = p \mid p$;
6 **end**

Algorithm 2. OBFUSCATOR

Data: The original CNF S_1,Husk CNF S_2,Husk result R_H
Result: The obfuscated CNF S, variable mapping M

1 **begin**
2     $mark(S_1)$;
3   **foreach** $c \in S_1$ **do**
4   lit =get literal $\in R_H$;
5        $c = c \cup \neg lit$;
6   **if** $c \in Key\ Clause\ Set\ KCS$ **then**
7           $nc = generate\_new\_clause(c, lit)$;
8           $S_2 = S_2 \cup nc$;
9   **end**
10   **end**
11   **foreach** $c \in S_1$ **do**
12       $averagelen = \frac{\sigma_{c' \in S_1} |c'|}{|S_1|}$;
13   **while** $|c| < averagelen$ **do**
14   lit =get literal $\in R_H$;
15   **while** $\neg lit \in c$ **do**
16   lit=get literal $\in R_H$;
17   **end**
18          $c = c \cup \neg lit$;
19   **end**
20   M =remap all variable in $S_1 \cup S_2$;
21   S =reorder all clause in $S_1 \cup S_2$;
22   **end**
23 **end**

Thus, S and S_1 can be solved with the same SAT solver, and the solution of S_1 can be extracted from that of S by projection on variables set of S_1. More details of OBUFSCATOR algorithm is presented in Algorithm 2.

In algorithm 2, Procedure **mark** at Line 2 marks key clauses and output variables of some kind of gate in CNF formula. Procedure **generate_new_clause** at Line 7 generates some new clauses matching the key clauses, such that identifying gates become more difficult.

Different mark algorithms are needed for different gate types with different CNF signatures. But their complexity are of the same. As all circuits can be represented by a combination of AND2 and INV, and the $mark$ algorithm for INV is trivial, so we only present the implementation of $mark$ for AND2 in Algorithm 3.Similarly, we also only present the implementation of **generate_new_clause** for AND2 in Algorithm 3.

4.3 Recover the Original Solution

The variables set of S is a superset of S_1. Therefore, to get solution of S_1, we need to filter assignment of variables in S_1 from R_O according to the variable name

---

Algorithm 3. mark and **generate_new_clause**

---

1 **mark**;
  Data: CNF formula S
  Result: marked S
2 **begin**
3   **foreach** $(C \in S)$ & $(|C| \equiv 3)$ **do**
4   **foreach** $l \in C$ **do**
5   **foreach** $(C_1 \in S)$ & $(\neg l \in C_1)$ & $(|C_1| \equiv 2)$ **do**
6   **foreach** $l_1 \in C_1$ **do**
7   **if** $(\neg l_1 \in C)$ & $(l_1 \neq l)$ **then**
8                    $match + + $;
9   **end**
10   **end**
11   **end**
12   **end**
13   **if** $match \equiv 2$ **then**
14   mark L as output literal ;
15   mark C as key clause ;
16   **end**
17 **end**
18 **generate_new_clause**;
  Data: key clause C in AND2, Husk literal lit
  Result: new clause C_1
19 **begin**
20     $olit$=Getting output literal from C ;
21     $C_1 = lit \cup \neg olit$;
22 **end**

---

mapping table M. R_O is the solution of S returned by SAT solver in Cloud. This is implemented by MAPPER in Algorithm 4. M[var].variable and M[var].formula At Line 4 and 5, represent the original variables of var and formula the var belongs to, formula can be S_1 or S_2.

5   Correctness, Effectiveness and Performance Analysis

5.1   Correctness Proof

The Correctness means that the algorithm should keep the solution space unchanged, that is, for CNF formulas S_1 and its obfuscated formula S:

1. If S_1 is unsatisfied iff S is unsatisfied. And the unsatisfied core of S_1 can be obtained from unsatisfied core of S by deleting literal in S_2;
2. If S_1 is satisfied iff S is satisfied. And the solution of S_1 can be obtained by projecting solution of S into variables set of S_1 .

Algorithm 4. MAPPER

Data: Obfuscated result R_O, variable mapping table M, Husk result R_H

Result: Result R

1 **begin**
2   **foreach** $lit \in R_O$ **do**
3          $var = abs(lit)$;
4          $rvar = M[var].variable$;
5   **if** $M[var].formula \equiv S_1$ **then**
6             $R[rvar] = lit > 0?rvar : \neg rvar$;
7   **else**
8             $Hlit = lit > 0?rvar : \neg rvar$;
9   **if** $Hr[rvar] \neq Hlit$ **then**
10   alert("something wrong with CloudSAT solver");
11   **end**
12   **end**
13 **end**

OBFUSCATOR algorithm can be simplified to the follows rules.

1. **Rule 1:** For each clause $c \in S_1$, we take one or more variables from S_2, and insert them into c according to the following rule: if variable is T in R_H, insert its negative literal; if variable is F in R_H, insert its positive literal. The resulting clause set is denoted as S_3.
2. **Rule 2:** generating new clauses with literals from R_H and output variables in S_1 according to the following rule: if variable is T in R_H, insert positive literal into clause; if variable is F in R_H, insert negative literal into clause; Literal of output variable is extracted directly from the key clause and inverted. New clauses set generated in this way is denoted as S_4.
3. **Rule 3:** Combining and randomly reordering S_2,S_3 and S_4 to produce S.

Definition 5 (Original Variable). *Variable and clause in original CNF formula is denoted* **original variable** *and* **original clause**. *Variable and clause in Husk formula is called* **Husk variable** *and* **Husk clause**. *Literal of Husk result is called* **Husk literal**.

Husk formula's solution is unique. Let the result be $R_H = \{b_1, b_2, \ldots, b_m\}$. S_2 can be satisfied only when assigning R_H to it.

According to Rule 1, each clause in S_3 can be expressed as $C = A \vee B$, where A is clause from S_1 , and $B = B_i$, $B_i = z_i \vee B_{i-1}$, $B_1 = z_1$. If $b_i \equiv F$, then $z_i = y_i$. if $b_i \equiv T$, then $z_i = \neg y_i$, while $y_i \in S_2$. With constrains of Husk clause, Husk variable must be assigned $R_H = \{b_1, b_2, \ldots, b_m\}$, then z_i must be F. So $B = B_i = z_i \vee z_{i-1} \vee \cdots \vee z_1$ must be F. This means $C = A \vee B$, whose value is decided by A.

Clauses in S_4 consist of Husk literals and output variables from S_1. These clauses can be represented as $C = z_i \vee A$. If $b_i \equiv F$, then $z_i = \neg y_i$, otherwise

$z_i = y_i$, while $y_i \in S_2$. With constraint from Husk clauses, Husk variables must be assigned as following: $R_H = \{b_1, b_2, \ldots, b_m\}$. Thus $z_i = T, C = z_i \lor A = T$. So clause C will not constrain the assignment of variables in A.

Theorem 1. *With the following hypothesis and facts:*

1. *Clause $A = a \lor X$ and clause $B = b$, while $b \notin X$;*
2. *Let $S_1 = A$, $S_2 = B$, then $R_H = \{T\}$;*
3. *With R_H and Rule 1, we have clause $C = A \lor \neg b$ and $S_3 = C$;*
4. *With R_H and Rule 2, with literal $a \in A$, we have clause $D = \neg a \lor b$ and $S_4 = D$;*
5. *let $S = S_2 \land S_3 \land S_4$ then $S = B \land C \land D$.*

we can prove $S = S_1 \land S_2$.

Proof. We have

$$
\begin{array}{lll}
S = B \land C \land D & B = b & \models \\
S = b \land C \land D & C = A \lor \neg b & \models \\
S = b \land (A \lor \neg b) \land D & & \models \\
S = b \land A \land D & D = \neg a \lor b & \models \\
S = b \land A \land (\neg a \lor b) & & \models \\
S = b \land A & B = b & \models \\
S = B \land A & S_1 = A,\ S_2 = B \models \\
S = S_1 \land S_2
\end{array}
\tag{3}
$$

In conclusion, obfuscated formula S is equivalent to $S_1 \land S_2$.

□

5.2   Effectiveness Analysis

According to Section 3, analyzing hyper-graph and bipartite graph is major approach to recover circuit structure. To show the effectiveness of our algorithm, we present qualitative analysis of its changes to hyper-graph. The changes to bipartite graphs is similar to hyper-graph.

Figure 2a) and 2b) shows the CNF signature and hyper-graph of two AND2 gate a and e. While their CNF signature and hypergraph after obfuscating are shown in Figure 2c) and 2d). There are four types of changes:

1. First, the length of key clauses c_1 and c_5 are changed from 3 to 4, this defeats structure detection techniques [9] based on key clause oriented pattern matching;
2. Second, CNF signatures of a and e are changed. This defeats structure detection techniques[8] based on sub-graph isomorphic;
3. Third, there are new clauses added in formula, such as c_4 and c_8, this change also defeats structure detection techniques[8] based on sub-graph matching;
4. Furthermore, by inserting proper literal in key clauses and generating new clause, CNF signature of instance e is changed from AND2 to AND3, shown in Figure 2b),2d). Husk variable E, which becomes a input variable of gate AND3, is indistinguishable with f and g, which are original input variables of AND2. this makes it impossible to distinguish AND2 and AND3.

5.3   Complexity of the Algorithm Analysis

The main procedure of OBFUSCATOR algorithm consists only one layer of loop. Although **mark** consists 4 layers of loop, the runtimes of the 3 inner loops are bounded by length of clauses. So the complexity of the OBFUSCATOR algorithm is $O(n)$. The complexity of the MAPPER algorithm is obvious $O(n)$.

6   Related Works

Secure Computation Outsourcing Based on Encryption: R. Gennaro et al.[13] presented the concept of verifiable computation scheme, which shows the secure computation outsourcing is viable in theory. But the extremely high complexity of FHE operation and the pessimistic circuit sizes make it impractical. Zvika et al.[12] constructed an obfuscated program for d-CNFs that preserves its functionality without revealing anything else. The construction is based on a generic multi-linear group model and graded encoding schemes, along with randomizing sub-assignments to enforce input consistency. But the scheme incurs large overhead caused by their fundamental primitives.

Secure Computation Outsourcing Based on Disguising: For linear algebra algorithms, Atallah et al. [15] multiplied data with random diagonal matrix before outsourcing. and recovered results by reversible matrix operations. Paper [16] discussed secure outsourcing of numerical and scientific computation, by disguising with a set of problem dependent techniques. C.Wang[5] presented securely outsourcing linear programming(LP) in Cloud, by explicitly decomposing LP computation into public LP solvers and private data, and provide a practical mechanism which fulfills input/output privacy, cheating resilience, and efficiency.

Verifiable Computation Delegation: Verifiable computation delegation is the technique to enable a computationally weak customer to verify the correctness of the delegated computation results from a powerful but untrusted server without investing too much resources. To prevent participants from keeping the rare events, Du. et al. [14] injected a number of chaff items into the workloads so as to confuse dishonest participants. Golle et al. [17] proposed to insert some

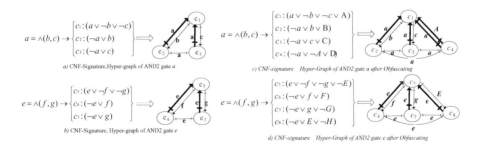

Fig. 2. CNF signature and Hyper-Graph of a and e before and after obfuscating

pre-computed results images of ringers into the computation workload to defeat untrusted or lazy workers. Szada et al. [18] extended the ringer scheme and propose methods to deal with cheating detection.

7   Experiments

Algorithms presented in this paper are implemented in language C. The experiments is conducted on a laptop with Intel Core(TM) i7-3667U CPU @ 2.00GHz, 8GB RAM. We unroll circuits in iscas89 benchmark into 30 times and transform them into CNF formulas, and generate Husk formula with variables number $vn = 675$ and clauses number $cn = 2309$, and then obfuscate the CNF formula by changing 2 input gate.we use MiniSat as solver. Figure 3 presents size of CNF formula after obfuscating with (vn/cn), number of 2 input gate in formula with $(Gate\ marked)$, SAT Solver time before and after obfuscating $(SAT\ Time\ before/SAT\ Time)$, obfuscating time $(Obfuscating\ Time)$,and solution recovery time $(Result\ time)$.

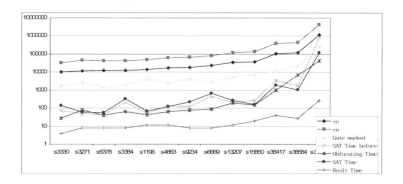

Fig. 3. Relationship between Runtime and Size of CNF formula

Experiments show that, SAT Solver time is likely to be the same between before and after obfuscating, since the original CNF formula is much larger than the Husk formula. At the same time,obfuscating time is linear with the gates marked in CNF formula, while solution recovering time increases linearly with the size of circuit vn and cn.

8   Conclusion

This paper propose the first Cloud SAT solver algorithm that can prevent the confidential information in the CNF formula from being recovered by untrusted Cloud server. Theoretical analysis and experimental result show that our algorithms can significantly improve security of the SAT solver with linear complexity while keeping its solution space unchanged.

Acknowledgements. This work was supported by the National High Technology Development 863 Program of China under Grant No.2013AA01A212.

References

1. Putnam, D.: A Computing Procedure for Quantification Theory. Journal of the ACM 7(3), 201
2. G. Hachtel, F. Somenzi. Logic synthesis and verification algorithms, I-XXIII, 1-564 (2006)
3. Clarke, E., Grumberg, O., Jha, S., Lu, Y., Veith, H.: Counterexample-Guided Abstraction Refinement. In: Emerson, E.A., Sistla, A.P. (eds.) CAV 2000. LNCS, vol. 1855, pp. 154–169. Springer, Heidelberg (2000)
4. Tseitin, G.: On the complexity of derivation in propositional calculus. Studies in Constr. Math. and Math. Logic. (1968)
5. Wang, C., Ren, K., Wang, J.: Secure and practical outsourcing of linear programming in Cloud computing. In: INFOCOM 2011, pp. 820-828 (2011)
6. Li, C.: Integrating equivalency reasoning into Davis-Putnam procedure. In: AAAI 2000, pp. 291-296 (2000)
7. Ostrowski, R.: Recovering and exploiting structural knowledge from cnf formulas, CP 2002, pp. 185-199 (2002)
8. Roy, J., Markov, I., Bertacco, V.: Restoring Circuit Structure from SAT Instances. In: IWLS 2004, pp. 361–368 (2004)
9. Fu, Z., Malik, S.: Extracting Logic Circuit Structure from Conjunctive Normal Form Descriptions. In: VLSI Design 2007, pp. 37–42 (2007)
10. MiniSat — SAT Algorithms and Applications Invited talk given by Niklas Sörensson at the CADE-20 workshop ESCAR, http://minisat.se/Papers.html
11. You, H.: Get Off of My Cloud: Exploring Information Leakage in Third-Party Compute Clouds. In: CCS 2009 (2009)
12. Brakerski, Z., Rothblum, G.: Black-Box Obfuscation for d-CNFs. In: ITCS 2014, pp. 235–250 (2014)
13. Gennaro, R., Gentry, C., Parno, B.: Non-interactive Verifiable Computing: Outsourcing Computation to Untrusted Workers. In: Rabin, T. (ed.) CRYPTO 2010. LNCS, vol. 6223, pp. 465–482. Springer, Heidelberg (2010)
14. Du, W., Goodrich, M.: Searching for High-Value Rare Events with Uncheatable Grid Computing (2004)
15. Atallah, M., Pantazopoulos, K., Rice, J., Spafford, E.: Secure outsourcing of scientific computations. Advances in Computers 54, 215–272 (2001)
16. Atallah, M., Li, J.: Secure outsourcing of sequence comparisons. Int. J. Inf. Sec. 4(4), 277–287 (2005)
17. Golle, P., Mironov, I.: Uncheatable distributed computations. In: Naccache, D. (ed.) CT-RSA 2001. LNCS, vol. 2020, pp. 425–440. Springer, Heidelberg (2001)
18. Szajda, D., Lawson, B.G., Owen, J.: Hardening functions for large scale distributed computations. In: ISSP 2003, pp. 216–224 (2003)
19. Paralleling OpenSMT Towards Cloud Computing, http://www.inf.usi.ch/urop-Tsitovich-2-127208.pdf?
20. Formal in the Cloud OneSpins New Spin on Cloud Computing, http://www.eejournal.com/archives/articles/20130627-onespin/?printView=true

The Improved AC High-Performance Pattern-Matching Algorithm for Intrusion Detection

Dongliang Xu, Hongli Zhang, and Miao Hou

School of Computer Science and Technology, Harbin Institute of Technology, Harbin, China
xudongliang@pact518.hit.edu.cn

Network Intrusion Detection Systems (NIDS) have become widely recognized as powerful tools for identifying, deterring and deflecting malicious attacks over the network. New generations of network intrusion detection systems create the need for advanced pattern-matching engines. This paper proposes an improved AC algorithm, called Semi-AC. We contribute modifications to the Aho-Corasick string-matching algorithm that drastically reduce the amount of memory required. Its efficiency is close to the standard AC, but the space is saved 50% or more.

Keywords: NIDS, attacks, AC, pattern-matching.

1   Introduction

String-matching is a fundamental problem in computer science that has resulted in a substantial number of publications in the past several decades. The heart of almost every modern intrusion detection system is a string-matching algorithm. String matching is computationally intensive. For example, the string matching routines in Snort account for up to 70% of total execution time and 80% of instructions executed on real traces [1].

String matching problem is defined as "to find all occurrences of pattern in text". Aho-Corasick[2], Knuth-Morris-Pratt[3], Boyer-Moore[1], etc. proposed the solutions of this problem. Aho and Corasick [1975] proposed an algorithm for concurrently matching multiple strings. Aho-Corasick (AC) algorithm used the structure of a finite automation that accepts all strings in the set. It is a very efficient multi-pattern matching algorithm. But the AC algorithm has some drawbacks. The focus of this paper is to improve the storage and search cost of NIDS string matching using Aho-Corasick trees.

The rest of this paper is organized as follows. Section two briefly describes the relevant prior work in pattern matching algorithms. Our advanced algorithm Semi-AC, which performs the multi-string matching, will be presented in section three. The forth section carries out various experiments and evaluation with Semi-AC algorithm. Finally, section five draws the conclusion and discusses future directions for our work.

2   Related Work

Arikawa and Shinohara[4] proposed the method of reducing the table size by dividing symbols. State transitions are two times as many as the original AC algorithm. Table size can be reduced to one-eighth.

W. Han et al. (Eds.): APWeb 2014 Workshops, LNCS 8710, pp. 200–213, 2014.
© Springer International Publishing Switzerland 2014

Snort [5] is one of the most popular public domain NIDS. The current implementation of snort uses the optimized version of the Aho-Corasick automaton. To reduce the memory requirement of the Aho-Corasick automaton, Tuck et al. [6] have proposed starting with the unoptimized Aho-Corasick automaton and using bitmaps and path compression.

Using Aho-Corasick's algorithms, the elapsed time to search patterns in a text depends heavily on the time of data transmission from a primary storage or a secondary storage. To reduce the cost of data transmission, data compression can be used. Fukamachi, et.al.[7] developed a pattern matching machine, based on Aho-Corasick's one, that scans a compressed text by Huffman Codes[8] without decoding. Since Huffman Codes are variable length, the Aho-Corasick pattern matching machine may cause false detections. To avoid false detections, Huffman Tree is attached to the pattern matching machine. The algorithm runs in linear time with respect to the total length of compressed patterns and the length of a compressed text.

3   Optimization and Implementation

A   AC Algorithm

Aho-Corasick proposed a multi-string matching algorithm, which is able to match strings in worst-case time linear in the size of the input. The AC algorithm consists of two parts: (1) constructing the pattern matching machine from given patterns and (2) making pattern matching machine run on text. The Aho-Corasick algorithm scans an input text T of length m and detects any exact occurrence of each of the patterns of a given dictionary, including partially and completely overlapping occurrences. A pattern is a sequence of symbols from an alphabet, and a dictionary a set of patterns $P=\{p_1,p_2,...,p_n\}$. The algorithm first constructs an automaton on the basis of the dictionary and then applies the automaton to the input text. The transition from state to state depends on the current input symbol. A pattern is identified when the automaton enters a state flagged as final. The complexities of constructing a pattern automaton and scanning a text are linear to the total length of given patterns and the length of a text, respectively. Given a set of patterns = {ACGATAT, ATATATA, ATATGC, TAGAT}, we show the completed pattern matching machine in Fig.1.

The Aho-Corasick finite state automaton for multi-string matching is widely used in NIDS. But the AC algorithm has a drawback which is the automaton need a lot of memory space. Assume that the alphabet size is 256 (ASCII characters). For a given set of patterns, a natural way to store the automaton is to represent each state of the automaton by node that has 256 success pointers, a failure pointer. If a pointer takes 4bytes and the rule list is simply pointed at by the node, the size of each state node is 1KB. So the automaton requires a very large memory when there are more than ten thousands patterns. As the AC automaton requires such a large amount of space, there are many improved methods based on compressing the space occupied by the automaton. However, most methods reduce the efficiency of the algorithm while compressing space. But the Semi-AC algorithm we proposed drastically reduce the amount of space, its efficiency is close to the original AC.

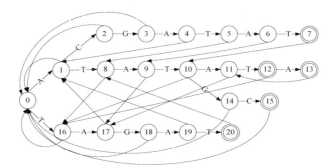

Fig. 1. The completed pattern Aho-Corasick automaton for the set{ACGATAT, ATATATA, ATATGC, TATAT}. Double-circled states are terminal.The dashed links represent the state-to-state supply function.

B   Experimental Statistics

We conducted an experiment on a pattern set containing 50 thousands patterns, which the min length of pattern is 4 characters and the max length is 249 characters. We statistic the frequency of the first n (n=2, 3, 4,$\cdots$, and n is less than the max length of pattern) characters of each pattern in the whole pattern set. The statistical results of the first 4 and 5 characters are showed in Fig.2 and Fig.3.

In Fig.2, the percentage of the first four characters which appear only once in the pattern set is 55%. The sum of the percentage of the first four characters which appear one, two, three and four times in the set is 83.85%.

In Fig.3, the percentage of the first five characters which appears only once in the pattern set is 69%. The sum of the percentage of the first five characters which appear one, two, three and four times in the set is 91.6%.

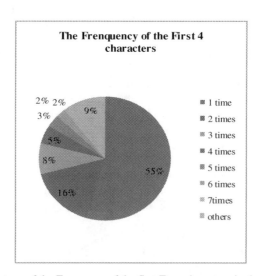

Fig. 2. The Percentage of the Frequency of the first Four characters in the whole pattern set

As shown in Fig.4, the sum of the percentage of the first n characters which appear one, two, three and four times in the pattern set increasing with n. we can understand that the percentage of the first n characters which appear more than five times is very small. For the efficient memory, we can build a semi-automatic machine with the first n characters of each pattern.

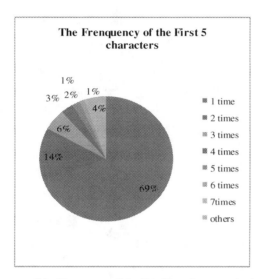

Fig. 3. The Percentage of the Frequency of the first Five characters in the whole pattern set

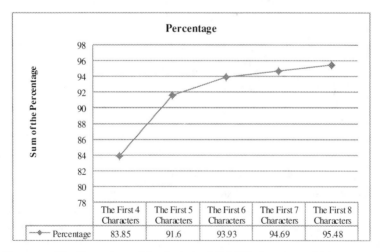

Fig. 4. The sum of the Percentage of the first n(n=4,5,6,7,8) characters which appear one, two, three and four times in the whole pattern set

C   Data Structure in Semi-AC

Because the size of each state node in original AC is 1KB, in the current conditions, there is not enough memory to build such a large automatic machine when there are more than ten thousands patterns. We have to improve the automatic machine. From the above experiments we can see that the repetition rate of the first n ($n>4$) characters is very low. In order to reduce space, we can build a semi-automatic machine with the first n characters of each pattern. Before describing Semi-AC automaton, we introduce a data structure on a set of patterns, called a trie. The trie of set $P=\{P_1,P_2,\cdots P_r\}$ is a rooted directed tree that represents the set P. Namely, every path starting from the root is labeled by one of the patterns P_i, and conversely, every pattern $p_i \in P$ labels a path from the root. We give the trie for $P=\{ACGATAT,$ $ATATATA, ATATGC, TAGAT\}$ in Fig.5.

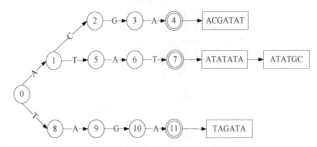

Fig. 5. The Semi-trie for the   set $P=\{ACGATAT, ATATATA, ATATGC, TAGAT\}$. The set Q of each terminal state is represented with squares.

First-n denote the first n characters used to construct the Semi-trie. $p_i^{c_i}$ is the ith pattern of set P, whose length is c_i. *Initstate*, *Curstate*, q and *Terstate* denote the initial state, current state, normal state and terminal state, respectively. $\delta(q,*)$ is the state transition function.

The build process is as follows: Firstly, the initial state *Initstate* (*Initstate*=0) is assigned to the current state *Curstate*. Secondly, get a pattern from the pattern set P and take the smaller of the *First-n* and c_i. Thirdly, for each state q of trie, we code $\delta(q,*)$ in a table of size $|\Sigma|$. The total size of the trie of a set P is worst-case $|\Sigma|*|P|$. It has the advantage of passing through a transition in constant time $O(1)$ by performing an access to a table. When the state q is the state of the forth character of the pattern or is a terminal state, a set $Q(q)$ points to a list of all the numbers of the patterns in P that correspond to q. Algorithm1 is the pseudo-code for the trie construction.

Algorithm 1. SemiTrie_Create($First\text{-}n$, P)

   Input: $First\text{-}n$, $P=\{P_1^{C_1}, P_2^{C_2}, \cdots, P_r^{C_r}\}$

Output: SemiTrie
Procedures :
01:    $Initstate \leftarrow 0$
02: **for** ($i \leftarrow 0$ to r)
03:    $Curstate \leftarrow Initstate$
04:    $j \leftarrow 0$
05:    $n \leftarrow \min\{First - n, c_i\}$
06:    **while** $j \leq n$ and $\delta(Curstate, p_i^j) \neq \theta$
07:        $Curstate \leftarrow \delta(Curstate, p_i^j)$
08:        $j \leftarrow j+1$
09:    **End while**
10:    **while** $j \leq n$
11:        Create a new normal state q
12:        $\delta(Curstate, p_i^j) \leftarrow q$
13:        $Curstate \leftarrow q$
14:        $j \leftarrow j+1$
15:    **End while**
16:    **if** ($Curstate$ has already been a terminal state)
17:        $Q(Curstate) \leftarrow Q(Curstate) \bigcup \{p_i\}$
18:    **else**
19:        mark $Curstate$ as a terminal state
20:        $Q(Curstate) \leftarrow \{p_i\}$
21:    **End if**
22: **End for**

The Semi-AC automaton built on P is the trie of P augmented with a "Failure function" *Failure*. The algorithm for building Semi-AC automaton is similar to building the standard AC automaton. The pseudo-code is shown in Algorithm 2.

The state *Curstate* goes in transversal order through the trie *Semi-trie* built on P. The state *Down* goes down the failure links from the parent of *Curstate*, looking for an outgoing transition labeled with the same character as between *Curstate* and its parent. $Q(Curstate)$ is initialized as empty when *Curstate* is first marked as terminal state.

Algorithm 2. SemiAC_Create(*First-n* , *P*)

Input: *First-n*, $P=\{P_1^{C_1}, P_2^{C_2}, \cdots, P_r^{C_r}\}$

Output: Semi-AC automaton

Procedures :

01: AC_Trie ⟵ SemiTrie_Create $(First - n, P = \{P_1^{C_1}, P_2^{C_2}, \cdots, P_r^{C_r}\})$

02: *Initstate* ⟵ root of AC_Trie

03: Failure(*Initstate*) ⟵ θ

04: **for**(*Curstate* in transversal order)

05:    Parent ⟵ parent of Curstate in AC_Trie

06:    $\sigma \leftarrow$ label of the transition from Parent to Curstate

07:    *Down* ⟵ Failure(*Initstate*)

08:    **while** *Down* ≠ θ **and** $\delta(Down, \sigma) = \theta$

09:        *Down* ⟵ Failure(*Down*)

10:    **End while**

11:    **if** *Down* ≠ θ

12:        Failure(*Curstate*) ⟵ $\delta(Down, \sigma)$

13:            **if** *Failure(Curstate)* is a terminal state

14:                mark *Curstate* as a terminal state

15:                $Q(Curstate)$ ⟵ $Q(Curstate) \bigcup Q(Failure(Curstate))$

16:            **End if**

17:        **else**

18:            (Failure(*Curstate*)) ⟵ *Initstate*

19:        **End if**

20: **End for**

The new algorithm can be illustrated with the help of Fig.6. In Fig.6, the Semi-AC automatic machine for the set *P*={ACGATAT, ATATATA, ATATGC, TAGAT} is constructed with the first four characters of each pattern. So the maximum depth of the automaton is 5. Each terminal state stores a pattern list.

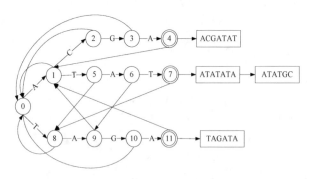

Fig. 6. The Semi-AC automaton for the set *P*={*ACGATAT, ATATATA, ATATGC, TAGAT*}. Double-circled states are terminal.The dashed links represent the state-to-state supply function.

With the data structure above, search of a pattern is easy. When the pointer reaches the terminal state, the text may be a pattern of the set. In that case, the text should be probed by comparing it with the pattern in the corresponding list, which determines if the text is indeed a pattern of the set or a false positive. The Algorithm3 summarizes the search algorithm on Semi-AC automaton.

Algorithm 3. SemiAC_Search(T, P)

Input: $T = t_1, t_2, \cdots, t_n$ $P = \{ P_1^{C_1}, P_2^{C_2}, \cdots, P_r^{C_r} \}$

Output: *match_num*   //the number of hit
Procedures :

01: Semi_AC $\leftarrow$ SemiAC_Create ($First - n, P = \{ P_1^{C_1}, P_2^{C_2}, \cdots, P_r^{C_r} \}$)
02: *Curstate* $\leftarrow$ Initstate of the automaton Semi_AC
03: **for**($j \leftarrow 1$ to n)
04:    **while**($\delta(Curstate, t_j) \neq \theta$ and Failure($Curstate$) $\neq \theta$)
05:       *Curstate* $\leftarrow$ Failure($Curstate$)
06:    **End while**
07:    **if** ($\delta(Curstate, t_j) \neq \theta$)
08:       $Curstate \leftarrow \delta(Curstate, t_j)$
09:    **else**
10:       **Curstate** $\leftarrow$ Initstate of Semi_AC
11:    **End if**
12:    **if** (*Curstate* is a terminal state)
13:       **while** ($Q(Curstate) \neq$ NULL)
14:          *pat* $\leftarrow$ get a pattern $P_1^{C_1}$ from $Q(Curstate)$
15:          **if** ((text $= t_{j-First_n}, \cdots t_{j+c_i}$) $=pat$)
16:             *match_num* ++
17:          **End if**
18:       **End while**
19:    **End if**

Table 1 is the representation of Semi-AC automaton in the memory. Each column of 256 ASCII value represent next state which we are from the current state to. The value which the terminal states correspond to in the column list is non-zero because each terminal state stores a pattern list.

Table 1. Semi-ac automaton in the memory

	ASCII value												
state	0	…	65(A)	66(B)	67(C)	…	103(G)	…	116(T)	…	255	list	fail
0	0	0	1	0	0	0	0	0	8	0	0	0	0
1	0	0	0	0	2	0	0	0	5	0	0	0	0
2	0	0	0	0	0	0	3	0	0	0	0	0	0
3	0	0	4	0	0	0	0	0	0	0	0	0	0
4	0	0	0	0	0	0	0	0	0	0	0	!0	1
5	0	0	6	0	0	0	0	0	0	0	0	0	8
6	0	0	0	0	0	0	0	0	7	0	0	0	9
7	0	0	0	0	0	0	0	0	0	0	0	!0	8
8	0	0	9	0	0	0	0	0	0	0	0	0	0
9	0	0	0	0	0	0	10	0	0	0	0	0	1
10	0	0	11	0	0	0	0	0	0	0	0	0	0
11	0	0	0	0	0	0	0	0	0	00	0	!0	1

4 Experimental Results and Evaluation

The experiments contain two parts. We compare the efficiency of Semi-AC with original AC algorithm by using equal length pattern set and unequal length pattern set, respectively.

A Experiments Environment

The overview of the experiments platforms are shown in Table2.

Table 2. Experiments Platforms

CPU	AMD Athlon II X4 2.8GHz
memory	2G
OS	Fedora 13

B Test Data Sets

We used two test data sets in experiment. The length of all patterns in the first data set is equal, which are 11 characters. The length of patterns in the second data set is unequal, which are 9-35 characters. Each data set contains 10 thousands patterns. The size of text being searched is 12.2MB.

Table 3. Data Set

	Number of patterns	Length of pattern	Size of text(MB)
Equal length pattern set	10000	11	12.2
Unequal length pattern set	10000	9-35	12.2

C   Results of Experiments

In this section, we explore the performance of our implementations.

1)   Compare Semi-AC with Original AC on Equal Length Pattern Set

Fig.7 shows the execution time of algorithm on equal length pattern set. Fig.8 shows the size of memory needed by automaton on equal length pattern set.

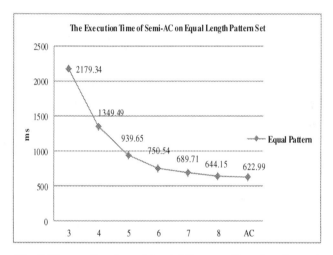

Fig. 7. The excution time of Semi-AC on equal length pattern set

In Fig.7, the horizontal axis represents the Semi-AC automaton of n (n=3, 4,⋯, 8) characters, the AC is the original AC. We can observe that the execution time of Smei-AC algorithm is reduced as the n is increasing. The efficiency of 8 characters Semi-AC automaton is very close to the original AC algorithm, which is 622.99ms.

In Fig.8, what the horizontal axis represents is the same as Fig.7. We can see that the memory size needed by Semi-AC automaton is larger as the n is larger. The automaton needs approximately 10MB memory for each additional character. The size of memory increases linearly with n.

Fig.9 shows the percentage of execution time reduce with the Semi-AC automaton increasing one character. The horizontal axis represents the Semi-AC automaton of n (n=3, 4,⋯, 8) characters. We can see that the percentage of reduction is up to more than 20% when n is 3, 4 and 5. The rate dropped to 8.1% when n is 6. The rate of last two automatons is lower. We can conclude that we should choose the Semi-AC automaton of six characters because the ratio of time and space is optimal.

2)   Compare Semi-AC with Original AC on Unqual Length Pattern Set

Fig.10 and 11 shows the execution time of algorithm and the size of memory needed by automaton on unequal length pattern set, respectively.

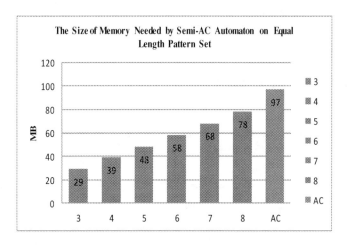

Fig. 8. Size of memory needed by automaton on equal length pattern set

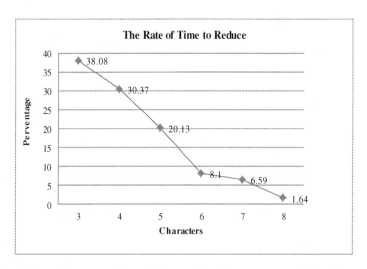

Fig. 9. The Rate of Time to reduce with increasing 1 character on equal length pattern set

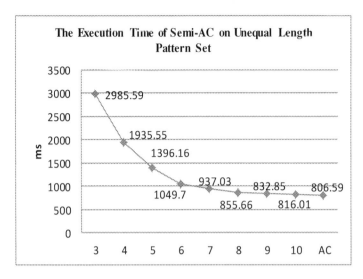

Fig. 10. The excution time of Semi-AC on unequal length pattern set

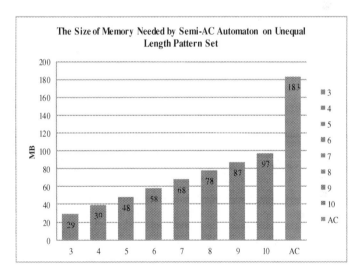

Fig. 11. The size of memory needed by automaton on unequal length pattern set

In Fig.10, what the horizontal axis represents is the same as Fig.7. We can see that the execution time of Semi-AC algorithm is also reduced while the n is increasing. It is smaller reduction in time when the number of character is more than 6. The efficiency of 10 characters Semi-AC automaton is very close to the original AC algorithm.

As a result of Fig.11, the automaton also needs approximately 10MB memory for each additional character. The efficiency of 10 characters Semi-AC automaton is very close to the original AC automaton, but the memory required is more than half of the original AC automaton. Space-saving effect of the Semi-AC algorithm on the unequal length pattern set is better than on the equal length pattern set.

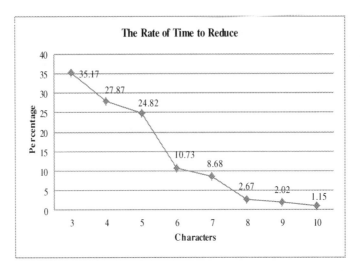

Fig. 12. The Rate of Time to reduce with increasing 1 character on unequal length pattern set

Fig.12 shows that the rate of execution time reduces with the Semi-AC automaton increasing one character on unequal length pattern set. We can see that the rate of reduction is up to more than 20% when n is 3, 4 and 5. The rate dropped to 10% when n is 6, but the efficiency of the algorithm is much lower than the original AC. The rate dropped to 2% and the efficiency is close to the original AC when n is 8. The rate of another two automatons is lower. We can conclude that we should choose the Semi-AC automaton of eight characters because the ratio of time and space is optimal.

5   Conclusion

In this paper, we proposed the Semi-AC algorithm, a multi-string matching algorithm. We contribute modifications to the Aho-Corasick string-matching algorithm that drastically reduce the amount of memory required. Its efficiency is close to the original AC, but the space save 50% or more. Although we get a markedly improvement in Semi-AC algorithm, there still is some enhancing space to improve. There may be any algorithm more efficient than Semi-AC. As future work we will do more research, and try to find more suitable applications on multi-pattern matching.

Acknowledgment. This work is supported by the National Basic Research Program of China under Grant No. 2011CB302605, the National High-Tech Development 863 Program of ChinA UNDER GRANT NO. 2010AA012504, 2011AA010705, 2012AA012506, the National Natural Science Foundation of China under Grant No. 61173145, 60903166.

References

1. Antonatos, S., Anagnostakis, K.G., Markatos, E.P.: Generating realistic workloads for network intrusion detection systems. In: ACM Workshop on Software and Performance (2004)
2. Aho, A., Corasick, M.: Efficient string matching: An aid to bibliographic search. CACM 18(6), 333–340 (1975)
3. Knuth, D.E., Moms, J.H., Pratt, V.R.: Fast pattern matching in strings. SIAMJ. Compt. 6(2), 323–350 (1977)
4. Arikawa, S., Shinohara, T.: A run-time efficient realization of aho-corasick pattern matching machines. New Generation Computing 2(2), 171–186 (1984)
5. Snort users manual 2.6.0 (2006)
6. Tuck, N., Sherwood, T., Calder, B., Varghese, G.: Deterministic memory-efficient string matching algorithms for intrusion detection. In: INFOCOM (2004)
7. Fukamachi, S., Shinohara, T., Takeda, M.: String pattem matching for compressed data using variable length codes(in japanese). Jouhougaku Symposium (1992)
8. Huffman, D.A.: A method for the construction of minimumredundancy codes. In: Proc. IRE, vol. 40, pp. 1098–1101 (1952)

The Research of Partitioning Model Based on Virtual Identity Data[*]

Lu Deng, Weihong Han, Aiping Li, and Yong Quan

Department of Computer Sciences, National University of Defense Technology,
Changsha, 410073, China
cool88220@126.com

Abstract. As the fast improvement of Internet, so do the data and information based on virtual identity. There are so many works on the data storage, whose main ideas are to store data in a distributed environment. And then a new issue is coming: how to decide the position of data efficiently. Because of the characteristics of the virtual identity data, they have their special pattern. In this paper, a partitioning model based on the virtual identity data is proposed. As an example, the Cassandra database is adopted to describe the model. The experiments are taken to test the feasibility of the model and the results show that the model can reduce the retrieval time efficiently.

Keywords: data partitioning, virtual identity data, Cassandra database.

1 Introduction

With the rapid development of Chinese Internet, the number of Chinese network's users has reached 618 million, e-commerce and online communication has become an important way of life, human activities are extended from the traditional physical space to the domain space. According to The 33nd China Internet Network Development Statistics Report which is announced by China Internet Network Information Center (CNNIC) [1]on January 16, 2014: To the end of December 2013, the Internet penetration rate was 45.8%, while the number of Chinese websites was 3.2 million which had an increase of 260 thousand in six months.

The situation mentioned above results in a tremendous growth in the application data based on virtual identity as websites basically need to store and record the static informations and dynamic interactive real-time informations. The amount of informations that Google[2] need to process increase from 100TB daily in 2004 to 20PB daily in 2008, in social networking services of Facebook [3], the number of page views has reached 1.5TB per day, In Yahoo's[4] Flickr photo-sharing service, the total storing amount of original images turn to 2PB, the number of read requests is 4 billion and requests of uploaded pictures is 0.4 billion per day. How to efficiently store and manage the virtual identity information has become an urgent problem need to be solved. because of the huge of virtual identity information which results in the solution that they cannot be stored

---

[*] This work was supported by NSFC (No.60933005, 91124002); 863 (No. 012505, 2011AA010702, 2012AA01A401, 2012AA01A402); 973 (No. 2013CB329601, 2013CB329602); 242 (No.2011A010); NSTM (No.2012BAH38B04, 2012BAH38B06).

W. Han et al. (Eds.): APWeb 2014 Workshops, LNCS 8710, pp. 214–224, 2014.

definitely in one or several machine nodes, data partitioning problem should be taken into account necessarily. While the virtual identity data have their own characteristics, the characteristics should be thought over in terms of data partitioning.

In this paper, the data partitioning methods of several databases and systems are listed firstly, and then the characteristics of virtual identity data are summarized, finally a data partitioning model that combines the machine node capabilities and virtual identity data characteristics is proposed based on Cassandra database which is taken as an example.

2   Related Work

Data partitioning is an important aspect of data organization and management and can be divided into horizontal partitioning and vertical partitioning. Horizontal partitioning can divide the relationship into multiple fragments and store them across multiple nodes, which takes advantage of the features of parallel and distributed nodes and reduces the transfer of data amount in the network and I /O. There are many kinds of horizontal partitioning as range algorithm, hash algorithm, mixing range algorithm, consistent hash algorithm, CMD algorithm. The main idea of vertical partitioning is to store the data in column and the main representative of the partitioning is columns database such as C-Store, MonetDB and so on. It can strengthen the performance of operations for the same column.

BigTable[5] and HBase[6] propose the concept of column family in the vertical direction that means to store data as a family which is formed by relevant properties, thus it can improve the aggregations between columns. In the horizontal direction, they take the range algorithm whose division is based on the Key's range in the table, so the data which are in the same column and in a specified range will be stored together.

PNUTS system[7] uses a relational database model, therefore, it does not do the partitioning in vertical direction. In the horizontal direction, the way it takes can support not only the range partitioning but also Hash partitioning.

Cassandra[8][9] also uses the concept of column family to do the partitioning in the vertical direction, but it takes improving consistent hash algorithm[10] in the horizontal direction as Dynamo system[11] which can reduce the amount of transfer that is resulted in by the Hash partitioning and range partitioning[12] when the situation that the machine node breaks down or a new machine node is coming happens..

The methods mentioned above are almost the partitioning[13] for one-dimensional Key which cannot satisfy the requests with the expansion of applications, so multidimensional data partitioning is needed. MD-HBase[14] is a multidimensional data partitioning[15] because of that it is designed for Location Based Services application whose location data are multi-dimensional data. If the partitioning takes the one-dimensional Key in NoSQL database[16], the data in the same position may be dispersed. It takes the KD-tree or Quad-tree algorithm to do the multidimensional partitioning, which can determine a defined location. After the multidimensional partitioning, the linearization technique is taken to reduce the dimensions of multidimensional data, such as Z-Order.

However, the partitioning which considers the machine node capabilities and the stored data's characteristics has been ignored in recent studies. In this paper, we propose a data partitioning model that think over the machine node capabilities and virtual identity data characteristics, which is based on the Cassandra database.

3   The Division Model of Virtual Identity Data

3.1   The Characteristics of Virtual Identity Data

The virtual identity data and informations are not only in large scales, but also in different types, they mainly have following feature:

1. All the data and informations are associated with a unique identifier which has contact with the real identity directly or indirectly.
2. The accounts of a virtual identity at different platform are set according to the user's own preferences, so they may be the same or have the similarity.
3. The virtual identity data and informations have a great relationship with the region, we can find that the users in similar areas may have similar hobbies, on-line time, more contacts and so on.

The analysis of the characteristics of virtual identity data is the promise of the discussion and research for partitioning about the virtual identity. If the characteristics are not recognized, the model proposed may not be suitable for virtual identity.

3.2   The Discussion of Partitioning Model

3.2.1   The Consideration of Machine Nodes' Capabilities

As we mentioned above, Data partitioning can be divided into two ways: vertical partitioning and horizontal partitioning. In this paper, the main research is based on Cassandra database, which use the concept of column family to divide the data in vertical partitioning and improving consistent hashing algorithm in the horizontal direction as Dynamo system. In the division of virtual identity database, the concept of column family is also adopted in vertical direction, but in the horizontal direction, a new weighted improving consistent hashing algorithm[17] which is based on improving consistent hashing algorithm is proposed.

It means taking the actual performance of the machine nodes (measured by resources) into account, which is called the weights.

Definition 31 (weights determined). *A represents the total resources of all machine nodes, a_i represents the resources capability of machine node i, Q_i represents the actual storage of machine node i, Q represents the storage of all machine nodes, then* $Q_i = Q * a_i / A$

Assume that hash ring is divided into 16 virtual nodes according to the improving consistent hashing algorithm, there are A, B, C, D 4 physical nodes at the initial states. According to the thought of weighted improving consistent hashing algorithm[18], the number of virtual nodes that every physical node stores should be different as a result of the difference of the resources performance of every physical node. Figure1(a) is the distribution situation of virtual nodes without the considering of the resources capabilities. It shows that every physical node have the same virtual nodes. Supposed that the proportion of resources performances for the physical nodes A, B, C, D are 2/16, 2/16, 4/16, 8/16, Figure1(b) shows the distribution situation of virtual nodes with considering of the resources capabilities, the numbers of virtual nodes that every physical node controls are 2, 2, 4, 8.

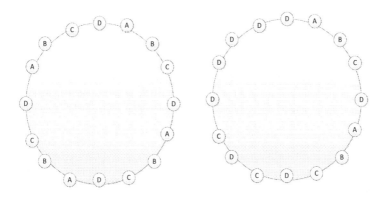

(a) improving consistent hashing algorithm   (b)weighted improving consistent hashing algorithm

Fig. 1. The division of virtual nodes

Discussing the situation of adding nodes[19][20], here is a new coming node E, which makes the proportion of resources capabilities for the physical nodes change from 2/16, 2/16, 4/16, 8/16 of A, B, C, D to 1/16, 1/16, 2/16, 4/16, 8/16 of A, B, C, D, E. It means that the reducing nodes of A, B, C, D are assigned to the physical node E and the new assignment should be done with the promise of making sure that the change of nodes A, B, C, D is the least. Figure2(a) shows the assignment before, Figure2(b) shows the assignment new.

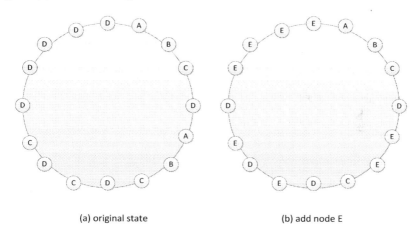

(a) original state   (b) add node E

Fig. 2. The node's adding in algorithm

And then considering the situation of decreasing node[21][22], the physical node D is cut down in the circumstance of four nodes A, B, C, D, which makes the proportion of resources performances for the physical nodes change from 2/16, 2/16, 4/16, 8/16 of A, B, C, D to 4/16, 4/16, 8/16 of A, B, C. Of course the virtual nodes' assignment is changed too. Because of the same size of virtual nodes, the extra nodes of D would be assigned to A, B,C in the proportion of 1:1:2. Figure3(a) shows the assignment before, Figure3(b) shows the assignment new.

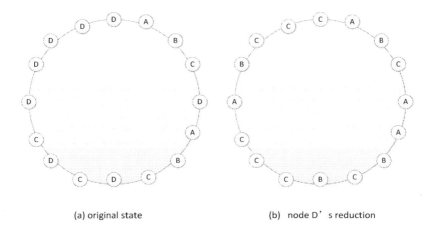

(a) original state   (b) node D's reduction

Fig. 3. The node's reduction in algorithm

3.2.2   The Consideration in One Machine Node

The discussion above is the wise storage size of every physical node, and the attention is focus on the storage in one node. According to the division's purpose, the data put in the same physical node should be related, which will not need too much span and save the query time. Therefore, how to judge the correlation between the two sets of data is the emphasis.

Firstly, considering the assignment of informations which are stored under the same unique identifier (stored by Column Family) and different platforms(stored by Key). For a user, the virtual accounts under different platforms are most likely the same or similar. The platforms whose attribute "virtual account" is follow the law mentioned above are stored together, which will be beneficial for the achieve of the query in accordance with the virtual account.

Figure4 (a) shows that the informations are arranged arbitrarily. After the observation, the law that the virtual account at platform Sina and Tencent are the same can be summarized. So with the consideration of the similarity of attribute "virtual account", the information of platform Sina and Tencent should be stored adjacently, which result in the platform Taobao stored after them. It can be shown in the Figure4 (b).

After that the assignment of informations is considered which are stored under the different unique identifier (stored by Column Family) in the same physical node. For the transfer of unique identifier, there is almost not query of range, so it will waste the advantage of order. According to the situation that virtual identities in the same place have more aspects in common and communicate with each other frequently, the informations under the different unique identifier are ordered by the attribute "region" that indicates the virtual identity's place, which will be benefit for the management of virtual identity, as it shows in Figure 5.

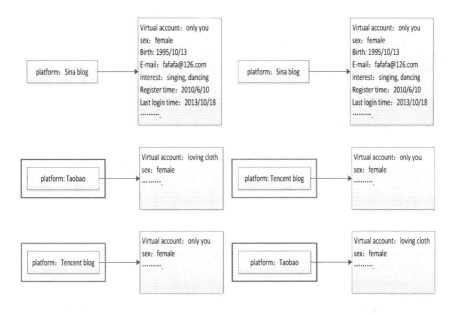

(a)storage not considering the correlation   (b)storage considering the correlation

Fig. 4. The correlation of virtual accounts

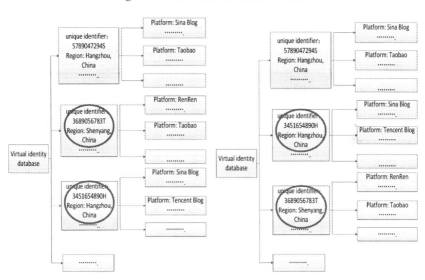

(a)Without considering the correlation in one unique identifier   (b)considering the correlation in one unique identifier

Fig. 5. The correlation of region

The whole thought of division is achieved with three important algorithm: Weighted-consistent-hash(B, M, m), InternalSort(E) and Insert(E,m).

Weighted-consistent-hash(B, M, m) is the achieve of weighted improving consistent hashing algorithm which contains the transfer of improving consistent hashing algorithm: Consistent-Hash(n_i). In the algorithm, B means the address array of all physical machine nodes, M means the whole data amount, m means the amount on the average of each virtual node. n is used to store the number of virtual node which is calculated by M/m.

Algorithm 1. Weighted-consistent-hash(B, M, m)

Input: the address array B of all physical machine nodes, the whole data amount M, the amount on the average m of each virtual node

Output: true or false, the success of distribution return true, otherwise return false

1: n<-M/m   //n is the number of virtual nodes
2: A<-0;   //A is the total resources of all machine nodes
3: for i<-0 to length[B]-1
4:        A<- A+a_i // a_i is the resources performance of machine node i
5:    for i<-0 to length[B]
6:        n_i=n*a_i/A
7: Consistent-Hash(n_i)//doing hashing algorithm based on the weights

---

The division of the machine nodes' performance is shown above, and the division in one machine nodes will be discussed. The algorithms are based on new information, which means the discussion is unfolded in the view of new coming record. Insertnewid(E, m) is the achieve of inserting a virtual identity that is never appeared before, which means the consideration is under the different unique identifier (stored by Column Family) in the same physical node. E means the data structure which store the virtual identity data, m means the new coming virtual identity.

Algorithm 2. Insertnewid(E, m)

Input: the data structure E which store the virtual identity data, the new coming virtual identity m

Output: true or false, the success of insert return true, otherwise return false

1: for j<-1 to length[E]-1
2:        diff<- E[j]
3:        same<- E[j-1]
4:        if(diff[region]!=m[region] and same[region]= =m[region])
5:             insert[m, j]
6:             return true
7:        else
8:                j<-j+1;
9: if (j==length[E])
10:      insert[m, j]
11:      return true
12: return

For a new coming record which means there exists the unique identifier in the database, what needed to do is querying the database to find the unique identifier in the new coming record, and insert the record to the appointed attribute. If the attribute did not exist, the attribute should be set up first, and then finish the insert. Insertinid(E, m) is the achieve of inserting a new record that the unique identifier already exists. E means the data structure which store the virtual identity data, m means the new coming record.

Algorithm 3. Insertinid(E, m)

Input: data structure E which store the virtual identity data, the new coming record m
Output: true or false, the success of insert return true, otherwise return false
1: for j<-0 to length[E]-1
2:      key<-E[j]
3:      if(key[id]=m[id])
4:          if(m[attribute] not in the key[attribute])
5:              create(key[attribute], m[attribute])
6:          insert(id, m[attribute], m[content])
7:          return true
8:      else
9:              j<-j+1
10: Insertnewid(E, m)
11: return

4   Experiments

Virtual identity data are collected from Sina, Alibaba and many other partners, so they are really true and reliable, and the huge amount of data can be definitely known as massive.

Cassandra cluster can be deployed from the single node to dozens or even more nodes as the cluster system. In the experiment, the total number of nodes in the cluster virtual identity system is set as 30, the specific configurations are shown as follows:

Table 1. Node Configuration

Group	Physical nodes' number	Main memory	CPU	System
1	10	4G	4 cores	Ubuntu12.0
2	10	4G	2 cores	Ubuntu12.0
3	10	2G	2 cores	Ubuntu12.0

In the division of data, Cassandra database adopt the improving consistent hashing algorithm to decide the node. In this paper, weighted improving consistent hashing algorithm is proposed with the consideration of the actual performances of machine nodes, which result in the balance of nodes. To test the difference performance between weighted improving consistent hashing algorithm and improving consistent hashing algorithm, the machine nodes are divided into three groups, the distribution is shown as follow: each group follows the distribution.

Table 2. The group of nodes

group	Main memory 4G, CPU 4 cores	Main memory 4G, CPU 2 cores	Main memory 2G, CPU 2 cores
1	0	0	10
2	0	5	5
3	4	3	3

Taking 50,000 query requests as an example, do the test among the machine configuration shown in Table 2. The horizontal axis represents the amount of database, here were taken up: 100GB, 200GB, 300GB, 400GB and 500GB, the vertical axis represents the total response time of 50,000 queries query. The results are shown in Figure 6:

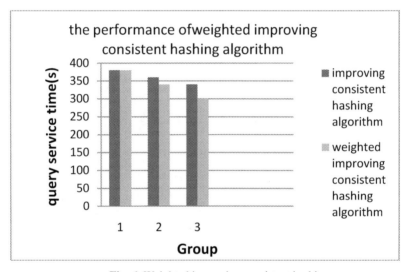

Fig. 6. Weighted improving consistent hashing

It can be concluded from the test that weighted improving consistent hashing algorithm has the same query service time as improving consistent hashing algorithm in group 1 because of there no difference performance of machine nodes, but the advantage of weighted improving consistent hashing algorithm can be seen from the group 2 and group 3 obviously.

Next, Test the storage of data in one machine node. The main idea is that data which have the same or similar attribute "virtual account" if they are under the same virtual identifier and virtual identifiers which have the same region should be stored adjacently. All the 30 machines are taken to store the database and test the performance. As the experiment above, the horizontal axis represents the amount of database, here were taken up: 100GB, 200GB, 300GB, 400GB and 500GB, the vertical axis represents the total response time of 50,000 queries query.

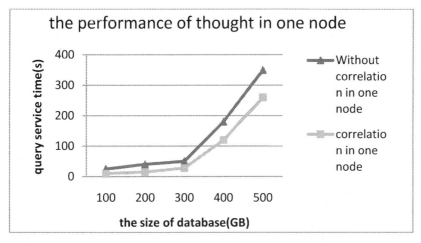

Fig. 7. Inside one node's consideration

The Figure 7 shows that the query response time have risen in both cases with the increase of databases' size, but the case which considered the internal relevant has a smaller rise. Therefore, taking virtual account and region similarity into account have influence on reducing query response time.

5   Conclusion

A partitioning model of the virtual identity data is proposed based on the Cassandra database in this paper, which takes the machine nodes' capabilities and virtual identity data characteristics into account. The research status are analyzed first, secondly virtual identity data characteristics and query requests are summarized, and then the model is proposed, at last, the experiments are taken to test the performance. The results show that the partitioning model can reduce the query time obviously.

References

1. China Internet Network Information Center, http://www.cnnic.net.cn/
2. Google, http://en.wikipedia.org/wiki/Goolge
3. Facebook, http://baike.soso.com/v214498.htm
4. yahoo, http://en.wikipedia.org/wiki/Yahoo
5. Chang, F., Dean, J., Ghemawat, S.: Bigtable: A distributed storage system for structured data. J. ACM Transactions on Computer Systems (TOCS) 26(2), 4 (2008)
6. George, L.: The hbase definition guide. O'Reilly Media, Inc. (2011)
7. PNUTS system, http://en.wikipedia.org/wiki/Pnuts
8. Lakshman, A., Malik, P.: Cassandra: a decentralized structured storage system. J. ACM SIGOPS Operating Systems Review 44(2), 35–40 (2010)
9. Lakshman, A., Malik, P.: Cassandra: structured storage system on a p2p network (2009)

10. Dong, L., Kang, C., Yunfei, C.: Research on the Reliable Replica Creation Technique Based on Virtual Ring Strategy. Proceedings of the World Congress on Engineering 2 (2013)

11. Crowley, C., Wolf, R.: Impact of the Neutral Wind Dynamo on the Development of the Region 2 Dynamo. J. Midlatitude Ionospheric Dynamics and Disturbances, 181–179 (2013)

12. Hu, L., Duan, J.: The Comparison of Two Kinds of Range Partitioning Method of Port Economic Hinterland: An Approach to Realize Two-type Development of Port Areas. J. Systems Engineering 2, 006 (2013)

13. Wang, S., Harvey, C.: Fracture Mode Partition Theory for One Dimensional Fractures. ECF18, Dresden 2010 (2012)

14. Nishimura, S., Das, S., Agrawal, D., Abbadi, A.: MD-HBase: design and implementation of an elastic data infrastructure for cloud-scale location services. J. Distributed and Parallel Databases 31, 289–319 (2013)

15. Yun, R., Jin, H.: Wavelength Division Multiplexing Subnet Partition Based on Clustering Algorithm in Multidimensional Space. J. Acta Optica Sinica 2, 004 (2013)

16. Celko, J.: Joe Celko's Complete Guide to NoSQL: What Every SQL Professional Needs to Know about Non-Relational Databases. Newnes (2013)

17. Jiakui, Z., Pingfei, Z., Lianghuai, Y.: Effective Data Localization Using Consistent Hashing in Cloud Time-Series Databases. J. Applied Mechanics and Materials 357, 2246–2251 (2013)

18. Lee, B., Jeong, Y., Song, H.: A scalable and highly available network management architecture on consistent hashing. In: Global Telecommunications Conference (GLOBECOM 2010), pp. 1–6 (2010)

19. Loesing, S., Hentschel, M., Kraska, T.: Stormy: an elastic and highly available streaming service in the cloud. In: Proceedings of the 2012 Joint EDBT/ICDT Workshops, pp. 55–60 (2012)

20. Jinbao, W., Sai, W., Hong, G.: Indexing multi-dimensional data in a cloud system. In: Proceedings of the 2010 ACM SIGMOD International Conference on Management of Data, pp. 591–602 (2010)

21. Garefalakis, P., Papadopoulos, P., Manousakis, I., Magoutis, K.: Strengthening consistency in the cassandra distributed key-value store. In: Dowling, J., Taïani, F. (eds.) DAIS 2013. LNCS, vol. 7891, pp. 193–198. Springer, Heidelberg (2013)

22. Raindel, S., Birk, Y.: Replicate and Bundle (RnB)–A Mechanism for Relieving Bottlenecks in Data Centers. In: Parallel Distributed Processing (IPDPS), pp. 601–610. IEEE (2013)

Implemention of Cyber Security Situation Awareness Based on Knowledge Discovery with Trusted Computer

Jiemei Zeng[1], Xuewei Feng[2], Dongxia Wang[2], and Lan Fang[2]

[1] Beijing Aerospace Control Center, Beijing 100296, China
[2] Beijing Institute of System Engineer,
National Key Laboratory of Science and Technology on Information System Security,
Beijing 100101, China
brafum@yeah.net

Abstract. Situation awareness aims to provide the global security views of the cyberspace for administrators. In this paper, a novel framework of cyber security situation awareness is proposed. The framework is based on a trusted engine, and can be viewed from two perspectives, one is data flow, which presents the abstracting of cyber data, and the other one is logic view, which presents the procedure of situation awareness. The framework's core component is a correlation state machine, which is an extension of state machine, and used to model attack scenarios. The correlation state machine is a data structure of situation awareness, and stored in a trusted computer in order to avoid being tampered. It is created based on the technology of knowledge discovery, and after being created, it can be used to assess and predict the threat situation. We conclude with an example of how the framework can be applied to real world to provide cyber security situation for administrators.

Keywords: cyber security, situation awareness, correlation state machine, threat assessment, trusted computing.

1 Introduction

With the dramatically increasing of cyber attacks during the last several years[1][2], cyber security situation awareness has become a major contributor for cyber security. Cyber security situation awareness belongs to the third generation of information security defense technology, and it aims to provide the cyberspace's global security views and states for administrators. Based on the comprehensive knowledge provided by situation awareness, administrators can make decisions timely and accurately, this will reduce the compromising considerably.

When attackers exploit the cyber infrastructure, security devices such as IDS (Intrusion Detection System) will generate hundreds of thousands of security events. It is impossible to recognize attack scenarios and be aware of security state of the cyberspace from these events by the manual way, so how to correlate these events to reconstruct attack scenarios and assess the attack activity's consequence automatically is the main goal of cyber security situation awareness.

W. Han et al. (Eds.): APWeb 2014 Workshops, LNCS 8710, pp. 225–234, 2014.

In this paper, after analyzing the current researches on situation awareness, we propose a framework of cyber security situation awareness based on state machine. It is more operable than the concept model provided by Bass [3][4], and can be used to guide the whole process of cyber security situation awareness. The framework can be viewed from two aspects. One is data flow, which presents the abstracting of cyber data, and is along with the processing of situation awareness. The quantity of cyber data is decreasing while the quality is increasing, data is abstracted to knowledge. The other one is logic view, which presents the procedure of situation awareness, that is methods or approaches can be acquired to processing cyber data to achieve cyber information and cyber knowledge.

The remainder of this paper is organized as follows: section 2 analyzes the exiting researches on cyber security situation awareness. Section 3 describes the structure of the situation awareness framework, and the main contents and methods of the framework are explained in detail in this section too. In section 4, we use an experiment to validate the technologies in the framework. At last we summarize this paper and suggest the future work in section 5.

2 Related Work

2.1 Models or Frameworks of Cyber Security Situation Awareness

Wang presented a framework for network security situation analysis in [14], which specializes in situation analysis of network, and doesn't comprise the situation prediction. The framework is derived from the data fusion model of Bass. The data fusion model [3][4] proposed by Bass in 1999 is the rudiment of network security situation awareness. The model includes five levels, and the objectives of each level are defined in this model. It mainly focuses on the concept of situation awareness, so in the practical environment, available technologies and methods for analyzing security situation are lacking.

Wang[5] proposes a hierarchical implementation model for network security situation awareness, but the situation information presented by this model only locate at security threat level, which can't indicate the whole situation of the network system. A security situation evaluation model for inter-domain routing system in the internet is proposed in [6], the model is a two tuples (TREE, EA), TREE is the routing tree of the routing system, and EA is the assessment algorithm. The ultimate results presented by this model are security metrics too. Otherwise, the EA's input is abnormal routing information, this information is always hard to acquire accurately. Amann presents a network intrusion detection method which is based on the real-time intelligence, such as black list. This method can improve the accuracy significantly while alert correlating[7]. Ahmed proposes to model the malicious executables based on the DFA, each state of the DFA indicates an action of a malicious executable. After being created, the DFA model represents causal knowledge, which can be used to correlate various alerts[8]. A hierarchical attack scenario reconstruction method is presented in [9], but the disadvantage of this method is that it must depend on the predefinition of the causal knowledge.

2.2   Methods or Technologies in Situation Awareness

In order to acquire the overall information of attack activities, such as time, place, multistage etc., Feng reconstruct attack scenarios from raw security events in [15][16] using the rudiment of correlation state machine, where the correlation state machine can not contribute to predict and assess cyber threat situation. Liu[10] presents a method of network security situation awareness based on artificial immunity system. The method uses network intrusion detection based on the theory of immunity as the base of situational awareness, to detect known and unknown intrusions with the help of biological technology. The goal of this method is to quantify the security situation of the network system. However, sometimes network security situation can't be boiled down to quantifiable information, such as attack scenarios. Gorodetsky[11] uses agents to detect and analyze network traffic, and then gets the security situation. The situation assessment method is based on asynchronous data flow, usually data flow can't represent all the basic security information in networks. Chen [12] gives a hierarchic assessment method to analyze and quantify the network security threats, but he does not give a practical algorithm when assessing the system level threat. Yegneswaran[13] uses IDS Bro to analyze network's activity information provided by honeynets, and then depict the curve of the security situation. However, the result will be intelligible only in the case of large scale attack happening.

2.3   Trusted Computing

The concept of "trusted" appeared firstly in 1970s, presented by J.PAnderson, trusted computing emphasizes on the controllability and the expectability of the computer system, it aims to acquire a secure and dependable computing environment through the integrity measurement, the hardware TPM/TCM is the basis of trusted computing[17]. So we can regard trusted computing as one approach to achieve information system's security.

Trust denotes a state of well-founded belief that a device or system will behave as expected, for a given purpose despite a level of physical or logical interference. The definition of trust emphasizes the consistency of behavior and expectation. What might be expected? We might require a device or system to perform its function whilst maintaining some general notion of security.

3   A Framework of Cyber Security Situation Awareness Based on State Machine

In order to achieve cyber security situation accurately and timely especially the threat situation, we present a framework based on correlation state machine, the framework is a successor of what was proposed in [14] by Wang. As shown in Fig.1, the framework is based on a trusted computer and comprises four parts mainly, namely database module, process module, correlation state machine and GUI. The database module is used to store sample/history data and attack scenarios which are discovered after mining. The process module comprises four processing components, namely Knowledge Mining, Correlation Analyzing, Threat Predicting and Threat Assessing. The correlation state machine is the core data structure, which is used to predict and assess cyber threat. GUI is an interactive interface between administrators and the framework. The green broken line in Fig.1 represents the data flow view while the black line represents the logic flow view, the two views are coordinated.

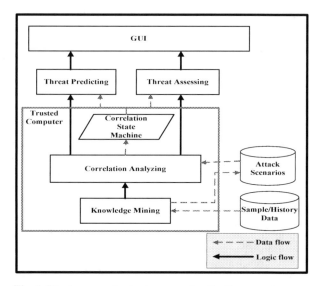

Fig. 1. The framework of cyber security situation awareness

Firstly, sample/history data is collected from experimental cyberspace or known datasets, and then stored in database. It is data source of Knowledge Mining and could be updated every now and then. After analyzing and discovering, Knowledge Mining generates various patterns of attack scenarios and stores them in databases. Secondly, the Correlation Analyzing component analyzes cyber security events at real-time based on the known attack patterns to reconstruct the occurrent attack scenarios. The occurent attack scenarios are represented by the correlation state machines, namely a correlation state machine is an attack scenario which is taking place in cyberspace at present. After being created, the correlation state machines can be input of the Threat Predicting and Threat Assessing components. The threat predicting component can deduce possible attack paths of an attack scenario based on the corresponding correlation state machine, and the Threat Assessing component assesses the consequence resulting from the attack scenario based on the corresponding correlation state machine too. These two components support administrators to make decisions directly. Finally, cyber security situation knowledge including the results of predicting and assessing is presented in a friendly way by the GUI, and the knowledge can be understood easily by administrators after being visualized. This can decrease administrators' cognitive burden considerably.

3.1   Discovering Attack Scenarios Based on Knowledge Discovery

Correlation state machine is the core of the framework, which aims to model and abstract the occurent attack activities. It is a dynamic data structure, and created based on the known patterns of attack scenarios. So, in order to achieve cyber security situation awareness, various patterns of attack scenarios must be discovered. We implemented this by data mining.

Firstly, we define a signature set, which is composed of characteristics of various attack activities, such as SYN-Scan, exploiting of certain vulnerabilities etc., this can be done referring to the existing projects or researches, such as Snort and so on. Every signature word in the set represents certain actions or steps of various attacks, and new words can be added to the set continually, that is the set should be expanded and updated continually in order to discover new attack activities. Secondly, the Knowledge Mining engine analyzes sample/history data to discover frequent itemsets which are composed of signature words. When the frequent itemsets are found out, correlation rules can be generated easily. Finally, if two rules have the same word, or they have the causal relationship, they are combined. After searching all of the rules in this way, various patterns of attack scenarios can be discovered and generated, which will be used when creating correlation state machine.

After generated, patterns of attack scenarios are formalized by XML. Specially, the sample/history data and signature set are critical to discover attack patterns, so administrators should collect useful information at daily work, but if administrators have known some patterns of attack scenarios, they can write the XML files directly, no need to mine the sample data.

3.2   Correlation Analysis of the Network Security Events

Correlation analysis of cyber security events is the foundation of acquiring high-level knowledge of cyber security situation, it distinguishes and standardizes raw security events generated by security devices, and then analyzes the logic relationships between these events based on attack patterns. When events are submitted, correlation engine will match security events with attack patterns discovered by the Knowledge Mining till finding out the corresponding attack pattern or patterns. After the pattern or patterns are found out, one or more correlation state machine will be created according to the patterns. Each attack pattern responses to one state machine.

Definition 1 Correlation State Machine. Correlation state machine is a data structure, which maps the XML file of corresponding attack pattern to memory. Each state of the machine is a tuple and derived from the corresponding rule node of the pattern, that is *state_i =(plugin_id, plugin_sid, src_ip, dst_ip, src_port, dst_port, protocal, timeout, occurrence, srcIP_record, dstIP_record, srcPort_record, dstPort_record, eventCounter, startTime)*. The first nine attributes of the state indicate the characteristics of the security events which can be processed by this state, and their meanings are the same as the rule presenting in [7]. *srcIP_record, dstIP_record, srcPort_record and dstPort_record* are used to record the characteristics of the events processed by this state, *eventCounter* is used to record the number of the events processed successfully and *startTime* indicates the start time that this state takes effect.

The correlation state machine is an intermediate process, it is used to track and record the process of matching between security events and attack scenario timely. *timeout* and *occurrence* are the two core concepts of the state machine. *timeout* indicates that how long the engine monitors one state, and it corresponds to the certain attack step's lasting time in multi-step attack. *occurrence* presents the number of the security events that the state can process, essentially, this attribute means to classify the similar security events, it reflects the idea of clustering analysis. *timeout* and *occurrence* work cooperatively, if the correlation engine processed "*occurrence*" security events successfully in the time limit of "*timeout*", the current states of the state machine will transfer, this reflects the development of the multi-step attack.

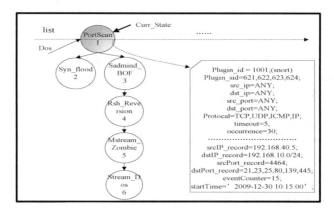

Fig. 2. The *Dos* state machine in the state machine queue

We illustrate an example of how a state machine working through the *Dos* attack scenario. In Fig. 2, the current state set includes one element, it is *State_1*. The correlation engine will monitor in this state for 5 seconds, if the number of the "*PortScan*" events processed by this state achieves 30 (that is *eventCounter* equals *occurrence*) in 5 seconds, the state will transfer. After transferring, the current state set of this state machine includes two elements, which are *State_2* and *State_3*.

When the current states process the matchable events, there is a question need to be paid more attention to. The states will record the events' characteristics which have been processed. For example, in Fig.2 *State_1* will record 30 events' characteristics, the attack type of these events is the same, but the addresses of these events may be different, how to record them in one state. Here we integrate the IP addresses and congregate the port addresses. Integrating of IP addresses can present the source area and the destination area of the attack activity; it is very useful in large-scale network intrusion detection, as *dstIP_record* of *State_1* in Fig.2.

3.3   Predicting the Threat Situation Using State Machine

We research on attack intent reasoning mainly in this paper, so in our framework, predicting is used to deduce the trend of attack scenarios, that is estimating the next step of attack scenarios, this is helpful to decrease the damage caused by attack activities.

When attackers launched exploitation, they always want to find out the optimized path to exploit a target system. In order to simulate this procedure, we use the algorithm of trace-back to find out the optimized path in a correlation state machine. This is similar to an attacker's exploiting in the real world. If we can find out the optimized path in each correlation state machine, we can predict the corresponding attack scenarios' trend easily, that is we can predict the threat situation in the whole cyberspace.

Predicting means to find out the optimized path in different paths using the algorithm of trace-back that is to find out the path whose costs is the lowest. After that, the most possible attack path that an attack will adopt is presented. But, there are many factors that should be considered when predicting cyber threat situation, for example defense strategies on target system, vulnerabilities, attackers' capacity and so on. So it is impossible that the predicting path is always the same as the real path. In order to improve the accuracy of predicting, we consider some weight factors that could affect attackers' decision when exploiting, such as defense strategies, vulnerabilities. Using these factors to modify the optimized path timely, and presenting the probability of each path in various correlation state machines, trends of various attack scenarios and the threat situation of cyberspace can be predicted.

3.4   Assessment of the Threat Situation

Administrators can be aware of the global security state and security level of cyberspace through assessment. After creating the correlation state machine and predicting the trends of various attack scenarios, we implements threat situation assessment through calculating the threat metric of cyberspace based on each correlation state machine. The whole algorithm of computing the cyberspace's threat metric can be illuminated by (1) and (2).

$$f_{machine_i}(opt\_path_i) = \sum_1^n state_j \quad : state_j \in opt\_path_i \tag{1}$$

$$F_{Threat\_Situation}(machine) = \sum_1^n f_{machine_i}(opt\_path_i) \tag{2}$$

Formula (1) presents how to quantify the threats caused by one correlation state machine, namely to quantify the harm degree derived from an attack scenario. (1) indicates that every state machine's threat $f_{machine_i}(\ )$ is only related to its own optimized attack path opt_path_i. When predicting component presents the optimized attack path of a correlation state machine, every state which is in this path will be set a quantitive metric, the metric is a contributor to the machine's threats. Then, all the metrics in the path are added together to compute the machine's threats. In this way, administrators will acquire a number of numerical values, which indicate the harm degree caused by various attack scenarios to cyber infrastructure. Especially, the quantitive metric of each state can be defined depending on the concrete needs. In our experiment, we use the variable of "*occurrence*" to represent the quantitive metric simply.

After acquiring the numerical threats of all the correlation state machines, we can use (2) to compute the global threats that the cyberspace suffers. The global threats $F_{Threat\_Situation}($ $)$ are determined by every state machine's threats $f_{machine_i}($ $)$. The assessment component adds up all the numerical threats to computer $F_{Threat\_Situation}($ $)$, which is an indicator of the whole cyberspace's security level.

Assessment of the threat situation should be executed periodically. In an assessment cycle, the component scans all the correlation state machines which is living in the memory, and computes the threats of the machines based on their optimized attack path. After adding up, the global threats in this cycle could be a point at y axis while time is the other axis. To link these points together in different cycles, a threat situation assessment curve will be presented to administrators.

4   Experiment Analysis

We implemented a prototype to validate the framework. The experiment procedure can be explained as follows. Firstly, Knowledge Mining analyzes the sample data of DARPA 2000 intrusion scenario specific data sets[18] to discover the pattern of Dos attack scenario, and after being generated, the pattern is stored in databases. Secondly, we replay the data sets in network 192.168.40.0/24. Security devices deployed in this network will generate a large number of events, and these events are sent to Analyzing Engine. Thirdly, after processing these events timely, Analyzing Engine generates a correlation state machine based on the pattern of Dos attack scenario, which can be illustrated by Fig.3.

The correlation state machine indicates that it is a distributed dos (DDos) attack activity, and the current state of the activity is "Mstream_Zombie", namely interacting between the controller node and zombie nodes. Attack path from State_1 to the current state State_5 presents the evolving procedure of the DDos activity and the possible attack path later. Especially, the predicting and assessing components are implemented in Analyzing Engine, and after being created, the correlation state machines can be used to assess the threat situation of network 192.168.40.0/24. Finally, administrators can view all the situation information through client, which provides a GUI.

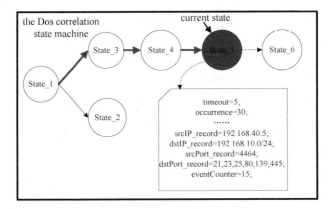

Fig. 3. The Dos correlation state machine

5   Conclusion

A framework of cyber security situation awareness based on knowledge discovery is proposed in this paper, which provides guidance and technical support for the whole situation awareness procedure, and is the foundation of the awareness work. There are four modules in the framework, database module, process module, correlation state machine and GUI. We deploy the engine and store the data structure in a trusted computer. After various attack patterns are discovered, attack scenario will be reconstructed through correlating security events, and the scenarios' abstraction can be used to predict and assess the threat situation in cyberspace, namely the correlation state machine.

One of the most important works in the future is to analyze and test the performance of the methods in real world, and we will also improve the predicting algorithm in the future.

Acknowledgment. The work of this paper is supported by the National Natural Science Foundation of China Project under grant No.61271252.

References

1. Ramaki, A.A., Ebrahimi, R., et al.: Enhancement Intrusion Detection using Alert Correlation in Co-operative Intrusion Detection Systems. Journal of Basic and Applied Scientific Research 3(6) (2013)
2. Blackhat Website: http://www.blackhat.com
3. Bass, T.: Multi-Sensor Data Fusion for next Generation Distributed Intrusion Detection Systems. In: 1999 IRIS National Symposium on Sensor and Data Fusion, Laurel, USA, vol. (1), pp. 24–27 (1999)
4. Bass, T.: Intrusion Detection Systems and Multi-Sensor Data Fusion: Creating Cyberspace Situation Awareness. Communications of the ACM 43(4), 99–105 (2000)

5. Huiqiang, W., Jibao, L., Mingming, H.: Research on the key implement technology of network security situation awareness. Geomatics and Information Science of Wuhan University 33(10) (2008) (in Chinese)
6. Xin, L., Xiaoqiang, W., Peidong, Z., Yuxing, P.: Security Evaluation for Inter-Domain Routing System in the Internet. Journal of Computer Research and Development 46(10), 1669–1677 (2009) (in Chinese)
7. Amann, B., Sommer, R., Sharma, A., Hall, S.: A Lone Wolf No More: Supporting Network Intrusion Detection with Real-Time Intelligence. In: Balzarotti, D., Stolfo, S.J., Cova, M. (eds.) RAID 2012. LNCS, vol. 7462, pp. 314–333. Springer, Heidelberg (2012)
8. Shosha, A.F., James, J.I., Liu, C.-C., Gladyshev, P.: Towards automated forensic event reconstruction of malicious code (Poster abstract). In: Balzarotti, D., Stolfo, S.J., Cova, M. (eds.) RAID 2012. LNCS, vol. 7462, pp. 388–389. Springer, Heidelberg (2012)
9. Xiao, F.U., Jin, S.H.I., Li, X.I.E.: Layered intrusion scenario reconstruction method for automated evidence analysis. Journal of Software 22(5), 996–1008 (2011) (in Chinese)
10. Nian, L., Sunjun, L., Yong, L., Hui, Z.: Method of Network Security Situation Awareness Based on Artificial Immunity System. Computer Science 37(1) (2010) (in Chinese)
11. Gorodetsky, V., Karsaev, O., Samoilov, V.: On-line update of situation assessment based on asynchronous data streams. In: Negoita, M.G., Howlett, R.J., Jain, L.C. (eds.) KES 2004. LNCS (LNAI), vol. 3213, pp. 1136–1142. Springer, Heidelberg (2004)
12. Xiuzhen, C., Qinghua, Z., Xiaohong, G., et al.: Quantitative hierarchical threat evaluation model for network security. Jouranl of Software 17(4), 885–897 (2006) (in Chinese)
13. Yegneswaran, V., Barford, P., Paxson, V.: Using Honeynets for Internet situation awareness [C/OL]. In: Pro of ACM/ USENIX Hotnets IV (2005),
    http://www.icir.org/vern/papers/
    sit-aware-hotnet05.pdf (January 12, 2008)
14. Dongxia, W., Xiaoyan, H., Lan, F., Xuewei, F.: Security Situation Awareness Information Model, Harsh Environment Resistance Annual Meeting of the computer (2010)
15. Xuewei, F., Dongxia, W.: Analyzing and Correlating Security Events Using State Machine. In: 2010 International Workshop on Frontiers of Secure Networks (2010)
16. Xuewei, F., Dongxia, W.: Research on the Key Technology of Reconstrucing Attack Scenario Based on State Machine. In: 2010 IEEE International Conference on Computer Science and Information Technology (2010)
17. Trusted Computing Group Website: http://www.trustedcomputinggroup.org
18. 2000 DARPA Intrusion Scenario Specific Data Sets[OL],
    http://www.ll.mit.edu/IST/ideval/data/2000/
    2000_data_index.html (January 24, 2008)

NNSDS: Network Nodes' Social Attributes Discovery System Based on Netflow

Shuo Mao[1], Zhigang Wu[1], Bo Sun[2], Shoufeng Cao[2],
Xiongjie Du[2], and Kaifeng Wang[3]

[1] Network Information Center, Beijing University of Posts and Telecommunications
100876 Beijing China
[2] National Computer Network Emergency Response Technical Team/
Coordination Center of China
100029 Beijing China
[3] Beijing TianYuanTeTong Technology Co., Ltd.
100029 Beijing China
maoshuo@bupt.edu.cn

Abstract. Currently, the most network traffic identification technologies focus on the applications of traffic, while ignoring the attributes of network terminal nodes which generate traffic. In this paper, we present a novel approach to identify the social attributes of network terminal nodes and design Netflow based network Nodes' Social attributes Discovery System(NNSDS).Firstly, we store the Netflow records using two hash tables to obtain the snapshots of the activity of the network. Then we discover the attributes of network nodes by the following elements: (1) social topology statistics, (2) social activity and (3) social roles of network nodes. We test our system on an IP backbone network. The experimental results show that our system can correctly identify various types of network nodes and the identification accuracy achieves 95%.

Keywords: Netflow, social attributes, Hadoop, terminal node.

1 Introduction

Network nodes are hosts appearing on the network with separate IP addresses. Social attributes of network nodes are the nodes' characteristic mined from the interconnection information between them. Social attributes include social topology statistics, social activity and social roles. Existing studies show that the network social attributes are important to assessing network security vulnerability [1] and managing the complexity of network [2]. Discovery of the social attributes of network nodes is quite important to meet new requirements in fine-grained classifying of terminal nodes, optimizing the configuration of network and managing network resources.

Recently, some novel approaches treat the problem of discovery of nodes' social attributes as a community ownership problem. These approaches classify nodes into different communities by spreading specific labels on network. However, nodes belonging to different communities from link topology may share similar social

W. Han et al. (Eds.): APWeb 2014 Workshops, LNCS 8710, pp. 235–245, 2014.

attributes. Concurrent processing on discovering large amount of nodes' social attributes brings new challenges in storage and computing. It is necessary to propose a method to obtain the network nodes' social attributes.

In this paper we design and implement the Netflow based network Nodes' Social attributes Discovery System (NNSDS) to analyze the social attributes of nodes in high-speed networks. NNSDS consists of two stages. Firstly, we propose a method to preprocess system inputs by producing snapshots aggregated from Netflow records. Secondly, we use Hadoop platform to mine social attributes from the aggregated data.We deploy NNSDS on an IP backbone network. The evaluation results indicate that NNSDS is able to discover social activity, social topology and social roles of network nodes by processing incremental Netflow data captured over the IP backbone network.

The rest of this paper is organized as follows: Section 2 introduces the related works of Netflow analysis and social attributes discovery technologies. Section 3 outlines the system architecture and function of NNSDS. Section 4 describes the implementation of NNSDS. Section 5 reports the experimental results conducted on an IP backbone network. Finally, we make conclusions in section 6

2   Related Works

2.1   Related works of Netflow Analysis

Netflow [3] is one of the key technologies used by network operators and administrators for monitoring large edge and core networks. It is widely used in flow measurement studies. However, there are several limitations of using Netflow technologies. Firstly, the number of concurrent stream of raw Netflow records captured over backbone network is up to 1 MB [4], which will cause space explosion. Secondly, operating on the raw records will reduce the performance of computing node. To overcome these limitations, the researchers have proposed two types of common solutions: reducing the amount of the raw data and increasing computing nodes [5].

Reduce Raw Data Scale. Stream sampling and similar stream merging are the two common solutions of reducing raw stream data. IPFIX [6] recommends the use of packet sampling, and the Sampled Netflow [7] has become a widely used monitoring solution among network operators. However, sampling has fundamental questions between accuracy and economy. Sampled Netflow suffers from limitations due to the static sampling method which cannot always ensure the accuracy of estimation [8].

Typical technology of data merging includes hierarchical merging [9], hash map based merging [10] and time granularity based merging. Hierarchical data merging can be computed with the optimal merge tree by using a dynamic program for a known set of arrival times. A significant problem of the dynamic program is that it requires time complexity $O(n+3)$ and space complexity $O(n+2)$[11]. HMJ is a typical Hash-merge join algorithm [12]. Specifying the optimal percentage of the number of flushed buckets to the total number of hash buckets is a key problem before hash map merging. The definition of similarity is important for similar data merging. Conflicts between similar data with different values are the most common problems need to be solved in data merging [13].

Increase Computing Nodes. Distributed flow data processing system deals with large amount of data by expanding the number of computing nodes [14]. However, existing distributed processing systems pay little attention on raw data compressing.

2.2   Related Works of Social Attributes Analysis

Most of the existing literatures of social attributes discovery are about mining the vulnerability of the network topology and discovering user communities on social network sites. There is rarely any paper focus on social attributes of terminal nodes on network. As network nodes share common features, studies on social attributes of network nodes still have reference values. There are three distinct approaches of nodes' social attributes discovery.

In topology Networks, assessments of social attributes such as connectivity, betweenness and the shortest path between nodes are important to maintain the network availability. Paxson V et al. used the central of the connectivity to measure the importance of the nodes [15].This paper defined the social attribute as connectivity between the nodes and the collection of their neighbors. It accurately figures out the network structure by spreading the labeled tag and detecting communities the nodes belonging to.

In the social network research filed, users' social attributes are divided into users' sex, age, educational background, profession, faith and so on [16]. Research on calculating the similarity of user nodes in social attributes is important to explain the spreading of the information online, discovery communities and assess the influence of the user nodes.

Karagiannis et al. proposed a novel approach named BLINC to mining the patterns of network terminal nodes [17]. BLINC defined the terminals' social attribute as the interaction communication degree between one node and the others. BLINC algorithm gives the statistics of the source IP and destination IP by counting the number of streams in a period of time. BLINC labels the terminal nodes with different social degree and labels the service nodes with the statistical results.

Discovering nodes' social attributes in network topology and social networks leads to additional consumption of resources used to construct and spread labeled information. BLINC can only decide on the role of a host after it gathering information from several flows, which prevents employing BLINC on real-time operation network.

We propose a system which can efficiently discover nodes' social attributes on real-time network based only on Netflow records. The system also proposes a proper data aggregation method which avoids the information loss and conflict problems.

3   System Architecture

NNSDS includes two phases: raw data preprocessing by generating snapshot of the Neflow records and social attributes discovering by analyzing snapshots stored in HDFS. The designation of NNSDS follows three considerations:

Decoupling modules: The boundary between origin data snapshotting module and follow-up analyzing module should be clear. Based on the consideration of system stability, we make the two parts relatively independent.

Standardized interaction: In order to pursue both processing efficiency and platform compatibility, NNSDS uses multi programming languages. We use data-interchange format such as Xml and JSON as the interactive data to ensure the correctness of data among modules.

Storing intermediate results: In order to do incremental processing as well as compare longitudinal data to observe system abnormalities, NNSDS stores the intermediate results in HDFS.

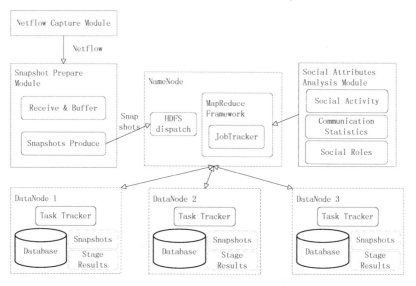

Fig. 1. Architecture of NNSDS

NNSDS includes three main functional modules: Netflow capture module, snapshots produce module and social attributes analysis module. Fig. 1 shows the architecture of NNSDS.

3.1   Netflow Capture Module

Netflow capture module is used to monitor specific gateway and produce Netflow records from the traffic flows on network.

Snapshot Produce Module
Snapshot produce module is the raw data preprocessing module which cane divided into two sub-modules. One is Netflow accepting part; it is responsible to ensure the data integrity by monitoring specific port number and receives the Neflow data. Another is the Netflow snapshotting part, which is used to map Netflow data into special format of snapshot according to the Neflow receipt time.

3.2   Social Attributes Analysis Module

This module is the core function module of the system. It applies different MapReduce tasks on snapshots stored in HDFS, calculates the network nodes' social attributes and stores intermediate analysis results into HDFS.

NNSDS can give three social attributes of Network nodes:

1）Communications statistics: The communications statistics includes the Session-In and Session-Out degree, up and down packages count, up and down bytes count of communications of the terminal nodes. This attribute reflects the real social connections of nodes and can be used to do deeper analysis of service types of the nodes.

2）Social Activity: The social activity of terminal nodes indicates the online frequency of special nodes. We calculate this attribute by counting the snapshots where specific addresses appear. Activity of nodes is the intuitional impression of social attributes of the real network. It provides evidence of the following service discovery operation. As fine-grained nodes' activity reflects nodes' real behavior better but is less convenient for observing than coarse-grained nodes' activity and different time granularities affect the discovery results with different intensity. The NNSDS is designed to get nodes' activity by period of month, day and hour, uses the fine-grained activity results as the input to calculate the coarse-grained one.

3）Social roles: The social roles of the terminal nodes means the service type of the nodes, we divide social roles into three categories: service provider, normal client nodes and NAT multiplexing nodes.

4   System Implementation

The implementation of NNSDS consists of four parts: Netflow capturing, Snapshots preparing, Mapreduce task issuing and Snapshots storing and analyzing. Fig. 2 shows the implementation flow chart of NNSDS. Netflow capture hosts are used to monitor the stream data in Ethernet and produce Netflow records which won't be discussed in this paper.

4.1   Snapshot Prepare

The task of snapshot prepare stage is to format origin input data of NNSDS. The main challenge it meets is to receive the complete data and do justifiable merge of the data. The implementations of the two steps are given below.

The original data received by NNSDS is in the form of Netflow, and high-speed receive rate will pressure the module used to produce snapshot afterwards. We open multi receive threads to store the in-coming data in ring buffers, then snapshot produce part reads data out from relative buffers. In addition, we do hash map during data cache storing the same source IP address in only one buffer to eliminate duplicate data operated afterwards.

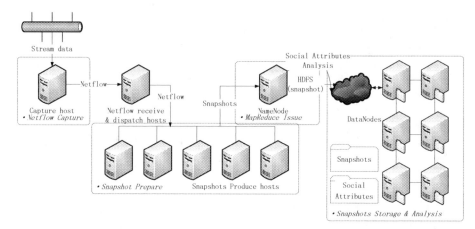

Fig. 2. Implementation of NNSDS

The snapshot producing methods are borrowed from paper [12], and we use hash table to locate data with the same source IP address to realize data merging. Snapshot includes count of destination IP addresses, count of ports used, count of protocol types of the special source IP address and so on. We use two hash tables called *conn_tab* and *ip_tab* to achieve the goal. Fig. 3depictsthe relationship between these two hash tables.

Conn_tab hash table is used to check whether the same 6-tuple Netflow has been stored and reduce operation complexity by avoiding to calculate the new-coming but already exists Netflow. *Conn_tab* use six tuples as hash key, they are source IP address (*src_IP*), destination IP address (*dst_IP*), source port (*src_Port*), destination port (*dst_Port*), protocol type of link layer (*proto_type*) and the gateway number that we get this Netflow (*fxo_id*). We store the origin Netflow information in *conn_tab* hash map with this hash key above and the hash value we stored is the information we needed from Netflow.

Ip_tab hash map is used to store the final statistical results of single source IP address. The hash key of *ip_tab* is source IP address appears in Netflow (*src_IP*).The statistical results include the total number of source ports the *src_IP* used till now, the number of destination IP addresses and the number of destination ports, the upstream and downstream degree of the *src_IP*, the total number of the upstream and downstream packets and bytes, the number of gateway we get this *src_IP* Netflow and the protocol count this *src_IP* used.

In order to gain the values stored in *ip_tab*,   we link the nodes sharing the same *src_IP* in *conn_tab* to *ip_tab* by storing pointers point to link list of the nodes. We use LRU (Least Recently Used) principle to ensure new added node is less possible to get out of the list for space storing hash maps is limited.

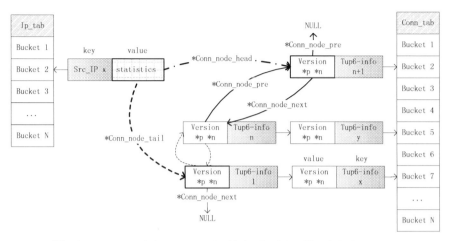

Fig. 3. The connectivity between double hash tables (IP_tab & Conn_tab)

There are nine possible situations when we deal with the newly captured Netflow data by checking both hash maps. Here we declare one of them to introduce our processing. When we get one Netflow whose *src_IP* is already stored in *ip_tab* but the 6-tuple is not stored in *conn_tab*, we deal with this Netflow as shown below.

Step 1. Add the information we gained (*src_IP, dst_IP, src_Port, dst_Port, Prototype, fxo_id, etc.*) from the Netflow and store it into *conn_tab*;
Step 2. Link this new added node to the node in *ip_tab*;
Step 3. Change the *conn_node_head* pointer of the list in *ip_tab*, and make it point to the newly added node;
Step 4. Traverse the newly updated list to check whether we should change the statistical results stored in *ip_tab*.
Step 5. Simultaneously with the Netflow operation, NNSDS outputs the communication statistic degrees from *ip_tab* hash table into snapshot files in form of JSON every 5 minutes.

4.2   Social Attributes Analysis

Social attributes of terminal nodes implemented currently include nodes' activity and attributes get from communication statistics. MapReduce is used to analyze the snapshot conveniently.

Attribute 1. Communications statistics
The Map function gets information of source IP address and communication statistical results, then output them in *<key(src_IP), value(communication values in each snapshot)>* pairs as intermediate results. The Reduce function calculates the communication statistical results of a longer period of time by adding the intermediate results of the same source IP address. After that in the Reduce function we give the Session-In and Session-Out degree, upstream and downstream packets count,

upstream and downstream bytes count, the ratio of bytes count and packets count and so on in form of *<key(src_IP), value(communication statistical value in 5 mins)>*which is the communication statistics of the terminal nodes.

Attribute 2. Social Activity

First of all, the Map function of MapReduce reads in the snapshots stored in HDFS in the<key, value>format, where the key means the Line ID of snapshot files and the value means the JSON format snapshot itself. Then in the Map function we get information of IP address and the snapshot generation time and output a pair of *<key(src_IP),value(appear time)>*as intermediate results. The key means source IP address and value means the time this address appeared. After that the Reduce function will read the intermediate results and make a bitmap string size of 24×12 to represent every 5 minutes of one day and fill the bitmap with 1 if the source IP appeared in the snapshot produced in that 5 minutes. In the end the Reduce function output source IP address and the bitmap in the form of *<key(src_IP), value(appear time periods)>* as the activity attributes of terminal nodes.

Attribute 3. Social roles

Social roles of network nodes are labeled using both nodes' activity and communication statistical results. The node which is always online and the communications degree to it is high is most likely to be a server host. The node which is used for less than half of the time and the communication degree to it is not high is most likely to be a client PC. Nodes communicating like client PCs but is used for almost all the time are most likely to be NAT nodes.

5   Experiment Results

We deploy our system in a backbone Network where the Network bandwidth is 10Gbps. The cluster used to run the NNSDS system includes 5 computers, two of them (Red Hat Enterprise Linux Server with 2.0GHz CPU and 128GB memory) are used to run the snapshot producing part, and three of them (Red Hat Enterprise Linux Server with 2.0GHz CPU and 128GB memory) are used to work as Hadoop platform.

NNSDS receives Netflow data at an average speed of 138K per second and produces snapshot every 5 minutes after operating origin data using double hash tables in form of JSON. We use historical data includes snapshots of 5 hours with about 4 million independent IP address, and then analyze and label them from terminal nodes' activity, communication statistical and social roles.

Attribute 1. Communication Degree Analysis of Network Nodes

The Session-In degree count results computed by NNSDS of the 5 hours data are depicted in Fig. 4. We can observe that the nodes' number decreases with the increasing of Session-In degree. 46.7842% terminal nodes of them show that the number of nodes linked to them is average zero per 5 minutes. The terminal nodes whose Session-in degree is above 500 network nodes per 5 minutes only occupy 0.0724%. We can speculate them as server nodes as their communication characteristics are similar.

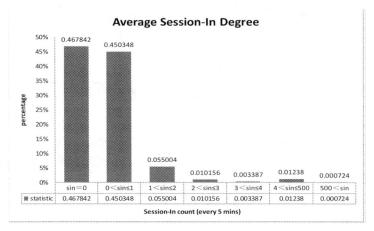

Fig. 4. Session-In degree of network nodes

Attribute 2. Activity of Network Nodes
We chose the network nodes whose Session-in degree above 500 every 5 minutes and check the social activity attribute of them.

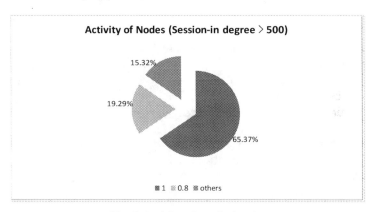

Fig. 5. Activity of terminal nodes

Results show that terminal nodes appeared all the time account for 65.37% of the whole data, nodes appeared for over 80% of the time account for 84.66%. Nodes whose activity is below 80% account for 15.32%.Thismay result from the network situation we choose. We can conclude the terminal nodes with all-time activity as server nodes.

Attribute 3. Social Roles of Network Nodes

According to the analysis results of nodes' activity and communication degree, we label nodes with above 500 accessing source nodes per 5 minutes and always stay active as service provider nodes. After examined manually, we verified that the accuracy of NNSDS is up to 95%. Table 1. gives several typical labeled network nodes along with their social attributes and the examination results.

Table 1. Typical Discovery Results

Heading level	Network Node's IP	Session-In degree	Activity degree	Examination Results
Server	180.149.132.99	2019	100%	http://yun.baidu.com/
Server	42.120.182.45	767	100%	http://www.taobao.com/
PC	106.9.255.105	3	30%	Hebei Hengshui Telecom

6   Conclusions and Future Work

Social attributes of network nodes are important for network configuration and management. However, existing approaches to discover social attributes of nodes focus on community ownership or have insignificance that cannot be employed in real-time operation network. To address these problems, we design and implement a Netflow based network nodes' social attributes discovery system (NNSDS). At first, NNSDS uses double hash tables to aggregate the raw data. Then obtains the social attributes of network nodes by using Hadoop to do incremental analysis. The social attributes include three elements: communication degree, nodes' activity and social service role. Experimental results indicate that NNSDS has good performance on the backbone networks.

NNSDS is now able to discover nodes' social attributes of communication statistics, activity and social roles. In the future, we aim to complete the implementation details by considering other social attributes.

Acknowledgement. Supported by the National Key Technologies Research and Development Program of the Ministry of Science and Technology of China under Grant No.2012BAH37B02; the National Information Security 242 Project of China under Grant No.2013A133.

References

1. Tan, Y., Wu, J., Deng, H.: Evaluation method for node importance based on node contraction in complex networks. Systems Engineering-Theory & Practice 11, 79–83 (2006)
2. Costa, L.F., Rodrigues, F.A., Travieso, G., et al.: Characterization of complex networks: A survey of measurements. Advances in Physics 56(1), 167–242 (2007)
3. Cisco I O S. Netflow introduction (September 2006), http://www.cisco.com

4. Estan, C., Varghese, G.: New directions in traffic measurement and accounting. ACM (2002)
5. Wang, C., Chen, W., Zhang, J.: Challenging Scientific Problems of Technology and Application of Big Data. Bulletin of National Natural Science Foundation of China (2014)
6. IPFIX. Internet Engineering Task Force, IP Flow Information Export Working Group, http://www.ietf.org/html.charters/ipfix-charter.html
7. Cisco Sampled Netflow, http://www.cisco.com
8. Choi, B.Y., Bhattacharyya, S.: On the accuracy and overhead of cisco sampled netflow. In: Proceedings of ACM SIGMETRICS Workshop on Large Scale Network Inference (LSNI) (2005)
9. Chen, N., Xu, T.: Study on NetFlow-based network traffic data collection and storage. Application Research of Computers 2, 75 (2008)
10. Li, M., Ye, J., Li, J.: Network Traffic Management Based on Hash Aggregation. Journal of Jiangxi Normal University (Natural Science) 35(3), 174–177 (2011)
11. Eager, D., Vernon, M., Zahorjan, J.: Optimal and efficient merging schedules for video-on-demand servers. In: Proceedings of the Seventh ACM International Conference on Multimedia (Part 1), pp. 199–202. ACM (1999)
12. Mokbel, M.F., Lu, M., Aref, W.G.: Hash-merge join: A non-blocking join algorithm for producing fast and early join results. In: Proceedings of the 20th International Conference on Data Engineering, pp. 251–262. IEEE (2004)
13. Naumann, F., Häussler, M.: Declarative data merging with conflict resolution (2002)
14. Lv, Z., Zheng, J., Huang, H.: A Distributed Real-Time Intrusion Detection System for High-Speed Network. Journal of Computer Research and Development 41(4), 667–673 (2004)
15. Paxson, V., Mahdavi, J., Mathis, M., et al.: Framework for IP performance metrics. Framework (1998)
16. Wang, J., Zhu, K., Wang, F.: Microblog Fans Network Evolving Model based on User Social Characteristics and Attractiveness of Behavior Properties. Journal of Computer Application 33(10), 2753–2756 (2013)
17. Karagiannis, T., Papagiannaki, K., Faloutsos, M.: BLINC: multilevel traffic classification in the dark. In: ACM SIGCOMM Computer Communication Review, vol. 35(4), pp. 229–240. ACM (2005)

A Survey of Network Attacks Based on Protocol Vulnerabilities

Gang Xiong[1], Jiayin Tong[1,2], Ye Xu[1,2], Hongliang Yu[3], and Yong Zhao[1]

[1] Institute of Information Engineering, Chinese Academy of Sciences, Beijing, China 100093
{xionggang,tongjiayin,xuye,zhaoyong}@iie.ac.cn
[2] University of Chinese Academy of Sciences, Beijing, China 100049
[3] Beijing University of Posts and Telecommunications, Beijing, China 100876
yuhongliang@bupt.edu.cn

Abstract. It has a long history to launch attacks using vulnerability in the network protocols. More and more researchers are attracted to the attack and defenses of network protocols. It will mitigate the severe consequences that attacks may lead to by maliciously using protocol vulnerability if we have reliable protocol design and prompt defenses. In this paper, we review the research progress about attacks based on protocol vulnerability. We take advantage of critical characteristics in information security to classify these attacks, namely confidentiality attack, integrity attack, availability attack. Some challenges confronted by the researchers are discussed in view of current researches. The prospect of this field in the future comes at last.

Keywords: network protocol, protocol vulnerability, network traffic, network attack, attack defenses.

1 Introduction

The Internet keeps developing rapidly in the recent years. Different network services based on various network protocols enrich the content on the Internet, satisfying diverse needs of users. However, the vulnerabilities in network protocols are likely to be taken advantage to launch network attacks. Some of these attacks aim at the resources of target hosts, such as consuming CPU resources and network bandwidth; some aim to crack encrypted traffic to steal the secret information; others masquerade as legitimate users to cheat victims. Though the attack targets and purposes are different from each other, they all relate to vulnerabilities in network protocols.

In order to carry out network attack successfully, it is necessary that the target network protocols suffer from vulnerabilities. These vulnerabilities mainly come from two aspects. One is the protocol design procedure, which will impact the nodes running that protocol when attack happens. The other is from the malicious use of legitimate protocol process. At this time, the attacked node might be any one linking to the Internet and the damage will be much more serious.

At present, vulnerabilities in the network protocols are included to the research field after some kind of attacks appear and lead to network destruction. Not satisfied

W. Han et al. (Eds.): APWeb 2014 Workshops, LNCS 8710, pp. 246–257, 2014.

with this passive method, more and more researchers start from different aspects gradually to analyze vulnerabilities in popular protocols, hoping to come up with solutions before massive attack occurs so as to narrow down the potential losses.

We discuss and summarize network attacks based on protocol vulnerabilities in this paper. We introduce the current state of related researches, study the latest research progress and prospect research directions in the future. The paper continues as follows: Section 2 presents a classification of some well-known protocol vulnerability attack cases from the perspective of the information security critical characteristics they break and the attack range they have. Section 3 focuses on both of attack and defense methods to provide research progress in related areas. The subsection about protection is divided by protocol layers. We debate future research directions in Section 4. The conclusion of this whole paper is shown in Section 5.

2   Network Attacks Classification Based on Protocol Vulnerabilities

Researches on network attacks based on protocol vulnerabilities are widespread. Classification of these researches will help to summarize current work, know the research state better and guide further study.

Traditional classification method generally considers the layer of protocols, using OSI model or TCP/IP model to do the classification. In this paper we will take advantage of a higher layer, the critical characteristics in information security, with the attack range to classify these network attacks, and then make a detailed discussion.

As early as in [1], information security golden triangle CIA model was referred and spread. This model contains three critical characteristics of information security, namely Confidentiality, Integrity and Availability. With the gradual development of information security subject, more and more attributes are put forward by researchers. These attributes extend the connotation of information security in different aspects, such as Controllability, Non-repudiation, and Authentication.

In the CIA triangle, Confidentiality means that, during the information transfer and storage, it could not be leaked to unauthorized entities, that is to say, only the authority could get the information. Integrity protects the information from being tampered by unauthorized entities during its transfer and storage, or discovers the tampered information in time. Availability represents that the authorized entities are able to use information and information systems normally, preventing denial-of-service attack or even compromise due to malicious attacks.

The network attacks based on protocol vulnerabilities discussed in this paper will first be classified into three categories according to the broken critical characteristic, naming Confidentiality attack, Integrity attack and Availability attack. For each category, in the light of attack range, we differentiate into two subclasses, only impacting the nodes running the protocol, or any nodes linked to the Internet. Then follow detailed discussion in every subclass. Because the scope of network attacks is really wide, this paper will focus on directions with latest progress. Some classical attack cases will also be included. The organization is shown in Fig. 1.

Classification Reasons		Typical Attack Cases	
		Protocol Nodes	Any Nodes
Information Security Attributes	Confidentiality	Crack Specific Protocols	Crack Encryption Algorithm
	Integrity	Man-in-the-Middle Attack	Service Hijacking
	Availability	Denial of Service	Reflection Attack

Fig. 1. Classification of typical network attacks based on protocol vulnerabilities

3   Survey of Research Progress

As shown in Fig. 1, this section will first review current status of researches on network attacks based on protocol vulnerabilities. Then discuss research progress about corresponding defense technologies.

3.1   Confidentiality Attack

Confidentiality attack refers to decryption and analysis on encrypted protocol to obtain transferred information, or identify the anonymous communication ends.

3.1.1   Attack on Protocol Nodes

Cracking the encrypted instant messaging protocol that used by large number of users is hot. Those protocols include popular VoIP software Skype and QQ. In addition, some researches about confidentiality attacks against mobile devices are carried out.

Through a dynamic analysis of Skype, Desclaux [2] found that the client would send some signaling data encrypted by RC4 after establishing TCP connections with other nodes. But two packets were the same. Because of exclusive or used in RC4, these two data packets could be decrypted.

Yu et al. [3] found that during QQ login, the key and the encrypted message were sent in the same datagram. Some data could be extracted using that key. Then making use of vulnerability in qqTEA encryption algorithm, combined with brute-force attack, they could decrypt all the data. The writers carried out the attack successfully.

Yi [4] further diminished the complexity of brute-force attack to QQ. He also discovered that the cycle in the pseudo-random number generator was so small that one could use exhaustive method in an acceptable time. Using this vulnerability, he accomplished the task of cracking a significant key during QQ login.

Lin et al. [5] studied the anonymity in Location Based Service. They presented a sending model for consecutive queries, pointing out the current k anonymity algorithm could not protect the privacy efficiently. Then they came up with query attack algorithm for two k anonymity algorithms, Clique Cloaking and Non-clique Cloaking. The anonymity set could not be appropriate to measure the query anonymity. So they suggested a new method based on entropy. The experiments showed that the query attack algorithm was efficient to identify the anonymous users.

Long ago, Lundberg [6] did researches on the security issues in Ad-hoc routing. He proposed several attacks including black hole attack, which referred that malicious node broadcasted itself to be the shortest path node and intercept all the packets.

For MANET, Ullah et al. [7] compared OSLR protocol and Ad-Hoc Demand Distance Vector protocol under black hole attack and evaluated the throughput, lag and network load using OPNET, showing their performances under different conditions.

3.1.2   Attack on Any Nodes

The encryption algorithms are widely used in the fields of identification verification, sensitive information protection and prevent data from leaking to illegal entities. Benefiting from the promotion of hardware capability and distributed computing, people are easy to get systems with powerful computing capability. Researchers are turning their eyes to crack encryption algorithms using high-performance computers. Zhang et al. took advantage of distributed computing, trying to crack block cipher algorithm [8] and stream cipher algorithm [9]. On the other hand, current HTTPS, FTPS, SSL VPN are all based on SSL/TLS protocol. Some researchers focus on the vulnerabilities in SSL/TLS to crack this protocol and achieve the encrypted information.

3.2   Integrity Attack

Integrity attack stands for tampering the transferring data by unauthorized entities. This kind of attack modifies the original content and affects the trust between nodes.

3.2.1   Attack on Protocol Nodes

Qian et al. [10] presented a new TCP sequence number inferring attack, which occurred in the middleware of firewalls. The attackers could get the cached data in the middleware to hijack TCP connections and inject malicious content.

Zhang [11] proposed two HTTPS attack methods. SSLSniff was on the basis of ARP spoof and certificate replacement, while SSLStrip on ARP spoof and content tampering. If a client lacked of checking identification or the validity of certificates, an attacker could use SSLSniff to hijack HTTPS sessions. Moreover, the security of HTTPS was tied to HTTP. Attackers might use SSLStrip to tamper content in HTTP.

Callegati et al. [12] also gave a man-in-the-middle attack in HTTPS. They first masqueraded as a default gateway in the LAN. Then they intercepted responses from the server and replaced the original certificate with a fake self-signed certificate.

Zhou [13] concluded some weaknesses in P2P stream, including data pollution, large volume, real time and closure, making it harder to keep P2P stream security.

3.2.2   Attack on Any Nodes

DNS hijacking is usually used by operators to spread advertisements. When a user surfs the Internet, he cannot refuse these extra content. This is certainly another kind of attack to integrity and it will affect all the users under such an operator.

3.3   Availability Attack

Availability attack mainly disturbs the normal nodes running via protocol vulnerability, making it unable to provide services, or even driving these valid nodes to attack any other hosts on the Internet. The former attack affects nodes that running that protocol, while the later could lead to wider impacts, and difficult to trace.

3.3.1   Attack on Protocol Nodes

ARP attack was raised as early as 1997. It took advantage of vulnerability in APR updating mechanism. A more famous attack is TCP SYN flooding attack. The half TCP connections consume many resources of servers, resulting in denial of service.

In 2003, Kuzmanovic et al. [14] first came up with Low-rate Denial-of-Service (LDoS) in TCP, which made use of TCP retransmission when timed out. It would impact on TCP throughput seriously. This vulnerability embeds in common self-adaption mechanism, thus harder to defense. Guirguis et al. [15,16] proposed RoQ attack, which in fact was also owing to vulnerabilities in TCP congestion control and routers management. He et al. [17] surveyed LDoS in detail, presenting different LDoS in TCP and LDoS to active queue management in routers. Kumar et al. [18] suggested a new attack based on LDoS. Different from former attacks, the malicious end was a TCP receiver. This malicious receiver might control the rate and mode of the sender remotely to exhaust its resources.

For P2P networks, in 2002, Douceur [19] presented a new attack method called the Sybil Attack. Malicious nodes used the vulnerability that creating new identifications cost nothing in the network. They masquerade as normal nodes to launch attacks, such as misleading nodes with incorrect routing tables, transferring unauthorized files. Since nodes in P2P network only need to send keep-alive messages at regular intervals to inform other nodes its status, the malicious nodes could also keep sending such messages to pollute neighbor nodes' routing tables, affecting the entire P2P network.

Wang et al. [20] put forward an asymmetric communication method using IP spoofing to defend internal attacks. The client sent requests through IM or email to the proxy. After receiving requests, the proxy sent the responses to the client using spoofing IP. This method could hide the IP address of the proxy, preventing other nodes to discover that proxy.

Wang et al. [21] researched SIP flooding attack in 3G network and shown related detection methods. SIP flooding was carried out by sending lots of SIP messages to proxy-session control function (P-CSCF) in IMS network via SIP ends. As a result, users in target region could not establish any calls, namely denial-of-service.

Related researches are increasing with the popularity of IPv6. Nakibly et al. [22] presented routing loop attack in IPv6 automatic tunnel and measures to release this problem. The vulnerability in IPv6-in-IPv4 automatic tunnel allowed attackers to form a routing loop to amplify traffic, triggering a DoS attack. Abley et al. [23] found a new worm which spread in IPv4-IPv6 protocol stack. This worm applied a two-layer scan mechanism. The simulation result showed the worm spread faster in IPv4-IPv6 stack, proving that two protocol stacks would affect the propagation of worms.

3.3.2   Attack on Any Nodes

The reflection method could be used to launch attacks against any nodes on the Internet, with the premise that a wide disparity between request and response data sizes.

Paxson [24] discussed different protocol vulnerabilities which could lead to reflection attacks in 2001. In this paper, we mainly focus on several reflection attacks with significant impact.

DNS reflection has a long history and is always used by attackers due to its remarkable effect. Though owning a shorter history, NTP reflection attack catches attackers' attention because of its as much as 200 times amplification factor. P2P networks, owing to its robustness, extensibility, high-performance, gradually replace traditional C/S networks. They are widely applied to services such as instant messaging, streaming media, and resource downloading, which in charge of massive online nodes. But their vulnerabilities could be maliciously used to launch network attacks.

3.3.2.1 DNS Reflection Attack Vaughn et al. [25] studied cases about DNS reflection attacks, finding that the attack changed from MX record responses of unauthoritative servers at early time to much larger responses of EDNS. The anti-spam organization Spamhaus met a 300Gbps DDoS attack in 2013. The attack employed large response packets of open recursive DNS servers and surely shocked the whole world.

3.3.2.2 NTP Reflection Attack In January 2014, CloudFlare, a cloud computing company providing DDoS attack prevention services, declared that it was attacked by more than 400Gbps DDoS traffic. This attack was based on servers running NTP (Network Time Protocol).

3.3.2.3 P2P Reflection Attack Naoumov et al. [26] studied Overnet protocol in eDonkey clients. Each eDonkey node manages its own index table and routing table using DHT. The online node will directly add new resource nodes and routing nodes to its index table without verification. They launched DDoS attacks by polluting P2P nodes indexes and routing tables.

Athanasopoulos et al. [27] discovered that in Gnutella network, the QueryHit response packet did not have a confirmation. So they forged some resource response packets to nodes in Gnutella, resulting in lots of nodes requesting the victim for resources or even downloading files from the victim.

Sia [28] analyzed BitTorrent protocol and pointed out that its tracker protocol had vulnerabilities. A DDoS attack could be triggered by sending some packets with spoofed IP and ports to online nodes, controlling them to request the victim. Defrawy et al. [29] also carried out DDoS attack using BitTorrent network. They made some fake torrent files, which contained a modified tracker server and address of the victim. When someone downloaded this torrent, the node would regard the victim as a tracker server, established TCP connections with it, thus consuming resources of the victim.

For Kad network, Steiner et al. [30] debated on possibilities of using Kad maliciously to launch attacks. Sun et al. [31] took advantage of masqueraded Kad response packets to attack nodes on the Internet. Yu et al. [32] indicated that the downstream data was much larger than the upstream and there was not an identification verification mechanism in Kad. Such vulnerabilities might be used by attackers.

In ESM [31], every node connects to another one aperiodically and sends current list of recorded nodes. The other will add new information to its own list without checking. Then it will send messages to these new nodes. The attackers would be able to launch attacks using this vulnerability.

In our previous work [33], after analyzing the packets, we found that Thunder also suffered from the pervasive problems in P2P protocol design. It lacks of procedures to verify node identification. We could easily amply the number of packets by 12 times.

The reason for the vulnerabilities mentioned above is that there is no necessary identification verification mechanism in the protocols. Ignoring identification verification could avoid a series of complex procedures, contributing to the improvement of the entire system throughput. However, it raises some security issues. Especially when the number of online nodes is tremendous, the vulnerability in protocol design could be amplified and lead to serious results. Attackers might forge some packets, masquerade as P2P nodes or servers to detect P2P network, confuse normal online nodes, and even drive them to attack victims.

3.4   Countermeasures

In massive applied network protocols, there exist potential vulnerabilities that could be used to launch attacks. Hence, researchers studied them to present corresponding countermeasures. We will discuss the progress according to different protocol layers.

3.4.1   Network Layer Countermeasures

Lv et al. [34] focused on countermeasures of IP spoofing. They first studied the defects of end-to-end IP verification mechanism, pointing out that it neglected damages from the retransmission of IP spoofing messages. Then they proposed ESP to improve IP spoofing protection. This mechanism combined open routers correlation technology with tags. It could support both dynamic routing and asymmetric routing.

Nakibly et al. [22] gave three methods to deal with routing loop attack in IPv6 automatic tunnel, including verifying nodes, filtering shared tunnel list and IP detection.

Yang et al. [35] presented a random marking method to trace the source of an IPv6 attack. Routers would extract packets passed by under a specific probability. In the meantime, the selected packets were marked with addresses of any two neighbor routers. Then the victim could recover attack path using these marks.

3.4.2   Transport Layer Countermeasures

Lemon [36] came up with SYN cache and SYN cookies to defense SYN flooding attack. The thought of SYN cache is that servers only assign least resources when receiving SYN packets and other resources are assigned until the connection is established. While the thought of SYN cookies is that sending an encrypted cookie to the original requester instead of status information. Terry et al. [37] also mentioned using

SYN cookies to prevent SYN flooding attack. The cookie here was a unique cookie for just one time, thus resisting retransmission. Using AES128 and private key, one could check if the address was real.

Bellovin et al. [38] proposed a method to defense TCP original sequence number attack. They designed an algorithm that using IP addresses, ports and a private key to generate sequence number.

To defense LDoS, many researchers put forward their schemes. Zhang et al. [39] improved traditional RED [40]. Their Robust RED algorithm increased the throughput of TCP. The basic thought was to detect and filter attack packets before deploying traditional RED. Xiang et al. [41] carried out a method using information matrix to detect and trace LDoS. They quantified probability distributions of different network flows through information matrix. Their method could detect early attacks in an extensive scale. Chang et al. [42] proposed a simple protection mechanism for LDoS naming SAP (Shrew Attack Protection). Different from two previous methods, it tried to select attack flows to trace attacks. This algorithm was based on destination port number, so easy to implement on current routing mechanisms.

3.4.3   Application Layer Countermeasures
There are so many different application layer protocols, so we classify them into four categories according to research progress, namely instant messaging, HTTP(S), P2P and others.

3.4.3.1 Instant Messaging Garfinkel [43] analyzed on Skype in detail, indicating that privacy, verification, availability, robustness, flexibility and integrity were significant factors to ensure the security of Skype.

Aiming at problems faced during QQ login, Yi [4] presented two methods to handle brute-force and pseudorandom number attacks. Increasing the times of TEA iteration and changing the last 7 bytes padding could defense brute-force attack. While a pseudorandom number generator with longer cycle would alleviate the attack. Extra MD5 operation could mitigate dependency on pseudorandom number.

3.4.3.2 HTTP(S) Huang et al. [44] did a security evaluation and behavior monitor about injection attacks on the Internet. They described a series of skills in software testing to apply them to Internet applications. They found that the performance of popular flaw scanning software WAVES was satisfactory.

Zhang [11] showed protection schemes to man-in-the-middle HTTPS attack. One was to strengthen verification of shared key, making it harder for attackers to get the master key. The other was using digital signature to encrypt sensitive information.

3.4.3.3 P2P Douceur [19] considered that the Sybil attack could be cleared only if a trust mechanism was adopted. The thought was to calculate a trust value for every node. Then online nodes would remove nodes with lower trust values to the edge of the network or even block them, thus diminishing the impact of malicious nodes.

Wang et al. [45] pointed out that streaming media P2P network were more likely to suffer DoS attack than resource sharing P2P network due to more bandwidth and real time. A DoS defense system was presented to move untrustworthy nodes away.

Zhou [13] came up with a security model for P2P streaming media based on content review and implemented a prototype system, which was verified to have the ability to defense content pollution attacks.

For eDonkey (Overnet protocol), Naoumov et al. [26] proposed an explicit identification verification method between online nodes. Through encryption and closed source, messages could only be published by the node itself, but it confronted reverse engineering. While using a centralized proxy server would consume more resources.

Athanasopoulos et al. [27] studied an algorithm for Gnutella. Each node brought in a safe list by checking resource nodes to alleviate the consequences of attacks.

For BitTorrent, Sia [28] modified the implementation of protocol to prevent attackers from forging IP addresses. While Defrawy et al. [29] referred that an inspection should be made to both clients and tracker servers, removing tracker servers without responses to avert attacks.

Sun et al. [46] investigated general defense methods against DDoS attack using P2P systems. Then they emphasized on the importance of verifying P2P nodes. At the same time, they designed an active detection method to validate online nodes, thereby reducing attack impacts. This method benefited from its adaptability and efficiency.

3.4.3.4 Others Wang et al. [21] found that due to multiple mechanisms 3G supporting, attackers could carry out attacks via illegal mobile SIP ends, making the attacks more difficult to detect. Regarding the difference between the number of initial REGISTER request messages during the sampling period and normal number of messages as the characteristic, they conducted the detection using an algorithm called DSMD.

4   Future Research Directions

Network attacks based on protocol vulnerabilities tend to be more violent, more diverse and more invisible. The corresponding defense technology is improving as well.

Violence, mainly refers that attackers could collect and make use of large scale of resources more easily to launch attacks due to the development of computing capability of hardware (e.g. GPU). This resource dissymmetry will bring disasters to common network users. The related defense methods need more innovations to guarantee efficient protection under a resource-constrained circumstance.

Diversity includes two aspects, namely device-diversity and protocol-diversity. With more users using smart devices and rapid development of mobile Internet, not only producers and developers benefit, researchers and attackers are also attracted to this area. It is quite frequent that attackers take advantage of weakness in mobile systems to attack mobile terminals directly. For researchers, how to find out and tackle with vulnerabilities in mobile hosts promptly is both an opportunity and a challenge. And in protocol-diversity, due to the exhaustion of IPv4 addresses, IPv6 appears as a next generation protocol, and it will surely attract more attentions. Attacks towards IPv6 are unavoidable. The battle between attackers and defenders will move on.

Invisibility is achieved through mixed use of multiple protocols, thus more difficult to trace the attackers, for example, P2P reflection attack mentioned above. For P2P protocols, it is necessary to implement identification verification mechanism, which could help to recognize illegal requests and connections from malicious users and protect nodes from detection, attack and exploitation. When designing and implementing P2P protocols, essential identification verification procedure is always omitted for many reasons. After P2P applications owning a large number of users, this weakness will lead to serious consequences. Though some protocols adopt encryption methods, the intensity is not enough to resist current crack and analysis technology. Stronger encryption algorithm stands for more resource consumption. The balance between protocol security and resource consumption is needed.

5   Conclusion

Various network applications provide different types of services for network users, meeting their different needs and helping promote the Internet. But the vulnerabilities in these network protocols have a chance to be maliciously used by attackers to launch network attacks. This paper gives a survey of network attacks based on protocol vulnerabilities and highlights research fields that make latest progress. We also present some expectations about the future researches.

Acknowledgements. This work is supported by the National High-Tech Research and Development Plan "863" of China (Grant No. 2011AA010703), the National Key Technology R&D Program (Grant No. 2012BAH46B02), the National Natural Science Foundation of China (NSFC) (Grant No. 61070184) and the "Strategic Priority Research Program" of the Chinese Academy of Sciences (Grant No. XDA06030200).

References

1. McCumber, J.: Information Systems Security: A Comprehensive Model (1991),
   http://cryptosmith.com/sites/default/
   files/docs/MccumberAx.pdf
2. Desclaux, P.B.F.: Silver Needle in the Skype (2006),
   https://www.blackhat.com/presentations/
   bh-europe-06/bh-eu-06-biondi/bh-eu-06-biondi-up.pdf
3. Yu, K., Zhang, Y., Wang, Y.: Research and Analysis on the Security of QQ Login Protocol (in Chinese). Netinfo Security 11, 55–57 (2008)
4. Yi, Z.: Research and Analysis of QQ Login Protocol and Improvement (in Chinese). Netinfo Security 6, 85–87 (2011)
5. Lin, X., Li, S., Yang, Z.: Attacking Algorithms against Continuous Queries in LBS and Anonymity Measurement. Journal of Software 20(4), 1058–1068 (2009) (in Chinese)
6. Routing Security in Ad hoc Networks,
   http://citeseer.nj.nec.com/400961.html
7. Ullah, I., Rehman, S.U.: Analysis of Black Hole Attack on MANETs Using Different MANET Routing Protocols. Blekinge Institute of Technology, Sweden (2010)

8. Zhang, L., Zhang, Y.: Brute Force Attack on Block Cipher Algorithm Based on Distributed Computation. Computer Engineering 34(13), 121–123 (2008) (in Chinese)
9. Zhang, L., Zhang, Y.: Brute Force Attack on the RC4 Encryption Algorithm Based on Distributed Computing. Computer Engineering and Science 30(7), 15–20 (2008) (in Chinese)
10. Qian, Z., Mao, Z.: Off-path TCP Sequence Number Inference Attack-How Firewall Middleboxes Reduce Security. In: IEEE Symposium on Security and Privacy (SP), pp. 347–361. IEEE Press, San Francisco (2012)
11. Zhang, H.: Security Analysis of HTTPS Protocol Based on MITM Attack. Shanghai Jiao Tong University, Shanghai (2009) (in Chinese)
12. Callegati, F., Cerroni, W., Ramilli, M.: Man-in-the-Middle Attack to the HTTPS Protocol. In: IEEE Symposium on Security and Privacy, pp. 78–81. IEEE Press, Oakland (2009)
13. Zhou, S.: P2P Streaming Media Security Research (in Chinese). Central South University, Hunan (2009)
14. Kuzmanovic, A., Knightly, E.W.: Low-Rate TCP-targeted Denial of Service Attacks: the Shrew vs. the Mice and Elephants. In: Proceedings of the 2003 Conference on Applications, Technologies, Architectures, and Protocols for Computer Communications, pp. 75–86. ACM, Karlsruhe (2003)
15. Guirguis, M., Bestavros, A., Matta, I.: Exploiting the Transients of Adaptation for RoQ Attacks on Internet Resources. In: Proceedings of the 12th IEEE International Conference on Network Protocols, pp. 184–195. IEEE Press, Berlin (2004)
16. Guirguis, M., Bestavros, A., Matta, I., Zhang, Y.T.: Reduction of Quality (RoQ) Attacks on Internet End-Systems. In: INFOCOM 2005, pp. 1362–1372. IEEE Press, Miami (2005)
17. He, Y., Liu, T., Cao, Q., Xiong, Q., Han, Y.: A Survey of Low-Rate Denial-of-Service Attacks. Journal of Frontiers of Computer Science and Technology 2(1), 1–19 (2008) (in Chinese)
18. Kumar, V.A., Jayalekshmy, P., Patra, G.K., Thangavelu, R.P.: On Remote Exploitation of TCP Sender for Low-Rate Flooding Denial-of-Service Attack. Communications Letters 13(1), 46–48 (2009)
19. Douceur, J.R.: The Sybil Attack. In: Druschel, P., Kaashoek, M.F., Rowstron, A. (eds.) IPTPS 2002. LNCS, vol. 2429, pp. 251–260. Springer, Heidelberg (2002)
20. Wang, Q., Gong, X., Nguyen, G.T.K., Houmansadr, A., Borisov, N.: CensorSpoofer: Asymmetric Communication Using IP Spoofing for Censorship-Resistant Web Browsing. In: Proceedings of the 2012 ACM Conference on Computer and Communications Security, pp. 121–132. ACM, New York (2012)
21. Wang, S., Sun, Q., Yang, F.: Detecting SIP Flooding Attacks against IMS Network. Journal of Software 22(4), 761–772 (2011) (in Chinese)
22. Nakibly, G., Templin, F.: Routing Loop Attack Using IPv6 Automatic Tunnels: Problem Statement and Proposed Mitigations (2011),
http://tools.ietf.org/search/rfc6324
23. Abley, J., Savola, P., Neville-Neil, G.: Deprecation of Type 0 Routing Headers in IPv6 (2007), http://www.ietf.org/rfc/rfc5095.txt
24. Paxson, V.: An analysis of Using Reflectors for Distributed Denial-of-Service Attacks. ACM SIGCOMM Computer Communication Review 31(3), 38–47 (2001)
25. DNS Amplification Attacks,
http://www.isotf.org/news/DNS-Amplification-Attacks.pdf
26. Naoumov, N., Ross, K.: Exploiting P2P Systems for DDoS Attacks. In: Proceedings of the 1st International Conference on Scalable Information Systems, pp. 47–52. ACM, New York (2006)

27. Athanasopoulos, E., Anagnostakis, K.G., Markatos, E.P.: Misusing Unstructured P2P Systems to Perform DoS Attacks: The Network That Never Forgets. In: Zhou, J., Yung, M., Bao, F. (eds.) ACNS 2006. LNCS, vol. 3989, pp. 130–145. Springer, Heidelberg (2006)
28. Sia, K.C.: DDoS Vulnerability Analysis of BitTorrent Protocol (2007), http://oak.cs.ucla.edu/~sia/pub/cs239spring06.pdf
29. El Defrawy, K., Gjoka, M., Markopoulou, A.: BotTorrent: Misusing BitTorrent to Launch DDoS Attacks. In: Proceedings of the 3rd USENIX Workshop on Steps to Reducing Unwanted Traffic on the Internet, pp. 1–6. USENIX Association, Santa Clara (2007)
30. Steiner, M., En-Najjary, T., Biersack, E.W.: Exploiting KAD: Possible Uses and Misuses. ACM SIGCOMM Computer Communication Review 37(5), 65–70 (2007)
31. Sun, X., Torres, R., Rao, S.: DDoS Attacks by Subverting Membership Management in P2P Systems. In: 3rd IEEE Workshop on Secure Network Protocols, pp. 1–6. IEEE Press, Beijing (2007)
32. Yu, J., Li, Z., Chen, X.: Misusing Kademlia Protocol to Perform DDoS Attacks. In: International Symposium on Parallel and Distributed Processing with Applications (ISPA 2008), pp. 80–86. IEEE Press (2008)
33. Tong, J., Xiong, G., Zhao, Y., Guo, L.: A Research on the Vulnerability in Popular P2P Protocols. In: 8th International Conference on Communications and Networking in China, pp. 405–409. IEEE Press, Guilin (2013)
34. Lv, G., Sun, Z., Lu, X.: Enhancing the Ability of Inter-Domain IP Spoofing Prevention. Journal of Software 21(7), 1704–1716 (2010) (in Chinese)
35. Yang, J., Wang, Z., Guo, H.: IPv6 Attack Source Traceback Scheme Based on Extension Header Probabilistic Marking. Application Research of Computers 27(6), 2335–2340 (2010)
36. Lemon, J.: Resisting SYN Flood DoS Attacks with a SYN Cache. In: Proceedings of the BSD Conference, pp. 89–97. USENIX Association, Berkeley (2002)
37. Terry, T., Yu, H., Yuan, X., Chu, B.: A Visualization Based Simulator for SYN Flood Attacks. In: Proceedings of the International Conference on Imaging Theory and Applications and International Conference on Information Visualization Theory and Applications, pp. 251–255. Elsevier (2011)
38. Bellovin, S., Gont, F.: Defending against Sequence Number Attacks (2012), http://tools.ietf.org/html/rfc6528
39. Zhang, C., Yin, J., Cai, Z., Chen, W.: RRED: Robust RED Algorithm to Counter Low-Rate DoS Attacks. IEEE Press Communications Letters 14(5), 489–491 (2010)
40. Floyd, S., Jacobson, V.: Random Early Detection Gateways for Congestion Avoidance. IEEE/ACM Transactions on Networking 1(4), 397–413 (1993)
41. Xiang, Y., Li, K., Zhou, W.: Low-Rate DDoS Attacks Detection and Traceback by Using New Information Metrics. IEEE Transactions on Information Forensics and Security 6(2), 426–437 (2011)
42. Chang, C., Lee, S., Lin, B., Wang, J.: The Taming of The Shrew: Mitigating Low-Rate TCP-Targeted Attack. IEEE Transactions on Network and Service Management 7(1), 1–13 (2010)
43. Garfinkel, S.L.: VoIP and Skype Security (2005), http://www.cs.columbia.edu/~salman/skype/SkypeSecurity_1_5_garfinkel.pdf
44. Huang, Y., Huang, S., Lin, T., Tsai, C.H.: Web Application Security Assessment by Fault Injection and Behavior Monitoring. In: Proceedings of the 12th International Conference on World Wide Web, pp. 148–159. ACM, New York (2003)
45. Wang, P., Sparks, S., Zou, C.: An Advanced Hybrid Peer-to-Peer Botnet. IEEE Transactions on Dependable and Secure Computing 7(2), 113–127 (2010)
46. Sun, X., Torres, R., Rao, S.: Preventing DDoS Attacks on Internet Servers Exploiting P2P Systems. Computer Networks 54(15), 2756–2774 (2010)

Evaluation Scheme for Traffic Classification Systems

Yong Zhao[1], Yuan Yuan[2], Yong Wang[2], Yao Yao[1], and Gang Xiong[1]

[1] Institute of Information Engineering, Chinese Academy of Sciences, Beijing, China
[2] National Computer Network Emergency Response Technical Team
Coordination Center of China

Abstract. Recent research on evaluation and comparison of traffic classification systems only used tagged offline dataset, thus the result can only reflect the performance of the classification systems on the network from which the offline dataset was collected. Besides, the difference of scopes and granularities of different traffic classification systems also render them not comparable. In this work, we propose a novel two-phased evaluation system which combines offline dataset evaluation and online evaluation. Our evaluation approach can help network manager pick the traffic classification system that fit their specific network most. In addition, we introduce three metrics corresponding to our evaluation scheme to do comprehensive evaluation and group applications according to their behaviors and functions to compare classification systems of different granularities.

Keywords: traffic classification, traffic classification evaluation, traffic classification comparison.

1   Introduction

With the development of the Internet, the volume of network traffic has increased a lot, and the content of the traffic has also changed dramatically. Years ago, network traffic was mainly comprised of packets of traditional protocols such as HTTP, FTP, etc, however, the ingredients of nowadays' network traffic are much more complicated. With the ratio of traditional protocols continue to shrink, traffic of numerous new applications are filling this gap, among which are P2P, VoIP, encrypted traffic, viruses and attack traffic. In response to the change of network traffic, a growing number of network operators start to use traffic classification technology and traffic classification devices to identify traffic flowing through their management node. By doing so, they can understand their network better and improve their service quality. Driven by intense demand of these network operators, more and more novel traffic classification methods, as well as traffic classification devices that use these new methods have been proposed and introduced to classify the complicated traffic in the recent years in this field.

While new classification methods and classification devices are emerging, technologies used to evaluate these methods and devices develop slowly. Due to different granularities and many other reasons, it is still difficult to evaluate different classification methods and classification devices comprehensively and systematically; existing

W. Han et al. (Eds.): APWeb 2014 Workshops, LNCS 8710, pp. 258–264, 2014.
© Springer International Publishing Switzerland 2014

research use offline dataset to do evaluation which is hard to reflect the results of online classification; besides metrics other than classification accuracy is in need to give an overall assessment for these methods and devices and compare their ad- vantages and disadvantages.

In this paper we first analyze the key issues prohibiting the development of traffic classification evaluation, then, we propose our novel two-phased evaluation scheme to solve these problems and introduce some metrics related to build a uniform evaluation system for traffic classification systems. The rest of the paper is organized as follows. In Section 2 we review some practical issues of building a traffic classification evaluation system. Then in the next 3 sectors, we propose our methods to solve these issues. In 3 we present our two-phased evaluation scheme which combines the offline dataset evaluation and the online evaluation together. In 4 we introduce three new metrics for a comprehensive evaluation. In 5 we try to unify the scopes and granularities of different traffic classification systems through grouping. In 6 we conclude our work.

2   Practical Issues

As traffic classification system may have very different performances on different dataset both in accuracy and speed, researchers have studied a lot on comparing and evaluating the performance of different traffic classification systems. However research in this field has encountered many tricky issues that may inhibit a better and overall comparison and evaluation, still the evaluation method many researchers used in their papers needs more polish. We conclude three key issues in this field as follows:

2.1   Offline Dataset.

Studies for evaluating traffic classification systems used offline dataset to do evaluation. For instance, Kim et al. [1] used packets collected from seven different backbones to do evaluation. They applied a payload based method to give each flow in the offline dataset an application-specific tag and used this tagged dataset to evaluate feature-based traffic classification using machine learning algorithms and host- behavior based traffic classification method. They find that SVM achieved the highest accuracy.

While the evaluation method of Kim which used tagged offline dataset can compare different traffic classification methods in accuracy and speed on the dataset they collected, it cannot reflect the performance of these classification methods on another dataset from a total different network, that is, SVM may achieve the highest accuracy on Kim's dataset, but it may not if the study is conducted on another dataset that is collected on another backbone trace.

2.2   Metrics

More useful metrics are in need to do overall evaluation for traffic classification systems. Dainotti[2] concluded the metrics used by researchers in this fields, four most frequently used metrics are accuracy, precision, recall and F-measure. These four metrics may be enough to only measure and compare the performance of different traffic classification systems in terms of accuracy, but to do a comprehensive evaluation, we need other metrics that can reflect the stability, the load capacity, speed and so on.

2.3   Scope and Granularity

Difference in scopes and granularities is another reason that makes it difficult to compare different traffic classification systems. Salgarelli et al. [3] indicated that many traffic classification methods have different granularities, for example, BLINC [4] attempts to identify peer-to-peer traffic irrespective of the specific underlying application, meanwhile, Bernaille et al.'s [5] methodology distinguishes between the various peer-to-peer applications such as Bittorrent, Emule, etc. in other words, Bernaille's method has a finer granularity of classification compared to BLINC method. Figure 1 shows the difference of granularities of these two methods which makes them not comparable to each other.

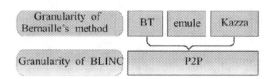

Fig. 1. Granularity of BLINC and Bernaille's classification algorithm

3   Online Evaluation

To address problem caused by using offline dataset, we propose an evaluation scheme of traffic classification systems that combine offline dataset evaluation and an online evaluation, our scheme can do real-time evaluation for different classification systems on a real traffic environment at the same time. The contribution of our work is that this novel evaluation scheme can help us find the best classification systems on a specific trace, in contrast to other evaluation method which can only find the best classification systems on the tagged offline dataset. The process of our evaluation scheme which consists of two parts is shown in Figure 2.

Offline dataset evaluation is conducted before the online evaluation. First, we have to collect enough packets as our offline dataset and then tag each flow with its application type which is used as the reference point to do the evaluation. Before sending these tagged packets to the traffic classification system we are about to evaluate, we should extract some packets from our offline dataset in a specific proportion and put them in a mixed order. By "in a specific proportion" we mean the proportion of packets of each application should be as approximate as possible to the proportion of

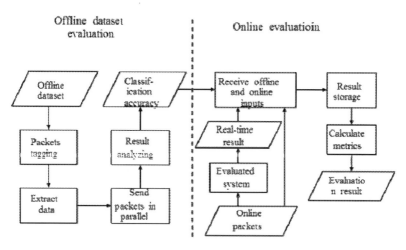

Fig. 2. Process of the two-phased evaluation scheme

packets of this application on the real network. By doing so, we can make the packets we send more similar to those on the real network in which we are going to evaluate in the online evaluation part. After sending these reconstructed offline data to classification systems to be evaluated, we receive the classification results of these systems through a uniform interface and calculate their classification accuracy of each application. The offline dataset evaluation in our two-phased scheme will then output this key information to the online evaluation as a benchmark to evaluate the real-time classification accuracy of the evaluated classification systems.

The online evaluation part of our scheme receives three inputs: the classification accuracy of each application from offline dataset evaluation as the benchmark, the real-time classification result from evaluated traffic classification systems and all the packets from the real network environment. When started, our online evaluation device will first read the benchmark information from the offline dataset evaluation. Then it receives all the packets from network and counts the byte number, packet number and flow number in a pre-defined time slot T, meanwhile it also counts the byte number and packet number of each flow, these numbers will be used at the end of each time slot T to calculate all the metrics we define in the next sector, and at the end of a time slot these numbers will be all cleared. When the online evaluation de vice receives the third input it needs, that is, a real-time classification result from the evaluated systems, it will decides if this result is accepted according to the benchmark information. Combining all these inputs together, online evaluation systems can calculate all the metrics we need and output them at the end of each time slot.

4   Metrics

As we mentioned in sector 2, metrics in terms of accuracy alone is far from enough to do comprehensive evaluation of traffic classification systems. Based on our two-phased evaluation scheme discussed before, we introduce three new metrics here to evaluate the real-time classification accuracy, the relative classification speed and the load capacity respectively.

4.1   Real-Time Classification Accuracy Estimation

This is the most important metric of our scheme because it can reflect the predictive accuracy performances of the evaluated classification systems on the exact network in which we conduct our online evaluation. The real-time classification accuracy estimation metric is calculated as follows:

Assume that we have n different types of application in our offline dataset numbered from 1 to n. in the offline dataset evaluation, S_{BDa} bytes of data tagged with numbered a have been send to the evaluated system D. By comparing the classification result with the tag, system D have correctly recognized R_{BDa} bytes of all the S_{BDa} bytes date of type a. So, we can get the accuracy for this application of system D we have $P_{BDa} = \dfrac{R_{BDa}}{S_{BDa}}$.

After started, online evaluation device first read the accuracy off all n types of applications from the offline dataset evaluation. In a complete time slot T, assume that classification system D have recognized S_{tBDa} bytes of data of type a, and the online evaluation device have received S_{tB} bytes of data. Then in this time slot T, we have the real-time classification byte accuracy estimation, $P_{tBD} = \dfrac{\sum_{i=1}^{n} P_{BDi} S_{tBDi}}{S_{tB}}$ .Similarly, we can also calculate the packet accuracy and the flow accuracy.

It is worth mentioning that the metric "real-time classification byte accuracy estimation" is an estimation of the real-time classification accuracy of the evaluated system, this is not tantamount to the real accuracy. However, as long as the data we extract and construct in the offline dataset evaluation process is similar enough to the data in the real network, then the classification accuracy of each application P_{BDa} will be close to P_{BDa_real}, which is the real accuracy of application a of evaluated system. Based on this, we have $P_{tBD\_real} \approx P_{tBD}$.

As we have mention before, "similar" means the proportion of packets of each application should be as approximate as possible to the proportion of packets of this application on the real network. Besides, the statistical feature should also be similar to those of a real network. In a word, the more alike, the better.

4.2   First-Recognizing Rate

This is a metric that can be calculated only when multiple traffic classification are evaluated at the same time. The first-recognizing rate of a given classification system is the proportion of the flow that is first recognized by this system to all the flows that have been recognized by more than one systems.

As we cannot figure out the algorithms applied by classification systems which is the business secret of the companies that sell these systems, we have to find another way other than analyzing the time complexity of the algorithm. Also, we are not sup- posed to use an absolute time metric here because the classification time depends largely on the bandwidth and the type of packets of the network, in this case, an absolute time metric can be misleading. Taking into consideration of these factors, we introduce such a relative metrics to compare the classification speed of different traffic classification systems. Though we cannot compare two classification systems evaluated in separate times using this metric, we can still get some useful information of the relative classification speed of the classification systems evaluated together.

4.3   Drop Rate

The drop rate of a given classification system is the proportion of the number of flows that are not output by the classification system to the total number of flows in the time slot T. it is worth mentioning that if the evaluated system receive the packets of a flow and fail to recognize its application type, it should also output the classification result for this flow claiming that this flow is not recognized. If the classification system does not output any information about such flows, then the packets of these flows are considered to have been dropped by the evaluated system because its limited load capacity on an high speed network environment.

Drop rate is introduced to reflect the process ability and load capacity of the evaluated system on a high speed network. The less the drop rate of the evaluated system the stronger its process ability is.

5   Unify the Scope

To address the scope and granularity issue we mentioned in sector 2, we may assign a unique application number to each type of application, and then group every application according to the behavior and function of the specific application. Groups may include P2P, audio and video, mail, database, game, network management etc. Table 1 shows an example of the mail and database group we create.

Table 1. Group example of mail and database

Group	Application	Num	Group	Application	Num
	POP3	242		CDDB	949
	SMTP	243		DB2	950
	IMAP	244		Informix	951
	Binkp	245		LDAP	952
MAIL	COTP	246	DATA BASE	MySQL	953
	LotusNotes	247		Oracle	954
	MAPI/OETP	248		PostgreSQL	955
	OWA	249		SQLServer	956
	X.400	250		Sybase	957

The application number is a fine granular way to distinguish different applications, and accordingly the group we establish is coarse way to distinguish applications.

It should be noted that the group we create may not be the best way to sort up all the applications, some of the applications may be grouped to the wrong class due to some misunderstanding, but this will not have major impact to our goal of grouping, since we only do this so that classification of different scope and granularity can be compared at the same premise and at the same granularity. For example, host- behavior based classification and feature-based classification using machine learning algorithms now can be compared at the same coarse granularity.

6   Summary

In this paper, we first review the key issues of recent works in traffic classification evaluation and traffic classification comparison, then, to address these issues, we proposed a novel two-phased evaluation scheme for traffic classification systems. By applying this scheme, we are able to conduct a comprehensive evaluation of multiple traffic classification systems of different scopes and granularities at the same time. The three new metrics we introduced help us to compare classification systems evaluated together and find out the classification system that fit the network best in which we do the online evaluation.

References

1. Kim, H., Claffy, K.C., Fomenkov, M., Barman, D., Faloutsos, M., Lee, K. (eds.): Internet traffic classification demystified: myths, caveats, and the best practices. Proceedings of the 2008 ACM CoNEXT Conference. ACM (2008)
2. Dainotti, A., Pescapè, A., Claffy, K.: Issues and future directions in traffic classification. IEEE Network 26(1), 35–40 (2012)
3. Salgarelli, L., Gringoli, F., Karagiannis, T.: Comparing traffic classifiers. ACM SIGCOMM Computer Communication Review 37(3), 65–68 (2007)
4. Karagiannis, T., Papagiannaki, K., Faloutsos, M.: BLINC: multilevel traffic classification in the dark. In: Proceedings of the 2005 ACM SIGCOMM, pp. 229–240 (2005)
5. Bernaille, L., Teixeira, R., Akodkenou, I., Soule, A., Salamatian, K.: Traffic classification on the fly. ACM SIGCOMM Computer Communication Review 36(2), 23–26 (2006)

Winnowing Multihashing Structure
with Wildcard Query[*]

Yichen Wei[1,2], Xu Fei[2], Xiaojun Chen[2], Jinqiao Shi[2], and Sihan Qing[1,2]

[1] School of Software and Microelectronics, Peking University, Beijing, China
cirda@sina.com
[2] Institute of Information Engineering, CAS, Beijing, China
xufei@iie.ac.cn

Abstract. Payload attribution is the process to identify source and destination of packets which appeared in the network and contained certain excerpt. Payload attribution structures process and store corresponding network traffic in order to support identification and analysis afterwards. The work of this paper is based on an existing payload attribution data structure which stores and processes network traffic based on Bloom Filters. We propose a novel data structure called Winnowing Multihashing structure with Wildcard Query (WMWQ). Our methods support wildcard queries efficiently and have higher data reduction ratio as well as lower false positive rate. In addition, we show that the time complexity of querying a WMWQ is shown to be constant in the number of inserted data elements. The proposed methods can be used for network forensics traffic processing in large scale networks and can improve the efficiency of network forensics processing and analysis.

Keywords: Network Forensics, Payload Attribution, Bloom Filter, Block Structure, False Positive.

1   Introduction

Nowadays network applications are spreading widely with the rapid development of network technology, and network information technology is affecting every aspect of people's life gradually. However, a direct consequence is a significant increase in cybercrime, which affects both society and individuals dramatically. It is impossible to prohibit cybercrime completely using technical methods due to complexity, uncertainty and diversity [1]. Therefore, the traceback, i.e. the process of tracking network data is playing more and more important roles. Thus, network forensics have received increased research interest. It is infeasible to record all the original traffic in today's network [2]. For example, the signature of new worm or the method of a new cybercrime is usually unknown beforehand [3], and we need to store several weeks or even months of traffic to support investigation timely. It is obviously impossible to store such huge quantity of data in today's era of data explosion. Additionally, recording the original network traffic, such as users' email and account [4], as well as

---

[*] Supported by the National Natural Science Foundation of China under Grant No. 61170282.

W. Han et al. (Eds.): APWeb 2014 Workshops, LNCS 8710, pp. 265–281, 2014.

commercial data of enterprises and confidential material of the government [5], exhibit severe privacy issues.

Payload refers to the actual cargo of a data transmission, that is, the actual data sent by the packets towards the destination. Payload is different from network traffic, according to the definition in Wikipedia [6], it is the part of the transmitted data which is the fundamental purpose of the transmission, to the exclusion of information sent with it (such as headers or metadata, sometimes referred to as overhead data) solely to facilitate delivery. In addition, payload attribution is a vital element in network forensics. Given the query demands of transmission history of packets or an excerpt of possible packet payload, a Payload Attribution System (PAS) is able to identify source, destination and the emerging times of all packets containing specific payload segment [2]. As one of the core modules of network forensics system, a PAS is able to investigate cybercrime through such as identifying who has received phishing email or finding which insider has permitted unauthorized disclosure of sensitive information [3]. Actually, attribution is trying to solve the problem of identifying the source or destination of some network traffic instances [7]. A PAS must have proper mechanism to guarantee users' privacy in the network deployed with PAS, and must have proper verification mechanism to ensure that information only can be disclosed to authorized subjects. Payload attribution is a procedure to identify the source and destination of all packets occuring in the network and containing specific excerpt. Payload attribution structure handles and stores corresponding network traffic in order to support identification and analysis afterwards. It is a valuable tool in helping determine attackers and victims of network incidents and analyze security incidents.

To investigate payload attribution, the collection and storage of a payload, as well as querying a payload are the most important tasks. The most straightforward method to solve these problems is blocking the network traffic and the most classical data structure is Bloom Filter [8]. Nevertheless, traditional Bloom Filter is used for determining whether one element belongs [9] to the set. It is a bit array with size of m and k random hash functions, due to its structure, it exists the possibility of false positives.

The research emphasis of this paper is to find an efficient data structure to deal with the payload [10] and compress it, then store the result in order to support excerpt query afterwards and give out fast and accurate reply. According to the current research situation, this paper utilized the most common block method to partition the payload, run Winnowing fingerprint algorithm which had more advantages than other fingerprint algorithms, then hashed the resulting blocks and inserted them into the Bloom Filters. Based on CMBF [3], our data structure well supports wildcard queries. It not only solves the first block offset problem, the alignment problem and the consecutiveness problem, but also improves data reduction and accelerates query speed under an acceptable false positive rate. Moreover, the traditional Bloom Filter does not support delete operations due to its structural characteristic. In fact, data in the network may need to be altered, that is, the actual storage structure needs to support delete operations. In order to support data deletion, we propose some ideas to extend the corresponding structure as well.

This article is organized as follows. In the next section, we review related prior work. Subsequently, we provide a detailed design description of our payload attribution structure Winnowing Multihashing with Wildcard Query. In Section 4, we conduct a theoretical evaluation of WMWQ, and finally we give out the conclusions in Section 5.

2   Related Work

Based on Bloom Filter, Kulesh Shanmugasundaram et al. proposed Block-Based Bloom Filter (BBF) [5] and Hierarchical Bloom Filter (HBF) [2], Chia Yuan et al. proposed Rolling Bloom Filter (RBF) [11], then Miroslav Ponec [2] et al. proposed Fixed Block Shingling (FBS), Variable Block Shingling (VBS), Enhanced Variable Block Shingling (EVBS), Multihashing (MH), Enhanced Multihashing (EMH), Winnowing Block Shingling (WBS), Winnowing Multihashing (WMH) [2] etc., however, these methods existed some problems, such as first block offset problem, alignment problem and consecutiveness problems, moreover, most of these did not support wildcard query. As an improvement, Mohammad Hashem Haghighat et al. proposed Character Dependent Multi-Bloom Filter (CMBF) [3]. Although it supported wildcard query, it still contained alignment problem.

2.1   Bloom Filter (BF)

Bloom Filter (BF) [12] is a well designed random data structure which is used to represent a set and support approximate member query. Bloom Filter was invented by Bloom in 1970s and had already been widely used in database application. Recently, Bloom Filter has been paid attention in the field of network forensic.

The reason for the widespread of Bloom Filter is its excellent arithmetic property. Besides the efficient compression of data, one distinguished advantage of Bloom Filter is that its query time does not depend on the number [13] of inserted elements and its query time complexity is O(1). Its principle shows in Figure 1. A standard Bloom Filter is a m bit array which is used for representing a set $S = \{x_1, x_2, ..., x_n\}$, in which all bits are set to 0 initially. It uses k independent hash functions $h_1, h_2, ..., h_k$, which values are integer and range from 1 to m. To simplicity, we suppose that the mapping of each element by these hash functions is subject to a random distribution and ranges from 1 to m. For each $x \in S$, the bit $h_i(x)$ mapped by hash function h_i is set to 1, where $1 \leq i \leq k$. A bit can be set to 1 multiple times, but only the first change is useful. In order to check whether an element y is in the set S, we need to check whether all $h_i(y)$ bit have been set to 1. If not, it is obvious that y is not the member of the set S. If $h_i(y)$ bit have been set to 1, we consider that y belongs to S, while it may make mistake by some probability. When we determine an element y is in the set S while it is not actually according to the above method, Bloom Filter produces a false positive.

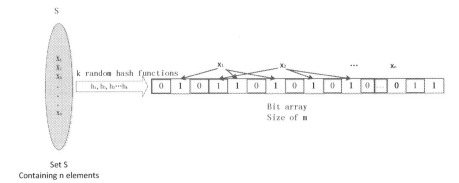

Fig. 1. The principle of Bloom Filter

False positive refers to affirmation by mistake, it is a kind of false. One example of false positive in Bloom Filter shows in Figure 2 [8]. The bit array of Bloom Filter is set to 0 initially, each element x_i in the set S is hashed k times and each time producing a bit position which is set to 1. If some bits are set to 1 multiple times, the later ones will not affect the former ones. For an element x_2 which is different from y_2, if we want to query whether it is in the Bloom Filter, we should follow the method by which elements are inserted into the Bloom Filter to find the k bits which are corresponding to y_2 being hashed k times, then check whether all these bits are set to 1. The verification in this situation shows that all these bits are set to 1, so it seems that y_2 is the element of the set. While, y_2 is not in the set actually, which is false positive, and the reason is hash collision. False positive rate refers to the ratio taken by the number of false positives upon the total number of queries.

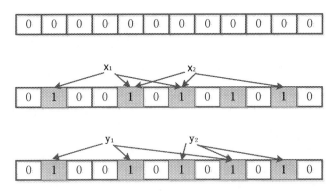

Fig. 2. False positive of Bloom Filter for instance

2.2   Rabin Fingerprint

Fingerprint is the short checksum of the string which has the feature that the probability of two different objects having the same fingerprint is tiny. Rabin defined a fingerprint policy on binary string based on polynomial [14]. Define a N-1 degree polynomial $S(x)$ for each binary string $S = (s_1,...,s_N)$, where $N \geq 1$:

$$S(x) = s_1 x^{N-1} + s_2 x^{N-2} + ... + s_N .$$

Then, we take a fixed irreducible polynomial $P(x)$ of degree K and define the fingerprint of S to be polynomial $f(S) = S(x) \mod P(x)$.

Rabin fingerprint has a good mathematical property, that is, the latter computing can be deducted by the former one, which can reduce the consumption of calculation greatly. It can be explained as follows:

Suppose there is a string $(s_1, s_2,..., s_i, s_{i+1},..., s_n)$, running with Rabin Fingerprint algorithm, the fingerprint of its substring $(s_1, s_2,..., s_i)$ is $f_1 = s_1 x^{i-1} + s_2 x^{i-2} + ... + s_i$, the fingerprint of the following substring $(s_2,..., s_{i+1})$ is $f_2 = s_2 x^{i-1} + s_3 x^{i-2} + ... + s_{i+1}$. It is obvious that f_2 can be derived from f_1, that is, $f_2 = x \times (f_1 - s_1 x^{i-1}) + s_{i+1}$. So it can reduce the calculation consumption and improve the operational performance.

2.3   Winnowing

Many applications show that Winnowing [2] has the best block result among fingerprint algorithms. Initially, it is used for generating the fingerprint of the documents, its principle is shown below:

Slide a window [15] of size k on the original text, compute a hash value for k characters each time, then store these generated hash values into an array in turn. Slide a window [16] of size w on the array of hash values, and choose the minimum value in each window. If there are more than one minimum values, choose the rightmost one, and all these values generating the fingerprint of the document. Figure 3 shows an example of generating the fingerprint of a document by Winnowing algorithm.

The red numbers in the figure is the fingerprint of the original document. This strategy ensure taking enough fingerprint information as well as generating fingerprint not too large. The red part is the fingerprint chosen by Winnowing, and we can also record the positions to track the positions where occur similar content. For example, [245,1] [63,5] [164,6] [384,8] [617,11] [339,14](Index starts from 0 and the second value shows the hash position in the original sequence). Fingerprints selected by winnowing are better for document fingerprinting than the subset of Rabin fingerprints, which contains hashes equal to 0 mod p, for some fixed p, because winnowing guarantees that in any window of size w there is at least one hash selected [2].

| Original Text: | **this is the WMWQ instance** |
| Delete unrelated content: | **thisistheWMWQinstance** |

Divide into 6-grams: **thisis hisist isisth sisthe istheW stheWM
theWMW heWMWQ eWMWQi WMWQin
MWQins WQinst Qinsta instan nstanc stance**

Hash 6-grams:

652	245	323	955	475	63	
164		478		384		728	
959		617	814	697	339	756

Slide a window
With size of 5:

（652,245,323,955,475） （245,323,955,475,63）
（323,955,475,63,164） （955,475,63,164,478）
（475,63,164,478,384） （63,164,478,384,728）
（164,478,384,728,959） （478,384,728,959,617）
（384,728,959,617,814） （728,959,617,814,697）
（959,617,814,697,339） （617,814,697,339,756）

Fig. 3. An example of Winnowing algorithm

3   Research on Winnowing Multihashing Structure with Wildcard Query (WMWQ)

A straightforward method to design a simple payload attribution system is to store all the packets payload. We could store the hash value of the payload instead of the actual payload in order to reduce the storage requirement [10] as well as provide privacy protection. This method reduces the data amount of each packet to 20 bytes (such as SHA-1), and the cost is false positive mistake brought by hash collision. We could save the space further through storing the payload into Bloom Filter. The false positive rate of a Bloom Filter depends on its providing data reduction ratio. The reason for which a Bloom Filter can protect the privacy is that we could only query whether a specific element is inserted and it could not reveal the list of stored elements. Even if someone try to query all the elements, the result will be useless because of false positive. Comparing with storing hash values directly, the advantages of using Bloom Filter are that storage space is saved and query speed is accelerated. Querying any packet's Bloom Filter will only take a short constant time.

However, it does not support payload excerpt query if inserting the whole payload into Bloom Filter. If not storing the whole payload, we can divide the payload into blocks according to some policies and insert them into Bloom Filter separately. This simple modification can support excerpt query, and in the query phase, we only have to divide the excerpt into blocks with the same strategy, then query them separately. Only if all the query results are positive, it seems that the excerpt belongs to the payload.

Wildcard is a kind of character, it can be represented by "?" and "*", and it is used for occupation and does not refer to any characters. The position where wildcard occurs can be replaced by any characters. Usually, "?" is used for matching 1 character and "*" is used for matching 0 or more arbitrary characters.

The existing technologies are limited to the query type they could respond [13]. Given the current network attack trend and the complex signature used for investigating these attacks, it is a vital issue. One common example is the growing of polymorphic worms. These worms change their representation as much as possible before they transmit themselves to the next victim. The changes in representation are often done by using a different key each time to encrypt malware for generating different malware bytes. We can extract some invariable parts of the payload of worms which will be separated by some random characters. Therefore, the signature of worms may be in the form of "A*B" [3], where "A" and "B" are two invariable strings which are separated by some unknown random strings. In this situation, we should find the block structure which supporting wildcard query and collect and store the payload in the network.

Compared to the existing structure, the advantage of our structure is supporting multiple unknown characters query in the excerpt, moreover, it improves data reduction ratio under the acceptable false positive rate.

3.1   WMWQ

Our structure bases on the algorithm of winnowing, it uses multiple instances and supports wildcard query. WMWQ is short for Winnowing Multihashing Structure with Wildcard Query, it uses 256 independent Bloom Filters which indexed from 0 to 255 and each block inserts into the corresponding Bloom Filter according its hash value. Each Bloom Filter uses only one hash function in order to speed up. The process of payload $\{c_1, c_2, ..., c_n\}$ contains following steps:

1) Slide a window of size k on the payload. The characters in the first window are $\{c_1, c_2, ..., c_k\}$, and the characters in the second window are $\{c_2, c_3, ..., c_{k+1}\}$ and so on, the characters in the i th window are $\{c_i, c_{i+1}, ..., c_{i+k-1}\}$, $1 \le i \le n-k+1$.

2) Compute a hash value $H(c_i, c_{i+1}, ..., c_{i+k-1})$ [3] for each window based on the formula:

$$H(c_i, c_{i+1}, ..., c_{i+k-1}) = (c_i \bmod q) \times p^{k-1} +$$
$$(c_{i+1} \bmod q) \times p^{k-2} + ... + (c_{i+k-1} \bmod q) \times p^0,$$

where p is a fixed prime and q is a constant, $q \le p < 256$. According to the property of the polynomial and in order to accelerate the processing speed, the hash value of the latter payload window can be computed from the former one [14], that is:

$$H(c_{i+1}, c_{i+2}, ..., c_{i+k}) = pH(c_i, c_{i+1}, ..., c_{i+k-1}) +$$
$$(c_{i+k} \bmod q) - (c_i \bmod q) \times p^k.$$

3) Recording the resulting hash values into an array $\{h_1, h_2, ..., h_i, ...\}$, where the
   i th item $h_i = H\left(c_i, c_{i+1}, ..., c_{i+k-1}\right)$.

4) Slide a window of size w on the hash array $\{h_1, h_2, ..., h_i, ...\}$ and select the minimum value in each window. If there are more than one minimum values, select the rightmost one. Insert a block boundary after the first character of the payload window corresponded to the selected hash value. For example, if the minimum hash value selected in one window is h_i, insert a block boundary after the character c_i in the payload, and the content between every two boundaries attaching the following o bytes generating the block content.

5) Inserting all the blocks into the corresponding Bloom Filters according to their hash values calculated by a random hash function.

6) Replacing the parameters in the above with different values and do the above operations several times (administrators can set the parameter t).

The query of excerpts is similar to the above steps, and the difference is in step 5, after finding the corresponding Bloom Filter, we check whether the corresponding mapping positions are set to 1 instead of inserting operation. If all the positions are filled with 1, it seems that the payload containing the excerpt.

In detail, we process string "thisistheWMWQinstance" as an illustration, shows in Figure 3.1. Suppose the size of payload window is k=6, then slide the window and generate 16 payload windows: "thisis","hisist","isisth","sisthe","istheW","stheWM", "theWMW","heWMWQ","eWMWQi","WMWQin","MWQins","WQinst","Qinsta", "instan","nstanc","stance". Compute the hash value of each window and generate an array { 652,245,323,955, 475,63,164,478,384,728,959,617,814,697,339,756 }. Slide a window of size w=5 on this array and generate 14 hash windows:{652,245,323,955, 475},{245,323,955,475,63},{323,955,475,63,164},{955,475,63,164,478},{475,63,16 4,478,384},{63,164,478,384,728},{ 164,478,384,728,95},{478,384,728,959,617},{38 4,728,959,617,814},{728,959,617,814,697},{959,617,814,697,339},{617,814,697,33 9,756}. The minimum hash values selected from these windows are 245,63,164,384,617,339, and they are corresponding to the characters in the payload which are indexed by 1,5,6,8,11,14, then the block boundaries are inserted after these characters. In addition, attaching the following o=2 characters, and the blocks generated in the first cycle are "isisth","the","heWM","WMWQi","Qinst". In the following, set new values for parameters k,w,p,q, and do the same operation again (t=2). The query of excerpts is similar to the process of payload, and we verify whether the corresponding mapping positions are set to 1 instead of the insertion operation in step 5. We consider that the excerpt belongs to the payload if all the answers are positive.

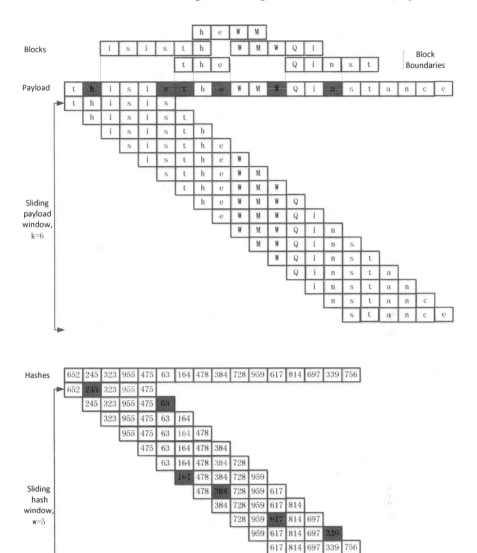

Fig. 4. Illustration of WMWQ

The block size is fixed in CMBF, in the excerpt query phase, we have to try all the possible positions because of the beginning position of the first matching block in the excerpt is unknown, which is so called alignment problem. If the fixed size is s, we have to try s-1 positions. The query times in the worst case is $v \times s$ [2] times than WBS, where v is the number of offsets. The alignment problem produces large consumption of the computing source and slows down the query speed. While, in WMWQ, the selection of block boundaries base on payload itself, which solves the alignment problem in CMBF effectively. In order to support wildcard query, we use q modulo in the hash computing, so we can map each character into one class between [0, q-1] and the query space is limited because the resulting hash values range from 0 to q-1 instead of all the possible values from 0 to 255. For example, we need to query the string "abcd?eghi" and the modulo q=4. The query can be divided into 4 substrings: "abcd0eghi","abcd1eghi","abcd2eghi","abcd3eghi".   We   query   each   substring independently, we thick the excerpt is contained in the payload if any substring is found. WMWQ can deal with more complex query as well. For example the query excerpt is "abcde[m-p]fghij" [3] and the modulo q=8. Then we map 4 possible unknown characters (m,n,o,p) into classes between 0 and 7 to create substring: "abcde5fghij", "abcde6fghij","abcde7fghij","abcde0fghij". We reduce the computing space and accelerate the process speed by modulo so that wildcard query with 7 unknown characters takes less than 1 second, while given the previous techniques, the same query takes about 4500 years to process [3].

Via the analysis of CMBF, we know that it is based on the fundamental structure FBS. Miroslav Ponec [2] had proved that WMH was better than FBS in almost every aspect, so it can be deduced that our WMWQ based on WMH is better than CMBF and the theoretical analysis is given in Section 4.

3.2   The Design Idea of Block Structure Supporting Deletion and Wildcard Query

The block structures mentioned above are all based on standard Bloom Filter, that is, we store the blocks into the standard Bloom Filter after dividing the payload by specific methods. The standard Bloom Filter is a bit array with size m, each bit can be set to 1multiple times and only the first change is useful. The standard Bloom Filter cannot be altered once elements are inserted. Actually, there will be situations in which the inserted elements have to be modified, so it is necessary to expand the structure of standard Bloom Filter to support the modification of elements (that is, deleting firstly then inserting afresh).

The most straightforward idea is that we replace each bit of the standard Bloom Filter with a counter [17] which is used for recording how many times each bit has been set to 1. If any element has to be altered, we should reduce the counters in the original mapping positions by 1 and increase the counters in the new mapping positions by 1.

We try to improve WMWQ to support payload deletion operation through expand the standard Bloom Filter, each Bloom Filter is no longer a bit array and is expanded to half-byte array, that is, we use 4 bits as a counter to record the times the corresponding position is set to 1, when it has to be deleted later, the corresponding counter will be reduced by 1. However, due to the introduction of counters, the data reduction ratio will be smaller and just reach 1/4 of WMWQ. What's more, the deletion operation on Bloom Filter is easy to bring in false negative problem. Because false negative is beyond our research scope, we ignore this problem in this paper. Figure 3.2 shows the situation in which the Bloom Filters in WMWQ is expanded into half-byte counters.

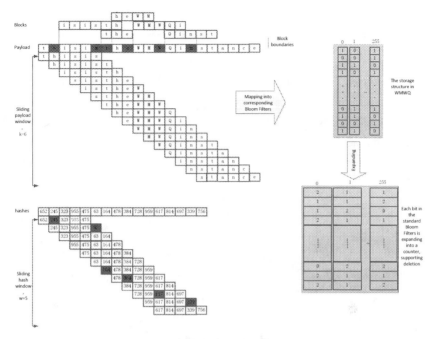

Fig. 5. The expanding Bloom Filters of WMWQ

This article has only proposed the idea of designing the block structure supporting wildcard query as well as deletion, the details and implementation method need to be studied further, which can be the next research direction.

4   Theoretical Analysis

4.1   False Positive Analysis of Standard Bloom Filter

1) False positive

According to Section 2.3, due to the structural property of the Bloom Filter, the possible hash collision [18] may bring about false positive. We could deduce the false positive rate of standard Bloom Filter based on the Knowledge of probability statistics,

that is, $FP = \left(1 - \left(1 - \dfrac{1}{m}\right)^{kn}\right)^{k}$ [8], where n is the number of inserted elements in

Bloom Filter, kn<m, with the premise that k hash functions in Bloom Filter is random completely. It can be proved as follows:

∵   k hash functions in Bloom Filter is random and the probability of any bit in the bit array with size m being set to 1 is equal.

$\therefore$ The probability of any bit being set to 1 is $\dfrac{1}{m}$, after all the elements being inserted, the probability of any bit still being set to 0 is $p' = \left(1 - \dfrac{1}{m}\right)^{kn}$.

$\because \lim\limits_{x \to \infty} (1 - \dfrac{1}{x})^{-x} = e$

$\therefore p' = \left(1 - \dfrac{1}{m}\right)^{kn} \approx e^{-\frac{kn}{m}}$

For clarity, let $p = e^{-\frac{kn}{m}} \approx p'$, ρ refers to the ratio of 0 in the bit array, then the mathematical expectation of ρ is $E(\rho) = p'$.

$\because$ The ratio of 1 in the bit array is $(1 - \rho)$.

$\therefore$ The false positive rate, that is, the probability of one element being not in the set while all the corresponding bits in the array being set to 1 is $FP = (1 - \rho)^k$.

$\because p'$ is the mathematical expectation of ρ, M. Mitzenmacher [17] had proved that the distribution of ρ is concentrated near its mathematical expectation.

$\therefore FP = (1 - \rho)^k \approx (1 - p')^k \approx (1 - p)^k$

Substituting p and p' separately, we have:

$f' = (1 - (1 - \dfrac{1}{m})^{kn})^k = (1 - p')^k$, $f = (1 - e^{-\frac{kn}{m}})^k = (1 - p)^k$. We often use p and f for convenience.

2) Optimization

The number of hash functions (k) in the Bloom Filter could not be neither too big nor too small. The reason is that, the probability we get 0 when query an element that is not in the set will be very large if k is very huge, otherwise, there will be too many 0 in the bit array if k is very small. We deduce the best number of hash functions (k) as follows [8]:

Let $g = k \ln(1 - e^{-\frac{kn}{m}})$

$\because f = \exp(k \ln(1 - e^{-\frac{kn}{m}}))$

$\therefore f$ is minimum when g is minimum

$\because p = e^{-\frac{kn}{m}}$

$\therefore g = -\dfrac{m}{n} \ln(p) \ln(1 - p)$

According to the law of symmetry, when $p = \dfrac{1}{2}$, that is $k = \ln 2 \cdot \dfrac{m}{n}$, g is

minimum, the minimum of false positive rate $f_{\min} = (\dfrac{1}{2})^k \approx 0.6185^{\frac{m}{n}}$.

Because p refers to the probability of one bit being 0, $p = \dfrac{1}{2}$ refers to 0 and 1 occupying the half of the array, that is, we should make the half of the array be empty if we want to get a low false positive rate.

Of course, the conclusion that false positive rate [8] is minimum when $p = \dfrac{1}{2}$ is not limited to approximation p and f. Because for f' = exp(k ln(1 − (1 − 1/m)^{kn})), g' = k ln(1 − (1 − 1/m)^{kn}) , p' = (1 − 1/m)^{kn} , we still have

$g' = \dfrac{1}{n \ln(1 - \dfrac{1}{m})} \ln(p') \ln(1 - p')$, according to the law of symmetry, g' is

minimum when $p' = \dfrac{1}{2}$.

4.2   False Positive Analysis of WMWQ

1) False positive
Based on the derivation of CMBF [3], it can be proved that the false positive produced by fingerprint algorithm is far less than the false positive produced by the structure of the Bloom Filter in WMWQ, for simplify, its false positive rate can be replaced with the false positive rate produced by the structure of the Bloom Filter.

Because WMWQ uses 256 Bloom Filters and each Bloom Filter uses only one hash function, assuming that characters of the string are uniformly distributed, the average size is s, the total number of inserted characters in the Bloom Filters is n, and the length of each Bloom Filter is m, then the insertion number of each Bloom Filter is

$N = \dfrac{n}{256s}$, k=1. According to the former derivation, the false positive rate of each

Bloom Filter is [3]:

$$\alpha = \left(1 - \left(1 - \dfrac{1}{m}\right)^N\right)^1 = \left(1 - \left(1 - \dfrac{1}{m}\right)^N\right) = \left(1 - \left(1 - \dfrac{1}{m}\right)^{\frac{n}{256s}}\right).$$

Assuming that the query excerpt is S and has not been inserted into WMWQ actually, it makes false positive due to the former inserted string S' which has the length of l. The reason for false positive may be the i different bytes between S and S'. So, all the possible values of i between 1 and l should be taken into consideration in the procedure of computing false positive rate. The i different characters between S and S' can generate at least $\left\lceil \dfrac{i}{s} \right\rceil$ blocks [3]. WMWQ will make false positive when all these blocks produce false positive. Because the false positive rate of each Bloom Filter (α) is less than 1, the upper bound of false positive rate produced by these blocks is $\alpha^{\left\lceil \frac{i}{s} \right\rceil} \leq \sqrt[s]{\alpha}$. Moreover, the probability of S and S' having i different bytes is $\dfrac{C_l^i \times 255^i}{256^l}$. Therefore, the false positive rate [3] due to the i different bytes between S and S' is:

$$FP_{BF_{n,i}} < \alpha^{\frac{i}{s}} \cdot \frac{C_l^i \times 255^i}{256^l} = \frac{C_l^i \times \left(255 \times \sqrt[s]{\alpha}\right)^i}{256^l}.$$

The false positive rate produced by Bloom Filters in WMWQ can be summarized by the sum of $FP_{BF_{n,i}}$ of all possible values of I, that is:

$$FP_{BF_n} = \sum_{i=1}^{l} FP_{BF_{n,i}} < \sum_{i=1}^{l} \frac{C_l^i \times \left(255 \times \sqrt[s]{\alpha}\right)^i}{256^l}$$

Overall,

$$= \frac{1}{256^l} \times \left(\left[\sum_{i=1}^{l} C_l^i \times \left(255 \times \sqrt[s]{\alpha}\right)^i \right] - 1 \right) = \frac{\left(1 + 255 \times \sqrt[s]{\alpha}\right)^l - 1}{256^l}$$

the upper bound of the false positive rate in WMWQ is $\dfrac{\left(1 + 255 \times \sqrt[s]{\alpha}\right)^l - 1}{256^l}$,

where s is the average value of block sizes.

2) Optimization

The reduction of s will cause more elements inserted into the Bloom Filters resulting for the increase of false positive rate. Similarly, the certainty of query is reduced when s increases, which resulting for higher false positive rate of Bloom Filter. To find the optimum value of s to make the false positive rate be minimum, we should find the root of the false positive. From the former derivation,

$$FP \approx FP_{BloomFilter} < \frac{(1 + 255 \times \sqrt[s]{\alpha})^l - 1}{256^l}, \text{ where } \alpha = \left(1 - \left(1 - \frac{1}{m}\right)^{\frac{n}{256s}}\right)^l, \text{ partial}$$

derivative with respect to s and let it to be 0 [3]:

$$\frac{\partial FP}{\partial s} = \frac{\partial\left(\frac{(1+255\times\sqrt[s]{\alpha})^l - 1}{256^l}\right)}{\partial s} = 0$$

$$\Rightarrow \frac{l}{256^l}\times(1+255\times\sqrt[s]{\alpha})^{l-1}\times\frac{\partial(1+255\times\sqrt[s]{\alpha})}{\partial s} = 0$$

$$\Rightarrow \frac{\partial(1+255\times\sqrt[s]{\alpha})}{\partial s} = 0$$

$$\Rightarrow \frac{\partial\left(\left(1-\left(1-\frac{1}{m}\right)^{\frac{n}{256s}}\right)^{\frac{1}{s}}\right)}{\partial s} = 0$$

Replacing s with $\frac{1}{k}$ [3], we can get the optimum s:

$$\Rightarrow \frac{-k^2\times\partial\left(\left(1-\left(1-\frac{1}{m}\right)^{\frac{nk}{256}}\right)^{k}\right)}{\partial k} = 0$$

$$\Rightarrow \frac{\partial\left(\left(1-\left(1-\frac{1}{m}\right)^{\frac{nk}{256}}\right)^{k}\right)}{\partial k} = 0$$

According to the derivation of optimum number of hash function (k), we have:

$$k = \ln 2 \cdot \frac{256\times m}{n}$$

So, we have the optimum value of s:

$$\Rightarrow s = \frac{n}{256\times m\times\ln 2}$$

Therefore, under the certain false positive, we can calculate the size of bit array (m) and the number of inserted elements (n) as follows:

$$m = \frac{-n \times \ln\left(\dfrac{\sqrt[l]{256^l \times FP + 1} - 1}{255}\right)}{256 \times (\ln 2)^2}$$

$$n = \frac{-256 \times m \times (\ln 2)^2}{\ln\left(\dfrac{\sqrt[l]{256^l \times FP + 1} - 1}{255}\right)}$$

Where, l is the length of excerpt and FP refers to the false positive rate.

Furthermore, we could calculate the data reduction ratio for WMWQ under certain false positive rate, using the optimum value of s. Because we use 256 independent Bloom Filters, which occupy $256 \times m$ bits of the storage to store n bytes. Therefore, the data reduction ratio of WMWQ is [3]:

$$DRR = \frac{8 \times n}{256 \times m} = \frac{-8 \times (\ln 2)^2}{\ln\left(\dfrac{\sqrt[l]{256^l \times FP + 1} - 1}{255}\right)}$$

5   Conclusion

This paper mainly focus on the collection of payload in the network traffic. The WMWQ data structure taking the idea of dividing the payload into blocks bases on Winnowing fingerprint algorithm, and it inserts the blocks generated by our strategy into 256 distinct Bloom Filters. WMWQ supports wildcard query and has higher data reduction ratio under acceptable false positive rate. Moreover, it has solved the alignment problem in CMBF and has accelerated the query speed of the excerpt. In addition, we have proposed an idea of the structure which can support deletion as well as wildcard query. We have provided the theoretical analysis of WMWQ and the following work is to do some experiments to verify the performance of WMWQ. As a part of the future work, we are going to implement a payload attribution system based on WMWQ as a core module of network forensic system.

References

[1] Sembiring, I., Istiyanto, J.E., Winarko, E., Ashari, A.: Payload Attribution Using Winnowing Multi Hashing Method. International Journal of Information & Network Security (IJINS) 2(5), 360–370 (2013) ISSN: 2089-3299
[2] Ponec, M., Giura, P., Wein, J., Bronnimann, H.: New Payload Attribution Methods for Network Forensic Investigations. ACM Transactions on Information and System Security 13(2), Article 15 Publication. date: (February 2010)
[3] Haghighat, M.H., Tavakoli, M., Kharrazi, M.: Payload Attribution via Character Dependent Multi-Bloom Filters. IEEE Transactions on Information Forensics and Security 8(5), 705–716 (2013)

[4]   Ranum, M.J.: Intrusion Detection and Network Forensics. Technical Report. Report from the Second USENIX symposium on Internet Technologies & Systems (USITS 1999), Boulder, Colorado, USA (1999)

[5]   Shanmugasundaram, K., Brönnimann, H., Memon, N.: Payload attribution via hierarchical bloom filters. In: Proc. 11th ACM Conf. Computer and Communications Security, pp. 31–41. ACM (2004)

[6]   Wiki. Payload (computing),
      `http://en.wikipedia.org/wiki/Payload_computing` (March 15, 2014)

[7]   Ponec, M., Giura, P., Brönnimann, H., Wein, J.: Highly efficient techniques for network forensics. In: Proc. 14th ACM Conf. Computer and Communications Security, pp. 150–160. ACM (2007)

[8]   Jiao, M.: The concept and theory of Bloom Filter,
      `http://blog.csdn.net/jiaomeng/article/`
      `details/1495500.2007-01-27`

[9]   Snoeren, A.: Hash-based ip traceback. Proc. ACM SIGCOMM Computer Communication Rev. 31(4), 3–14 (2001)

[10] Demir, O., Ji, P., Kim, J.: Session based logging (sbl) for ip-traceback on network forensics. In: Proc. 2006 Int. Conf. Security and Management, pp. 233–239 (2006)

[11] Cho, C., Lee, S., Tan, C., Tan, Y.: Network forensics on packet fingerprints. In: Fischer-Hübner, S., Rannenberg, K., Yngström, L., Lindskog, S. (eds.) Security and Privacy in Dynamic Environments. IFIP AICT, vol. 201, pp. 401–412. Springer US (2006)

[12] Bloom, B.: Space/time tradeoffs in in hash coding with allowable errors. In: CACM, pp. 422–426 (1970)

[13] Broder, A., Mitzenmacher, M.: Network Applications of Bloom Filters: A Survey. Internet Mathematics 1(4), 485–509 (2004)

[14] Rabin, M.O.: Fingerprinting by random polynomials. Center for Research in Computing Technology, Harvard University, Report TR-15-81 (1981)

[15] Babcock, B., Datar, M., Motwani, R.: Sampling from a moving window over streaming data. In: Proceedings of 13th Annual ACM-SIAM Symposium on Discrete Algorithms (2002)

[16] Datar, M., Gionis, A., Indyk, P., Motwani, R.: Maintaining stream statistics over sliding windows. In: ACM Symposium on Discrete Algorithms, pp. 635–644 (2001)

[17] Mitzenmacher, M.: Compressed Bloom Filters. IEEE/ACM Transactions on Networking 10(5), 604–612 (2002)

[18] Garfinkel, S.: Network Forensics: Tapping the internet

Efficient IEEE 802.15.4 AHB Slave of Security Accelerator in Wireless Senor Networks

Rui Chen, Lan Chen, and Ying Li

Common Technology Research Department,
Institute of Microelectronics, Chinese Academy of Sciences,
Beijing, China
{chenrui,chenlan,liying1}@ime.ac.cn

Abstract. We present a power and area efficient AHB slave IP for IEEE 802.15.4 security suite. The full AES-CCM* operations are implemented by the proposed new scheme of reusing only one compact Advanced Encryption Standard (AES) core. Without common low-power optimization such as clock gating, the resulting circuits offer both lower power consumption and smaller die size in 0.13um technology compared to valuable references.

1   Introduction

1.1   Background

IEEE 802.15.4 (WPAN: Wireless Personal Area Network) [1] is a well-known standard developed as potentially low cost solutions to a variety of real-world Wireless Senor Networks (WSN) applications.

IEEE 802.15.4 chooses AES-CCM* operations as its security suites, which is the CCM*(CTR & CBC-MAC) operations based on the Advanced Encryption Standard (AES) with 128 bits key. CCM* is a generic combined encryption and authentication block cipher mode, it coincides with the original specification for the combined CTR (counter mode encryption) with CBC-MAC (cipher block chaining message authentication code) mode of operation for messages that require authentication and, possibly, encryption, and also supports encryption only messages.

1.2   Related Work

Since AES was announced in 2001, many hardware implementations of AES for different characteristic design have been proposed. Timing, area, and power properties of AES vary from different implementations by up to orders of magnitude [2]. M. Feldhofer's work with only 3400 gates "serves as a benchmark for future hardware implementations of the AES algorithm that are optimized for low-resource conditions" [3]. Even though their hardware takes over 1000 clock periods to finish the encryption of 128-bit data, which might be too slow to support the data rate of WSN, the thought of their work is remarkable and useful.

W. Han et al. (Eds.): APWeb 2014 Workshops, LNCS 8710, pp. 282–290, 2014.

The implementation of AES-CCM* (or AES-CCM) is not widely published as AES core. Huai's AES-CCM claims to be very energy-efficient [4]. Unfortunately, their AES-CCM is designed as an independent module, disregarding the inner-connection with other layers or modules in the WSN. Song's work presents an efficient architecture of security accelerator satisfying the IEEE 802.15.4 specifications [5]. But it cannot be easily applied since its interface is declared on their own rules.

1.3   Organization

There has been a growing interest in applying WSN to heterogeneous scenes, which can be greatly benefit from node design with flexible architecture, like AMBA-BUS. Therefore, in this paper, we propose an easily reusable IP as an AHB slave with power and area saving design, which is compatible with the AES-CCM* security suite of IEEE 802.15.4.

The rest of the paper is organized as follows:

Section 2 describes an 8-bit compact AES encryption core designed in both low power and area consideration. Section 3 presents the detailed mechanism of AES-CCM* with 3 kinds of security operation sharing one compact AES core. The architecture of a reusable AHB AES-CCM* slave and the interfaces are provided in Section 4. And Section 5 gives the conclusion.

1.4   Aes Core Design

The AES is a symmetric block cipher algorithm and can process 128 bits data blocks, using cipher keys with lengths of 128, 192, and 256 bits. It operates in a certain number of rounds varies between 10, 12, and 14 depending on the size of key [6]. IEEE802.15.4 defines the length of key as 128 bits [1], which indicates 10 rounds in one operation.

After the initial operation, the following 9 rounds are composed of a sequence of four transformations and last round contents only 3 transformations, as the Fig. 1.

1.5   Architecture Consideration

The 128 bits data are stored and transferred as 16 bytes. The 16 bytes are presented as a 4x4 matrix in Fig. 1. The width of dataflow determines the structure of algorithm which affects the size of design. The most significant difference between different data width is the number of S-Box, which is a critical module in AES coreused in SubByte and KeyExpansion as a non-linear byte substitution table.

The number of S-Box in AES core directly determines the size of design [2]. For instance, if we design the dataflow in 128-bit, 16 S-Box for SubByte and 4 S-Box for KeyExpansion should be used to finish the 128 bits data transformation in one clock period. In the same manner, a design of 32 bits needs 4 S-Box for SubByte and 1 S-Box for KeyExpansion to finish the 128 bits data transformation in 4 clock period.

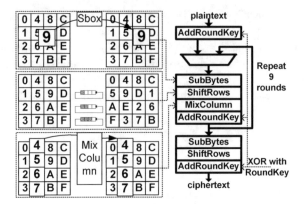

Fig. 1. Algorithm flow of AES

Even though the scheme of reducing S-Box would add extra control logic, it saves more resource in hardware. The throughput of AES also fall down with the scheme reducing of S-Box, the 128 bits AES has a throughput over Gbps and the 32 bits has over 500Mbps according to our design experiment.

As the data rate of WPAN is only 256kbps, we prefer an 8 bits structure. The result shows the throughput reaches 24Mbps at 32MHz, which is the frequency of AMBA in our system.

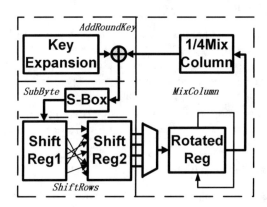

Fig. 2. Architecture of AES core

We use one S-Box for the SubByte and one S-Box for the KeyExpansion. With control logic, a 128-bit block consumes 170 clock periods to finish the encryption, 17 clock periods for each round. The architecture of our AES core is shown in Fig. 2. The four transformations of a round are pointed out in it.

Either the shift registers or multiplexer/de-multiplexer can be option to form the control logic, but shift registers saves more power [7] and easy for the operation: ShiftRows and optimization of MixColumn. The shift registers run for 16 clocks to register all the input 128 bits data and shift out a byte data for the operation: SubByte and AddRoundKey.

1.6   S-Box Implementation

The further consideration about S-Box is how to make the design efficiently. As mention before, S-box is a non-linear byte substitution table which could be implemented as a 256-byte ROM or look-up table. In the design with ROM, most of the area of AES core is occupied by the ROMs. An efficient implementation is to transform the original field $GF(2^8)$ into a smaller composite field $GF(2^4)^2$ [8], which is facile in hardware implementations and has smaller size. According to the security rules in WPAN, only the encryption of AES is needed. Result in DC shows that the area of S-Box designed by $GF(2^4)$ is less than 1/10 of the S-Box made by ROM.

1.7   MixColumn

AddRoundKey and ShiftRows are comparative easily to implement and optimize in the serial architecture. The rest optimization is MixColume.

The formula of MixColume differs from the input sequence but has similar architecture. If we rotate the 4 bytes data of a column for 4 times and send them into the same structure at every time. We could get the result of a column's MixColume operation.

We register a column of data in to a four byte rotated shifter register, from the Shift Reg2 in the Fig. 2, which is the input of 1/4MixColumn combination logic.

1.8   Result

The result of our design compared with other reference is shown below. Our design achieves the low power and low number of logic gates at a suitable timing property.

Table 1. AES core comparison

	Ref [3]	Ref [4]	Ref [9]	Ours
Tech	0.35um	0.18um	0.13um	0.13um
Area	3400gates	4023gates	3200gates	3761gates
periods	1016	160	160	170
power	4.5uW@1 00kHz,1.5v	1.06mW @10MHz	30uW @1MHz	19uW @1MHz 6.0uW @100kHz

Comparing to [9], our design achieves the low power because of 2 reasons. First, the S-Box structure in [9] is tree structure but chain structure in ours due to the more expansion of $GF(2^8)$ in [9]. Tree structure has higher switching rate than chain structure [10]. Second, [9] uses multiplexers for "ShiftRows" instead of some registers for small area optimization. However, register has lower input capacity than multiplexer which means lower power. On the other side, additional registers in ours switch only at proper clock period.

2   IEEE 802.15.4 Security Suite

AES-CCM* is the combination of CTR, CBC-MAC and CCM defined in IEEE 802.15.4 (2006).

In this section, formula O=AES(Key,I) means the encryption operation on the data I, with the encrypt key Key. The output is O.

2.1   CTR and CBC-MAC

The CTR is a confidential mode that features the application of the forward cipher to a set of input blocks, called counters, to produce a sequence of output blocks which are XORed with the plaintext to produce the ciphertext, and vice versa. The sequence of counters must have the property that each block in the sequence is different from others. Given a sequence of counters, T_1, $T_2...T_n$, the CTR mode is defined as follows:

$$C_i = P_i \oplus AES(Key, T_i) \quad i = 1 \text{ to } n \tag{1}$$

$$P_i = C_i \oplus AES(Key, T_i) \quad i = 1 \text{ to } n \tag{2}$$

Where C_i is the cipher block, P_i is the payload of MAC frame (FRAME) grouped into contiguous blocks. Encryption and decryption operation have the same architecture.

The CBC-MAC algorithm makes use of an underlying block cipher to provide data integrity on input data. The data to be authenticated is grouped into contiguous blocks, B_1, $B_2....B_n$, which is the additional authentication message (AUTH). The calculation of the MIC is given by the following equations:

$$O_{i+1} = AES(Key, O_i \oplus B_i) \quad i = 1 \text{ to } n \tag{3}$$

Where O_0=128'b0. The MIC is selected from O_{i+1}.

CCM is the combination of encryption and authentication. The T_x node uses CBC-MAC to get the MIC first, then uses CTR to encrypt the plaintext followed by MIC. The receive node uses CTR to get the plain and the authentication message from the received message. Then CBC-MAC calculates the authentication message from the decrypted plain and compare with the authentication message decrypted [1].

2.2   Implenmentation of AES-CCM*

Every mode in AES-CCM* needs the AES core. In general, the AES-CCM* module uses two AES modules in the parallel structure. One AES module is used for MIC calculation in CBC-MAC mode and the other is used for data encryption in CTR mode [5]. In [4], CCM is implemented in the parallel structure which declared to cut down the duty cycles of the whole processing. This scheme is not power-efficient enough and it was only useful at the CCM mode. However, according to the IEEE 802.15.4-2006, CCM mode only used in MAC command frame. Large amount of data frame only need encryption in CTR mode, so the parallel structure of 2 AES cores is wasted for most frames.

We find that almost the same structure can be applied to both CBC-MAC and CTR. Hence we design the security suite hardware with only one AES core. The dataflow is controlled by a FSM to cover both CTR and CBC-MAC, as the Fig. 3. The CCM is operated as CTR mode followed by CBC-MAC mode. As the CCM is rarely used, the extra time extension of CCM will not affect the performance of the WSN node.

In our design, the dataflow transfers via the Mux and DeMux as the Fig. 3. The path "1" (CBC-MAC Feedback in figure) is the data path for CBC-MAC while path "0" (Cnt/B0 and the path from DeMux to Contrl Logic in figure) is for CTR mode and the initial and last cycle of CBC-MAC. The two paths share a single AES core. The Mux and DeMux are controlled by FSM. The transition between states is determined by the mode signal from AHB and the flag generated from RAM R/W address.

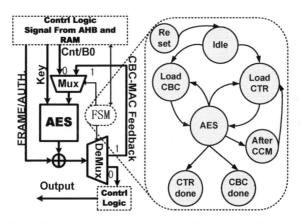

Fig. 3. Architecture of security operation and state transmission graph

At the beginning, when the first time the state turns into LoadCBC or LoadCTR (depends on the security mode), the plaintext (Cnt in CTR mode or B0 in CBC-MAC mode) of AES and "Key" is loaded into AES module. Next, the state turns into AES and the AES module operates the encryption for 170 cycles. When AES is done, state turns to LoadCBC or LoadCTR. The additional authentication message (AUTH) or the payload of MAC frame (FRAME) are read from RAM. AUTH XORs with the output of AES module. The result feeds back to the input plaintext of AES module in state LoadCBC. FRAME XORs with the output of AES module which is the result of CTR mode of the current block. If all the FRAME or AUTH in the RAM has been managed, the state turns into CTRdone or CBCdone if the mode is not CCM. Otherwise, the state will be AfterCCM and return to the LoadCTR for the encryption operation in CCM mode. All the transactions to Idle or Reset state are omitted in the Fig. 3.

2.3   Result

Our design saves large area due to the shared AES core, which achieves equally 11156 gates. The most complex operation CCM for 16bytes data takes 400 clock cycles, which means 80000 blocks of data can be managed per second in our design. It is enough for the narrow band sensor node. Our design consumes few power compare to other design shown in Table 2.

Table 2. AES core comparison

	Ref [4]	Ref [5]	Ours
power/MHz	0.04mW[a]	0.44mW	0.03mW

a. Estimated from FPGA,1/10 of the original result

3   AHB Slave

3.1   AHB Slave Structure

The structure is shown in the Fig. 4. The AES_SLAVE interface (AES_AHBIF) exchanges signals from AHB and stores them into controllable memory (RAM_I) and registers in AES-CCM*. There are four kinds of input data which are used in the AES-CCM* operation, payload of MAC frame (FRAME), additional authentication message (AUTH), initial vector (IV) for counter and the key (KEY) of AES. The last 2 kinds of data are sent directly into the registers of module AES-CCM*, due to their short length. The payload and additional message are buffered into the controllable RAM. As AHB is recommended to be at least 32-bit width [11], which is most efficient for the bus, an input buffer is designed to transfer 32 bits wide data into 8 bits.

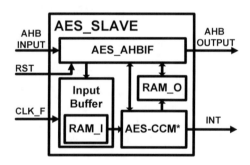

Fig. 4. Structure of AHB Slave

3.2   AHB Interface

All of the important registers in the interface are 32 bits, each of which has an address for access:

1. Reg_DataIn: The 4 kinds of data mentioned above are written into the address of Reg_DataIn. The data will be used by InputBuffer(RAM_I) or registers in AES-CCM*. To distinguish between each kind, we must use the Reg_Config.
   (a) Reg_Config: The register stores:

(b) mode(Reg_Config[5:3]): The security level which determines the function of AES-CCM* e.g. CTR, CBC-MAC or CCM.

(c) DataType(Reg_Config[2:0]): type of the data begins to transfer from the next clock. Four types of data are AUTH, FRAME, KEY and IV, as mentioned in 2.1.

2. Reg_ ReadData: The Master will read the data from the address of Reg_ReadData to get the result of AES-CCM*. The register fetches data from RAM_O continuously when AHB read signal is valid.

3.3   AHB Transfer Sequence

Here we focus on the registers mentioned in 3.2, and omit the description about control signals based on AMBA.

After the transfer is prepared, the first data should be written into Reg_Config with the correct mode of the security operation require and keep to the end. At the same time, begin to transfer "KEY", which means the DataType in Reg_Config should be written as KEY, and in next cycle, the KEY is written into Reg_DataIn.

The following transfer data differs from the "mode".

- CCM: (Key->)AUTH->FRAME->IV
- CBC-MAC: (Key->)AUTH ->IV
- CTR: (Key->) IV->FRAME

It should be pointed out that the CCM is the operation in transmission mode, the CCM in receiving mode is replace by CTR plus CBC-MAC.

In our scheme, when the AES_SLAVE is running the security operation, the bus is released and available to other tasks. The AES-CCM* module generates an INT signal to MCU when it has written the result into RAM_O after the security operation. MCU then responses to the INT request at the proper time by reading data from Reg_ReadData at the valid AHB read signal.

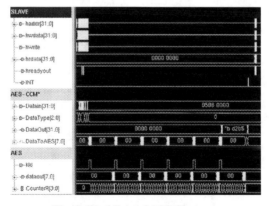

Fig. 5. Workflow of AES-Slave

3.4   Result

The slave's area is equal to 14000 gates. The Fig 5 shows the workflow of AES-Slave. The control signals and data are given by AMBA interface. AES-CCM* handles the security request by using AES module. At the end, the security result is read by master via AMBA. All the system with AMBA could easily take advantage of our design.

4   Conclusion

We present an efficient implantation for IEEE 802.15.4 security suite and make it reusable as an AHB slave IP. Only one compact AES core is needed to cover several security modes. Both the AES core and AES-CCM* module achieve competitive power and area results compare with former valuable references under 0.13um technology. The total power is 0.6mW at 32MHz after clock-gating optimization. Furthermore, the access for the security material may also be accomplished in our IP instead of MCU in our future research.

References

1. IEEE, IEEE standard 802.15.4-2006, Part 15.4: Wireless Medium Access Control (MAC) and Physical Layer (PHY) Specifications for Low-Rate Wireless Personal Area Networks (WPANs) (2006)
2. Tillich, S., Feldhofer, M., Popp, T., Großschädl, J.: Area, Delay, and Power Characteristics of Standard-Cell Implementations of the AES S-Box. Journal of Signal Processing Systems 50(2), 251–261 (2008)
3. Feldhofer, M., Wolkerstorfer, J., Rijmen, V.: AES implementation on a grain of sand. IEE Proc. Inf. Secur. 152, 13–20 (2005)
4. Huai, L., Zou, X., Liu, Z.L., Han, Y.: An Energy-Efficient AES-CCM Implementation for IEEE802.15.4 Wireless Sensor Networks. In: Networks Security, Wireless Communications and Trusted Computing (NSWCTC), vol. 2, pp. 394–397 (April 2009)
5. Song, O., Kim, J.: An Efficient Design of Security Accelerator for IEEE 802.15.4 Wireless Senor Networks. In: Consumer Communications and Networking Conference (CCNC), Las Vegas, Nevada USA (2010)
6. National Institute of Standards and Technology (NIST) FIPS-197: advanced encryption standard (November 2001), http://www.itl.nist.gov/fipspubs/
7. Chang, C.J., Huang, C.W., Tai, H.Y., Lin, M.Y.: 8-bit AES Implementation in FPGA by Multiplexing 32-bit AES Operation. In: Data, Privacy, and E-Commerce (ISDPE), Chengdu, China (2007)
8. Wolkerstorfer, J., Oswald, E., Lamberger, M.: An ASIC Implementation of the AES SBoxes. In: Preneel, B. (ed.) CT-RSA 2002. LNCS, vol. 2271, pp. 67–78. Springer, Heidelberg (2002)
9. Hamalainen, P., Alho, T., Hannikainen, M., Hamalainen, T.D.: Design and Implementation of Low-Area and Low-Power AES Encryption Hardware Core. In: Dubrovnik, C. (ed.) Digital System Design: Architectures, Methods and Tools (DSD) (2006)
10. Rabaey, J.M., Chandrakasan, A., Nikolic, B.: Digital Integrated Circuits: A Design Perspective, 2nd edn. Prentice Hall (2003)
11. AMBA^{TM} Specification (rev2. 0), http://www.arm.com

A Survey of Role Mining Methods
in Role-Based Access Control System

Liang Fang[1,2] and Yunchuan Guo[2,⋆]

[1] Beijing University of Posts and Telecommunications, Beijing 100876, China
[2] Institute of Information Engineering, CAS, Beijing 100093, China
fangliang@nelmail.iie.ac.cn

Abstract. Dealing with the risks in information sharing technology, role based access control(RBAC) mechanism has more advantages than traditional access control mechanism like DAC and MAC. Role takes core position in both building, maintaining the architecture of the RBAC system and migrating the non-RBAC system to the RBAC system. Then role engineering is proposed to find the appropriate roles for the RBAC system. Role mining problem, as an automatic way to find the roles, has been a hotspot for the role engineering. In this paper, we briefly introduce the basic definition of RBAC. A contribution of this paper is to classify some of the exist method into clustering methods and Binary Matrix Decomposition Method. We also evaluate some of these methods on both real-world data and experimental data. At last, we analyze the results to find out what differences may appear using different method and dataset.

Keywords: RBAC, role engineering, role mining, evaluation, overview.

1   Introduction

The rapid development of network technology has brought us more conveniences and efficiencies sharing information resources. However, the unauthorized access to sensitive information may bring tremendous risks. Researchers have put their concern on how to prevent the unauthorized access and thus reduce the system risks.One way is to focus on the security policies.For example, Bao et al.[1] present a method that automatically transform a given security policy into a compliant one.Another important way to prevent these risks is access control mechanism.It is designed to fight the threats caused by the unauthorized access operations. But with the rapid growing number of users and increasingly complicated and diversified information, the great management resource consumption that brought by the classic DAC and MAC has made it difficult to meet the

⋆ This work was supported by the National Natural Science Fundation of China (61100181,61100186), the National High Technology Research and Development Program of China (863) (2013AA014002) and Beijing Key Laboratory of IOT Information Security(Institute of Information Engineering, CAS).

W. Han et al. (Eds.): APWeb 2014 Workshops, LNCS 8710, pp. 291–300, 2014.

increasing security requirements. This needs us to find some new access control methods.

Role-based access control (RBAC) has strong flexibility and easy operational that can be applied to manage the authorization for large systems. Ferraiolo and Kuhn[2] first proposed the concept of role-based access control model. They introduced the concept of role and then established the users-roles-permissions relationship to flexible represent the access and accessed relationship between users and resources. After their work, Pro.R.Sandhu[3] proposed a more complete RBAC basic framework and the classical RBAC96 model presented by him establishes the foundation for the development of RBAC.

RBAC associates the permissions with the users by using the role as an intermediary. Moreover, the user can get the appropriate permissions by assigned to a corresponding role. The relationship between users and roles is a many-to-many relation, which means a user can have multiple roles and a role can be assigned to multiple users at the same time. Relationship between roles and permissions is also a many-to-many relation. In this way, we can realize the separation between the users and permissions logically so that we can achieve the goal of having a better security protection system. By this logically separation, there is no need to have a fine-grained management directly to every user-permission when the system administrators manage the user accesses. In a RBAC system,users just need to be reassigned the corresponding role to realize the change of authority when the permissions changed.And we just need to modify the corresponding role when the organization changes so that we can implement the system update. RBAC not only reduces the complexity of the system maintenance but also reduces the complexity of permission management.

How to find appropriate roles to implement the corresponding user-role assignment and role-permission assignment takes the core position in both the problem of building and maintaining the architecture of the RBAC system and the problem of migrating the non-RBAC system to the RBAC system. Therefore, Role Engineering technology has been put forward and has gotten the attention of researchers both at home and abroad. Role engineering formulates the role according to the security requirements of the enterprise and assigns the corresponding access control permissions to roles. Then assign a corresponding role to the user according to the user's security needs of the business will. We can create a complete role-based access control system according to the two steps above. According to different role generation methods, the role engineering can be divided into two categories: Top Down engineering method and Bottom-Up engineering method, namely the Role Mining (Role Mining) technology.

Based on the business logic in enterprises, the top-down approach analyses the relations between the specific functional requirements of enterprises and business logic between users.Then using expert knowledge to construct reasonable allocations between roles and permissions and guarantee the functionality and security requirements of the system. Coyne[4] first proposed the definition of the top-down approach. Fernandez[5] then put forward a top-down method based on use cases. By using the enterprise use cases and use case sequence, we may capture the permissions required by the use cases and build corresponding roles to describe the use case.

However, for companies which have large numbers of users and a large number of business logics, the implementation of the top-down approach requires a lot of human work and unable to realize automated building of RBAC system. Therefore, how to find an automated analysis method to generate the appropriate user-role-permission relationship has become the focus of the research, we call this kind of automated method as bottom-up engineering technology, in another way-Role Ming.

2   Definition of the Role Mining Problem

When representing a complete state of RBAC in RBAC system,$\rho =< U, P, UP >$ is expressed as input,$\gamma =< R, UA, PA, RH, DUPA >$ is expressed as a complete output state .The output state includes role inheritance relationship between the state. The meanings of symbols are as shown below:

U: the user set of the RBAC

P: the permission set of the RBAC

R: the role set of the RBAC

UP: the user-permission relation

$UA \subseteq U \times R$: assignment between the users and permissions

$PA \subseteq R \times P$: assignment between the users and permissions

$RH \subseteq R \times R$: hierarchical relationships between characters

$DUPA \subseteq U \times P$: relation between the allocation of all users and permissions

As we mentioned in the previous section, the role engineering technology can be divided into two categories: Top-down method and Bottom-Up method (also called the Role Mining Problem (RMP)). We will introduce a detailed introduction and definition of the RMP in the following sections.

2.1   Role Mining Problem

Role Mining is an automatic or semi-automatic method that transforms the initial access control mechanism into a RBAC mechanism . It can realize the configuration and updating of RBAC system. Another purpose of the RMP is to provide an efficient and secure access control strategy by effectively detecting configuration errors, such as unreasonable for information system. Kuhlmann[6] proposed the concept of role mining for the first time in 2003. Vaidya[7] then proposed the formalized definition of RMP as follows:

Definition 1. *Basic Role Mining Problem (RMP): Given a set of users U , a set of permissionsPRMS, and a user permission assignment UPA, and a set of roles, ROLES, a user-to-role assignment UA, and a role-to-permission assignment PA 0-consistent with UPA and minimizing the number of roles, k.*

Definition 2. *δ-approx RMP: Given a set of users U, a set of permissions PRMS, a user-permission assignment UPA, and a threshold δ, and a set of*

roles, $ROLES$, a user-to-role assignment UA, and a role-to-permission assignment PA, δ-consistent with UPA and minimizing the number of roles, k.

Definition 3. *Minimal Noise RMP: Given a set of users U, a set of permissions $PRMS$, a user-permission assignment UPA, and the number of roles k, and a set of k roles, $ROLES$, a user-to-role assignment UA, and a role-to-permission assignment PA, minimizing*

$$||M(UA) \otimes M(PA) - M(UPA)||_1$$

where $M(UA), M(PA)$, and $M(UPA)$ denote the matrix representation of UA, PA and UPA respectively.

Different from the notion of noise defined in the Definition 3, I.Molly[8] gives a different noise data definition, namely:

Definition 4. *Correctness Noise. This kind of noise can be divided into two categories.*

Type I errors (false positives): A user has been over-assigned permissions and has gained additional privileges.

Type II errors (false negatives):User is not assigned all the permissions required for their job when he joins the organization or a new project, division of the company, etc.

Definition 5. *Applicability Noise. RBAC is not a panacea for access control, and it may not be suitable for all access control needs, such as when sharing is low. How to identify assignments that are applicable for roles in RBAC is a problem.*

Meanwhile, constraint is an important part of the access control system. Traditional role engineering technologies do not take into account how to generate a series of constraints such as common exclusive role constraints, constraints based on the set cardinality, etc. Finding out these reasonable constraints can satisfy the function and safety requirements of the enterprises in the process of constructing a secure RBAC system.

In addition, user attributes, permissions weights, interpretability of the role and secure business logic information have been introduced into the role-mining algorithm. These kinds of information can help us to enhance accuracy of the role-mining algorithm. At the same time, it can improve the conciseness and interpretability of RBAC system by optimizing the role set or the complete state of RBAC which are generated by different role-mining methods

3   Classification of Role Mining Methods

Generally, a role mining problem aims to find out a role set which decouple the relation between the user set and the permission set from the given user-permission assignment (UPA). It seems that we should find out the UPA first. However, both the Access Control Lists(ACLs) and the Access Control Matrix(ACM) can describe the given the UPA. So the question turns out to be how to find out a method that can generate an appropriate role set. In the rest part of this chapter, we will introduce some of these role mining methods.

3.1   Role Mining Methods Based on Clustering Methods

The clustering methods use the clustering technology to generate the whole RBAC structure of user-role-permission and give out the candidate set of roles. Schlegelmilch[9] first proposed a method called ORAC that combines the ORCA tools and hierarchical clustering method to visualize the clustering results. The ORCA method first use hierarchical clustering method to analyze the user-permissions relation. Then generate a complete state of RBAC. ORAC method clusters the users who have the same permissions together. Then select and merge the permissions clusters that have maximum user intersection to form a new cluster. At last repeat this process until there is no permissions cluster that can be merge. Eventually, permissions are assigned to just a role. But this method has obvious shortcomings, permissions in the generated role sets cannot be overlapped. This means permissions in a role cannot assign to other roles. It is obviously disobeys the requirement of RBAC. Aiming at this problem, Vaidya[10] put forward a new method called Complete Miner (CM), which generate new candidate roles and sort the roles by priority. The main idea of this method is to use the thought of enumeration. It computes all possible combinations sets of the initial roles. Then generate the permissions intersections between these combinations and generate permissions intersections between intersections. Eventually enumerate all the intersection. The specific process can be divided into three steps: 1) preprocess the original data to select the initial sets 2) enumerate the permission set and generate new permission sets on the basis of the original set 3) calculate the number of generated users in the new sets. This method has a very high complexity whose time complexity is exponential in the size of original sets. Therefore, in order to reduce computing consumption, the authors put forward an improved scheme called Fast Miner (FM). FM simplifies the process of enumeration all initial role sets. It only enumerated the intersections between the two initial role, so its computation complexity is reduced to $O(n^2)$. The process of calculating the number of users in the intersections is done at the same time. After generate candidate role in CM and FM quantify the priority of these roles.

HP Role Minimization (HPr)[11] is an algorithm that aims at finding a minimal set of roles to cover the original upa relation. A user u's permission set is denoted as $P(u)$. All users that have all of user-permissions form the set $U(u)$. Similarly for each permission p, the set of users that have p is denoted as $U(p)$ and the set of per-missions that are assigned to all users in $U(p)$ is denoted as $P(p)$. In each step, the algorithm selects a user u(or a permission u) and finds a pair $<U(u),P(u)>$ (or $<U(p),P(p)>$) which forms a role. All user-permission assignments between $U(u)$(or$U(p)$) and$P(u)$(or$P(p)$) are then removed and the remaining user-permission assignments will be considered in subsequent iterations. Consider of the complexity of the HPr algorithm, it takes at most$O(min(m,n))$iterations for the number of roles is at most $min(m,n)$. Each iteration takes $O(m + n)$ to scan all users and permissions and select the next user (or permission). The total running time of HPr is $O(mn)$.

Secure business logic information such as use frequency etc. has been introduced into the role mining algorithm to help enhancing the role mining algorithm accuracy. Based on weighted structural complexity (WSC), I.Molly[12] proposed a method aims at optimize the role semantic and system complexity of role mining algorithm. Moreover, they have proved that we can maintain a very good role inheritance while mining good roles. But according to the WSC definition, the algorithm allow direct user-permissions assignment that may increase the man-made factors which violated the original intention of RBAC. Ma[13] first introduced the Jaccard coefficient to calculate the initial similarity between user-user relation and permissions-permissions relation and used the matrix form to denote the similarity. Then he used the iterative calculation method to strengthen the similarity matrix between the user-user and permissions-permissions. This improves the accuracy and effectiveness of similarity permissions-permissions relation.

Using these methods mentioned above, the role sets might be just simple numerical sets. There is a lack of semantic meanings for the mined role sets.If we can understand these roles easier, we can achieve a better management of the access control system.

3.2   Role Mining Methods Based on Binary Matrix Decomposition

Besides these clustering methods mentioned in the previous section, Vaidya and Lu[14] introduces the notion of matrix decomposition into the role mining problem. They decomposed the original user-permission matrix into a user-role matrix and a role-permission matrix. The authors also introduced an edge-based role mining problem based on the analysis of the different roles mining problems.Meanwhile,they use the matrix form to denote all kinds of role mining problem.

Then Vaidya[15] mapped the RMP to Minimum Tiling Problem and Discrete Basis Problem by showing that RMP and δ–RMP is NPC problem.

Geerts[16] first proposed the method using tiles to extract information from the database which contain only 0/1 data. Let an itemset I denote a collection (subset) of the 0/1. Then a tile t corresponding to an itemset I consists of the columns in itemset I as well as all the rows that have 1s in all the columns in I . Minimum Tiling Problem method can solve the basic problem of RMP properly.

Definition 6. *Minimum Tiling Problem. Given a Boolean matrix, find a tiling of the matrix with area equal to the total number of 1s in the matrix and consisting of the least possible number of tiles.*

The Discrete Basis Problem is proposed by Miettinen[17]. This technique simplifies a dataset by reducing multidimensional datasets to lower dimensions for summarization, analysis and/or compression. But the Discrete Basis Problem just considers the Boolean data and find Boolean bases.

Definition 7. *Definition 3.2 Discrete Basis Problem. Given a matrix $C \in \{0,1\}^{n*d}$ and a positive integer $k \leq min\{n, d\}$, find a matrix $B \in \{0,1\}^{n*d}$ minimizing*

$$l_\otimes(C, B) = min_{S \in \{0,1\}^{n*k}} ||C - S \otimes B||$$

In addition, Frank[18] introduced the notion of business-roles and technical-roles into the standard RBAC model by analyzing the correlation between different business information. A business-role is a set of users and a technical-role is a set of permissions. In this algorithm, a user belongs to just only one business-role and permission belongs to just only one technical-role. He clusters users into business-roles and permissions into technical-roles using the infinite relational model (IRM). Then the author proposed Disjoint Decomposition model (DDM) to calculate the probability of getting actual user-permission assignment x using given business-roles and technical-roles. The equation of DDM is given as follows:

$$p(x|z, y, \beta) = \prod_{k,l}[1 - \beta_{kl}]^{n_{kl}^{(1)}}[\beta_{kl}]^{n_{kl}^{(0)}}$$

z : represents business-role

y: represents technical-role

β : for parameter ,using the Gibbs sampling to randomly sample assignments from a Dirichlet process mixture model.

$n_{kl}^{(1)}$ represents the number of assignments ($n_{kl}^{(0)}$ represents missing assignments) for each (business-role, technical-role) pair .

In the access control systems, permissions belonging to one role may never be assigned to some users. These negative assignments can bring some advantages for the access control system. To deal with this kind of role mining problem, Lu[19] introduced a new model EBMD that extends the concept of BMD by allowing negative elements in one of the decomposed matrices. A good EBMD decomposition solution can discover a set of roles, which well constitutes the original permission assignments and reveals embedded SoD or exception constraints to better meet the business needs of an organization.

4   Comparison Experiments

Most of the methods based on probability need a manual configuration for the access control system. For example the method proposed by Frank needs to assign the permissions to the technical-group and the users to the business-group manually. This requires the system administrators having some prior knowledge. And the goodness of the result depend on their knowledge. We don't discuss these methods in this section.

In this section, we test some clustering methods through both real-world datasets and experimental datasets(shown in Table.1). We implement these clustering methods on an Intel (R) Core (TM)i3-2120 3.3G machine with 4 GB memory to evaluate their performance. The running platform is Windows7.

Table 1. Information of DataSets

	$domino$	$healthcare$	$firewall$	$emea$	$testdata$
num of users	3477	46	365	35	1000
num of permissions	1587	46	709	3046	25

1. number of roles

   In Table.2, we show the number of roles that generated by different methods. Conforming to the ideas of the HPr algorithm,We find that this method generate the least number of roles. Because using the enumeration method, FM methods first find out the combination of existing permissions list and then according to the combinations generate the intersection of the combinations. This leads to a result that the number of roles is very large that in the worst case, it can reach a number of 2^n.From the last column of Table.2,we can see that the FM generated 282489 roles. This is even larger than the users number. The Tiling method only performed well when the number of permissions is small. The growing number of permissions leads to a massive calculation complexity growth to generate the roles. So this method is not very applicable for the systems that have plenty of permissions.

Table 2. Number of Generated roles

	$domino$	$healthcare$	$firewall$	$emea$	$testdata$
HPr	158	13	59	33	15
GO	263	18	92	48	1369
FM	1778	29	266	242	282489
Tile	*	*	*	*	53

*:the calculation time of Tile method for these methods is too long, so we don't give out these results.

2. coverage of upa

   From Fig.1.(a) and Fig.1.(b) we observed that FM and Tiling method achieved a good coverage rate of the original UPA at a very low level of role numbers (less than 5 roles). This phenomenon indicates that the generated roles have a lot of redundancies and repeatability. Meanwhile, a role may contain numbers of permissions that lead to a concentration of permissions and increase the risk of the system. The GO method performed inconsistently. Coverage of the original UPA grows relatively smooth with the growth of the role numbers. Though under this circumstance the redundancy is lower, the needed number of the role to achieve a good coverage may be larger. The permissions have a relatively average assignment to different roles and can help to reduce the excessive assignment to users. But management consumption could be much higher for the increasing number

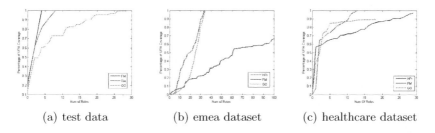

(a) test data   (b) emea dataset   (c) healthcare dataset

Fig. 1. We calculate the coverage of different number of roles for (a)test data(b)emea dataset(c)healthcare dataset

of roles. Therefore, an important point in the future research is to find out a certain method to optimize the existing method to get a reasonable coverage rate with a reasonable number of roles.

Different with the former two figures, we found that in Fig.1.(c) the FM algorithm got a relatively smooth growth while the other two methods achieved a very good coverage rate quickly. It indicates that there are some certain correlations between the original data and results. When the original data is different, we may get different results. In the future, we can study the possible different mining results brought by the original data.

5   Conclusion

In this paper, we introduced some role mining methods that constructing RBAC system. Though this methods has had great advantages in solve the role mining problem, there is much to be improved. Most current methods do not take the semantic meaning for the generated roles into consideration. This may causes a large management consumption and leads to a serious security risk because the system administrators may not understand what the role represent and assign the inappropriate role to the user. How to find out roles with semantic meaning is a research point that worth studying.

The technology of Internet of Things has a rapid development in recent years. It extends the communication between people and people to people and things. meanwhile, like distributed environment and wireless network ,the IOT systems are often complex and changeable. The user's behavior in the IOT system has a strong uncertainty and each domain has their own access control policy.The IOT system may also contain user's sensitive information such as location,user path etc. and the user number could be very large. These lead to risks that there may be disclosure of user's privacy.How to construct a appropriate RBAC system for this dynamic, interactive, uncertain and multi-domain environment while protecting the user's privacy is a big challenge in the future work.

Beside these, there are still some interesting problems for future research for the access control system. And we should not only pay attention to the RBAC system, but may take other access control mechanism into consideration.

References

1. Yibao, B., Lihua, Y., Binxing, F., Li, G.: A novel logic-based automatic approach to constructing compliant security policies. Science in China (Series F) 55(1), 149–164 (2012)
2. Ferraiolo, D., Kuhn, D.R., Chandramouli, R.: Role-based access control. Artech House (2003)
3. Sandhu, R.S., Coynek, E.J., Feinsteink, H.L., Youmank, C.E.: Role-based access control models yz. IEEE Computer 29(2), 38–47 (1996)
4. Coyne, E.J.: Role engineering (1996)
5. Fernandez, E.B., Hawkins, J.C.: Determining role rights from use cases. In: Proceedings of the Second ACM Workshop on Role-Based Access Control, pp. 121–125. ACM
6. Kuhlmann, M., Shohat, D., Schimpf, G.: Role mining-revealing business roles for security administration using data mining technology. In: Proceedings of the Eighth ACM Symposium on Access Control Models and Technologies, pp. 179–186. ACM
7. Vaidya, J., Atluri, V., Guo, Q.: The role mining problem: finding a minimal descriptive set of roles. In: Proceedings of the 12th ACM Symposium on Access Control Models and Technologies, pp. 175–184. ACM
8. Molloy, I., Li, N., Qi, Y.A., Lobo, J., Dickens, L.: Mining roles with noisy data. In: Proceedings of the 15th ACM Symposium on Access Control Models and Technologies, pp. 45–54. ACM
9. Schlegelmilch, J., Steffens, U.: Role mining with orca. In: Proceedings of the Tenth ACM Symposium on Access Control Models and Technologies, pp. 168–176. ACM
10. Vaidya, J., Atluri, V., Warner, J.: Roleminer: mining roles using subset enumeration. In: Proceedings of the 13th ACM Conference on Computer and Communications Security, pp. 144–153. ACM
11. Milosavljevic, N., Rao, P., Schreiber, R., Ene, A., Horne, W., Tarjan, R.E.: Fast exact and heuristic methods for role minimization problems (2008)
12. Molloy, I., Chen, H., Li, T., Wang, Q., Li, N., Bertino, E., Calo, S., Lobo, J.: Mining roles with semantic meanings. In: Proceedings of the 13th ACM Symposium on Access Control Models and Technologies, pp. 21–30. ACM
13. Ma, X., Li, R., Lu, Z.: Role mining based on weights. In: Proceedings of the 15th ACM Symposium on Access Control Models and Technologies, pp. 65–74. ACM
14. Lu, H., Vaidya, J., Atluri, V.: Optimal boolean matrix decomposition: Application to role engineering. In: IEEE 24th International Conference on Data Engineering, ICDE 2008, pp. 297–306. IEEE (2008)
15. Vaidya, J., Atluri, V., Guop, Q.: The role mining problem: A formal perspective. ACM Transactions on Information and System Security (TISSEC) 13(3), 27 (2010)
16. Geerts, F., Goethals, B., Mielikäinen, T.: Tiling databases. In: Suzuki, E., Arikawa, S. (eds.) DS 2004. LNCS (LNAI), vol. 3245, pp. 278–289. Springer, Heidelberg (2004)
17. Miettinen, P., Mielikainen, T., Gionis, A., Das, G., Mannila, H.: The discrete basis problem. IEEE Transactions on Knowledge and Data Engineering 20(10), 1348–1362 (2008)
18. Frank, M., Streich, A.P., Basin, D., Buhmann, J.M.: A probabilistic approach to hybrid role mining. In: Proceedings of the 16th ACM Conference on Computer and Communications Security, pp. 101–111. ACM
19. Lu, H., Vaidya, J., Atluri, V., Hong, Y.: Constraint-aware role mining via extended boolean matrix decomposition. IEEE Transactions on Dependable and Secure Computing 9(5), 655–669 (2012)

A Modified-k-Anonymity Towards Spatial-Temporal Historical Data in Location-Based Social Network Service

Chao Lee[1], Li-Hua Yin[1], and Lan Dong[2]

[1] Institute of Information Engineering,
Chinese Academy of Sciences Beijing, China
lichao@nelmail.ac.cn
[2] School of Computer and Information Technology,
Beijing Jiaotong University Beijing, China

Abstract. The paper introduces a modified k-anonymity spatial-temporal cloaking model on location based social network service. It makes a location historical record cannot distinguish from other $k - 1$'s records spatially,temporally. We propose an algorithm to generate the extended k-anonymity model cloaking region. Experiments show that the model is available and the algorithm efficient.

Keywords: location privacy, location-based social network service, diversity, k-anonymity.

1 Introduction

The Internet of Things (IoTs) has trended to growing use in commercial areas. It has been paid more and more attentions[1, 2]. In IoTs computing area, location-based services are most popular IoTs applications. Traditionally, the location-based service (LBS) is a service that utilizes GPS, WLAN and cellular network etc. techniques to obtain the location information of mobile terminals, and provide location information based services for mobile terminals through the wireless network[3]. On the other hand, Facebook, Twitter, Flickr and other online social networks are becoming more and more popular. Users can exchange information, share blog, video, image, location and other information in online social networks.With the development of online social networks and the popularity of mobile positioning devices, location-based services and online social networks has integrated to form a new service called *location-based social network services* (LBSNS).

LBSNS is based on the location information that build mobile service for the purpose of social interaction and sharing content via mobile devices. It has various applications, including the platforms for making friends, such as Loopt and Google Buzz, the check-in applications such as Foursquare and Gowalla, the applications of content sharing that combines location information such as Twitter and Weibo. These services can be utilized to find nearby friends, to

W. Han et al. (Eds.): APWeb 2014 Workshops, LNCS 8710, pp. 301–311, 2014.

promote interaction between them, but also allow users to share photos, videos or blog with location information indicating where the events happen, to help other users understanding the shared content more intuitively. LBSNS is the extension and continuity on the traditional social network service in a mobile environment. It is widely regarded as an important trend for the future development of social network service. According to the prediction of ABI Research, in 2103, the amount of LBSNS users will reach 82 million and the total revenue will reach 3.3 billon dollars.

However, user's privacy will inevitably exposure at the time the location information they share. Private information may be used by malicious users resulting in the unpredictable risk. Two surveys reported in July 2010 found that 55% of LBS users show concern about their loss of location privacy [4] and 50% of U.S. residents who have a profile on a social networking site are concerned about their privacy [5]. The results of these surveys confirm that location privacy is one of the key obstacles for the success of location-dependent services. In fact, there are many real-life scenarios where perpetrators abuse location-detection technologies to gain access to private location information about victims[6, 7].Therefore, research on the issues related to location privacy protection to LBSNS is important.

In fact, since LBS concept is proposed, location privacy problem has been a hot academic issue, the researchers also proposed various location privacy preserving approaches including false identification and location, spatial and temporal anonymous[8, 9], k-anonymity[10–12] etc.

However, LBSNS as a special case of LBS, its location information privacy protection requirements are also different.From the protection of the purpose, the goal of location privacy protection model for traditional LBS is to prevent attackers from identifying the users via location information. This means that an attacker who obtains some position information, but still not knows which users use the service at the position.In contrast, LBSNS location privacy protection model is to prevent user privacy place (for example, the exact location of the user's home) captured by those who might be malicious attackers or third party service providers.On other hand, in LBSNS, users share location information voluntarily. Therefore, we have to study new privacy protection strategy based on the existing privacy protection approach, to avoid the user privacy exposed security problems when providing services.

In this paper, we propose an extended k-anonymity Spatial-Temporal privacy protection approach to address the problem. We proposed spatio-temporal varied-k-anonymity and annotation l-diversity (VKL) model, which can makes a historical location record cannot distinguish from other $k-1$'s records spatially,temporally, with l different location annotations. We propose an algorithm to generate the extended k-anonymity model cloaking region. Experiments show that the model is available.

The rest of this paper is organized as follows: Section 2 presents the related work. In section 3, we describe the preliminary and some assumption of adversary user. Our proposed model is described in Section 4. Section 5 provides the achievement scheme of the model. Section 6 is the experiment of our proposed scheme. Finally, the conclusion can be found in Section 7.

2   Related Work

Our work on location privacy preserving studies on historical data in LBSNS is closely related to the traditional location privacy protection on LBS.There are several papers in the fields. According to Krumm's work[13], we abstract these approaches into two classifications: *anonymity* and *obfuscation*.

The basic idea of *anonymity* approaches utilizes a pseudonym and creates ambiguity by grouping with other people. These methods can be classified by: *pseudonym* and *k-anonymity*.For pseudonym, the identity of an event is altered in order to break the link between a user and his/her events.Using renaming function, the real identity of a user on each event is replaced by another pseudonym, which makes the attacker cannot observe the event of the actual user directly[14]. For k-anonymity, the person reports the region containing $k-1$ other users. Person's identity is indistinguishable from the location information of at least $k-1$ other subjects[9].

The basic idea of *obfuscation* approaches is used to protect user's location privacy through degrading the quality of information of user's location data, and at the same time, the degrading location data must be satisfied to service quality which user accepted. This is achieved by adding noise to the location and/or time-stamp of the events or by coarse graining them. Obfuscation is achieved mostly through perturbation or generalization algorithms. They can be can be divided by:hide location, adding noise, reducing precision. The distortion-based approach requires querying users to report their locations to the location anonymizer, but it also considers their movement directions and velocities to minimize cloaked spatial regions [15].

3   Problem Statement

3.1   Historical Event Model

Mobile users in LBSNS send messages through the wireless infrastructure to social network service providers. These messages can be the check-in data and micro blogs etc. containing the time stamps and locations. These messages are stored on servers that help users to review their lives. As we concentrate on the location privacy of these historical data, we ignore the content of the message, and only take the user, time stamp, location and location annotation into consideration.

We denote $\mathcal{U}$ as the set of users who are members of the location-based social network. Users are considered to be able to communicate by wireless network. Further, a user has a set of properties, such as name, gender, occupational, etc. We call the properties as *attributes* of the user. These attributes describe a user's profile.

In LBSNS, users report their locations when they post the message to provider server. We denote $\mathcal{L}$ the set of locations of users. Location $l \in \mathcal{L} = \{l_1, l_2, \ldots, l_n\}$ can be a two-dimensional point which represents a GPS coordinates, or a region that containing the point. We denote $\mathcal{T} = \{t_1, t_2, \ldots, t_n\}$ the set of discrete times that when users send the message. Besides, for each location, different users may have different annotation types. For example, given location l, user u_i and u_j, l may be u_i's home address, which may also be u_j's work address. It is easy to see that the location annotation is sensitive to user's privacy. We denote $\mathcal{A} = \{a_1, a_2, \ldots, a_n\}$ as the set of users's location annotation types.

In order to represent the case that a user $u_i \in \mathcal{U}$ had sent message to server at the time $t_i \in \mathcal{T}$, in location $l_i \in \mathcal{L}$, which labeled by $a_i \in \mathcal{A}$, we introduce the notion of *historical event* or *event* for short.

Definition 1. *An* event *is a triple* $e = \langle u, l, t, a \rangle$, *where* $u \in \mathcal{U}$ *is a user,* $l \in \mathcal{L}$ *is a location,* $t \in \mathcal{T}$ *is a discrete time, and* a *is the annotation of the location .*

3.2   Adversary Assumption

Since the event contains user's identification and the spatial-temporal information, user's privacy could be leaked easily. The traditional LBS attack model refers to the way to make the privacy protection model failure through mining and associate the user identity by looking for the inherent defects of location privacy protection model.The common LBS attack models including the joint attack model and the continuous query attack model.But unlike the LBS attack model, LBSNS attack model is to obtain the user's sensitive position, rather than the identity of the user. In this section, we propose a sensitive semantic location attack model.

Table 1. The example of location semantic annotation attack

pseudonym	Timestamp Region	Location Region	Location Annotation
u_1	$[t_1, t_2, \ldots, t_5]$	$[l_1, l_2, \ldots, l_5]$	*home*
u_2	$[t_1, t_2, \ldots, t_5]$	$[l_1, l_2, \ldots, l_5]$	*office*
u_3	$[t_1, t_2, \ldots, t_5]$	$[l_1, l_2, \ldots, l_5]$	*office*
u_4	$[t_1, t_2, \ldots, t_5]$	$[l_1, l_2, \ldots, l_5]$	*office*
u_5	$[t_1, t_2, \ldots, t_5]$	$[l_1, l_2, \ldots, l_5]$	*office*

For example, a location-based social network service provider release a historical dataset to a third-party research institute for analysis, as shown in Table 1, where the table columns represent the user identification, timestamp , obfuscated region and semantic annotation of location specifically. It is easy to see that the released dataset does not contain any explicit identifiers, such as user names, and the dataset is protected by k-anonymity approach where $k = 5$. A malicious user cannot distinguish the identity of the user through the first three fields of the table. However, when semantic annotation types of anonymity region are not diverse enough, even contain only one type, sensitive locations of the user is likely to be leaked. Table 1 shows that the anonymous region only contains one semantic annotation type of location. In this case, the location privacy protection model has failed. We assume the adversary user have the inference ability to distinguish the user probabilistically.

In traditional k-anonymity model, each user's query is regarded as an isolated event. It means that when each user position updates, it will calculate a new obfuscated region corresponding to the new position. In other words, the cloaking region of a position in two different timestamp may be different.When these cloaking information is stored, user sensitive position is likely to be leaked.

Take Fig. 1 as an example, the k-anonymity approach (where $k = 5$) produces two different regions in each containing 5 location points (users) specifically, when user visit sensitive position l at two different times.

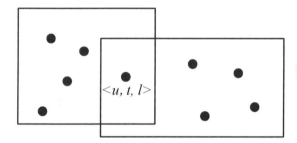

Fig. 1. An example of location continuous attack

An attacker makes an intersection of the two cloaking region, user u's sensitive location could be limited in a very small scope. In this way, user u's location privacy may be revealed. We assume the adversary user have the ability to launch the attack.

4   Proposed Model

To enable location privacy-preserving of LBSNS, our approach should achieve the following security guarantee: 1) location related attributes are all required for protection, including the time, location, location annotation. Protection of location sensitive semantic annotation could prevent a malicious user from deriving the user identity if he/she knows some external background knowledge about the annotation;2)when a user updates the cloaking region information of the same sensitive location at two different time, the information should be kept consistent spatially and annotation semantically as far as possible.

To meet the design goal, we proposed an varied k-anonymity and l-diversity location privacy preserving model towards spatial-temporal history data in LB-SNS.

Definition 2 (k-Anonymity Spatial Region). *k anonymity spatial region is a minimum boundary rectangle that contains at least k mobile users,denoted by $R_S = (l_{x-}, l_{y-}, l_{x+}, l_{y+})$, where (l_{x-}, l_{y-}) and (l_{x+}, l_{y+}) is the lower left corner and the upper right corner of the k users's locations specifically.*

Definition 3 (k-Anonymity Temporal Interval). *k-anonymity temporal interval is an interval of timestamp that contains at least k mobile users,denoted by $T_I = [t_{min}, \ldots, t_{max})]$, where $t_{min}, \ldots, t_{max}$ is the minimum and the maximum timestamp of k users specifically.*

To express intuitively, in the following we introduce the function $R : \mathcal{L} \rightarrow R_S$ to describe the cloaked region of the location set. For example, given $l_1, \ldots, l_k$ that are the k locations of k users, we use $R(l_1, \ldots, l_n)$ to denote the clocked region $(l_{x-}, l_{y-}, l_{x+}, l_{y+})$. We use $L(R^k)$ to denote the k locations in the cloaked region.

Similarly, we introduce the function $T : \mathcal{T} \rightarrow T_l$. It is intuitive to use $T(t_1, \ldots, t_k)$ to express $[t_{min}, \ldots, t_{max}]$.

In LBSNS, users are always annotate the locations they stayed. The annotation have the semantic information about the location of a user for a given time. It is possible to leak the user's interest or sensitive location.

Definition 4 (l-Diversity Annotation). *A cloaked region is l-diversity annotation if and only if in this region, the amount of the location type is greater and equal to l.*

It is easy to notice that for a same cloaked region, the greater the l is, the more uncertain about the interest of the users in the cloaked region. We use $A(a_1, \ldots, a_l)$ to express the annotation set of the cloaked region.

If it is not ambiguous, we use $T(t)$, $R(l)$,$A(a)$ to express the $T(t_1, \ldots, t_k)$, $R(l_1, \ldots, l_n)$ and $A(a_1, \ldots, a_l)$ specifically.

Definition 5 (Varied k-Anonymity Region). *Given k-anonymity spatial region $R_{u_1}^{k_1=k}(l_1)$ and $R_{u_1}^{k_2=k}(l_2)$ to denote user u_1's cloaked region at l_1 and l_2 with k-anonymity specifically. If $R_{u_1}^{k_1=k}(l_1) \cap R_{u_1}^{k_2=k}(l_2) \neq \emptyset$,then $R_{u_1}^{k_2=k}(l_2) \leftarrow R_{u_1}^{k_1=k}(l_l)$. We call $R(L(R(l_l)^{k_2=k})\backslash l_1)$ the varied k-anonymity region.*

For example, if a user u's two k-anonymity regions($k = 5$) about l_1 and l_2 are intersected shown in Fig. 2(a), according to Definition 5, user u's cloaked region at l_2 should be keep the same as to the region at l_1, and the locations except l_2 compose a new cloaked region R^v, as shown in Fig.2(b).

Based on the definitions from Definition 2 to Definition 5, we propose a spatio-temporal varied-k-anonymity and annotation l-diversity (*VKL*) model.

Definition 6 (*VKL* Model). *Given a set of event $E = \{e_1, \ldots, e_n\}$, for any event $e = \langle u, t, l, a \rangle$, e can be obfuscated to the following form*

$$AS(e) = \langle u, T(t), R^v(l), A(a) \rangle$$

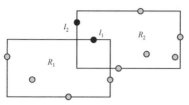

(a) user u's k-anonymity regions at location l_1 and l_2

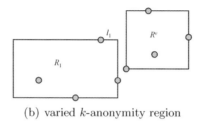

(b) varied k-anonymity region

Fig. 2. Illustration on when and how to compute varied-k-region

5   *VKL* Schema

In this section, we present an algorithm to generate the cloaked region satisfied the *VKL* model.

We give the definition of *spatial-temporal similarity* to decide whether need to process the algorithm.

Definition 7 (Spatial-Temporal Similarity). *We use* $sim(e_i, e_j)$ *to denote the spatial-temporal similarity of the event* e_i *and* e_j.

$$sim(e_i, e_j) = 1 - p_t \times sim(e.t_i, e.t_j) - p_l \times sim(e.l_i, e.l_j)$$

where $sim(e.t_i, e.t_j)$ *is the temporal similarity of time* t_i *and* t_j, *which can be computed by*

$$sim(t_i, t_j) = \sigma_t \times \frac{|t_i - t_j|}{\sqrt{t_i^2 + t_j^2}}$$

$sim(e.l_i, e.l_j)$ *is the spatial similarity of location* l_i *and* l_j, *which can be computed by*

$$sim(l_i, l_j) = \sigma_l \times \frac{dist(l_i, l_j)}{\sqrt{l_i^2 + l_j^2}}$$

If an event e and an event e' of user u's spatial-temporal similarity is smaller than a threshold, the same cloaked region could be used. Otherwise a new region need to be computed by the algorithm 1. The parameter p_t and p_l is used to

---

Algorithm 1. VKL cloaked region generation algorithm

Data: $E = \{e_1, \ldots, e_n\}$:location related history dataset, $e_i = \langle u_i, l_i, t_i, a_i \rangle$
Result: $AS(e)$:anonymized set of e

```
 1 begin
 2 │   foreach e ∈ E do
   │   │   /* if there have been existed a cloaked region for user u of
   │   │      event e                                                    */
 3 │   │   if IsExsitedRegion(e.u) == FALSE then
 4 │   │   │   computeCloakRegion(e.u);
 5 │   │   end
 6 │   │   else
   │   │   │   /* check the spatial-temporal similarity between the e and
   │   │   │      the e' of e.u in the cloaked region                    */
 7 │   │   │   if Sim(e, e') < δ then
 8 │   │   │   │   return AS(e) ← AS(e');
 9 │   │   │   end
10 │   │   │   else
11 │   │   │   │   R(e) = computeCloakRegion(e.u);
12 │   │   │   │   if R(e) ∩ R(e') = ∅ then
13 │   │   │   │   │   return R(e);
14 │   │   │   │   end
15 │   │   │   │   else
16 │   │   │   │   │   return AS(e) ← AS(e') ;
17 │   │   │   │   end
18 │   │   │   end
19 │   │   end
20 │   end
21 end
```

---

denote the weight on temporal and spatial to compute the whole similarity. σ_t and σ_l give the granularity to compute each similarity specifically.

Algorithm 1 takes the historical event record as input, and computes the cloaked region for each event record satisfied the VKL model. Firstly, for each event e, it checks whether e has some cloaked region at the interval time to decide compute the region or not. If at the location l, there not existed a cloaked region $R(l)$, then a new region need to be computed. Otherwise, e need to compare the spatio-temporal similarity with e', which generates the $R(l)$ at first. If they are similar, then e and e' are thought to be the same event,therefore e uses the same cloaked region. Otherwise, generate a new cloaked region. If the generated region $R(e)$ have intersection on $R(e')$, which are not satisfied to the adversary model discussed in section 3.2. Therefore, e uses $R(e')$ to be the cloaked region, as describe in definition 5.

6   Experiment

In order to evaluate the effectiveness of VKL model , this section designs a series of experiments.

The event data is generated by the following ways. We use Brinkhoff data generator[18] to generate the user locations on the map of Oldenburg, Germany. The outputs of the data generator are saved into files. We use a location annotation types from [19]. Then the generated locations choose the location type randomly to annotation the location semantics.

The simulation runs on a PC with Intel i3 3.40GHz CPU and 4G memory, running a Ubuntu 14 operation system. In the experiments, we choose 100000 events for study. The experiment evaluates the speeding time of each history event for different k and the spatio-temporal threshold.

Table 2. Parameters

parameters	values
number of events	100000
k	{5,10,15}
p_t	0.5
p_l	0.5
σ_t	{0.01,0.02,0.03.0.04,0.05}
σ_l	{0.01,0.03}
1	$\sigma_l = 0.01, k = 5$
2	$\sigma_l = 0.01, k = 10$
3	$\sigma_l = 0.01, k = 15$
4	$\sigma_l = 0.02, k = 5$
5	$\sigma_l = 0.02, k = 10$
6	$\sigma_l = 0.02, k = 15$
7	$\sigma_l = 0.03, k = 5$
8	$\sigma_l = 0.03, k = 10$
9	$\sigma_l = 0.03, k = 15$

Table 2 illustrate the experiment parameters, and the different parameters of our nigh groups experiments used.

The evaluation of the experiments is shown in Fig 3. From the Figure, we can notice that, the σ_t and k is not the key factors of the efficiency. When σ_l grows, the spend times increase quickly. Through the algorithm is $sigma_l$-sensitive, in a real situation, we can choose a small $sigma_l$ to keep the algorithm efficiency.

7   Summary

Location privacy has been heavily studied in LBS privacy-preserving algorithm such as anonymity and obfuscation. Krumm [13] has made a comprehensive survey. However, there is lack of a location privacy approach on LBSNS. Therefore,

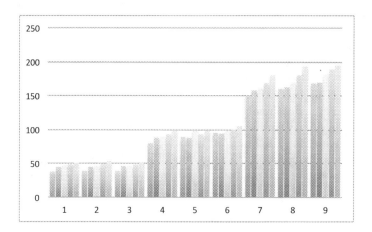

Fig. 3. The evaluation of our extended k-spatio-tempoal-anonymity approach

our work is focused on this area. Location privacy on LBSNS has some different feature of LBS that users are willing to share location information, which always has semantic annotations about the interest. This may cause inference attack on user location semantic information if attackers have external background knowledge. Meanwhile, the social data with the location information are stored on the service provider server, if attacker is capable of hacking the data, user's sensitive location information may be leaked. These historical data on server has the requirement to be privacy preserved. Therefore, we extended the traditional k-anonymity model that takes the spatio-temporal location annotation into consideration, and help avoid the continuous data association attack faced by the traditional k-anonymity model. We give the algorithm to compute the cloaked region, and make a simulation experiment to show the availability of the model.

Acknowledgment. This work was supported by the National Natural Science Fundation of China (61100181), the National High Technology Research and Development Program of China (863) (2013AA014002),the Fundamental Research Funds for the Central Universities Contract No. 2012JBM030, and Beijing Key Laboratory of IOT Information Security(Institute of Information Engineering, CAS).

References

1. Atzori, L., Iera, A., Morabito, G.: The Internet of Things: A survey. Computer Networks 54, 2787–2805 (2010)
2. Weber, R.H.: Internet of Things - New security and privacy challenges. Computer Law & Security Review 26, 23–30 (2010)
3. Virrantaus, K., et al.: Developing GIS-supported location-based services. In: Proceedings of the Second International Conference on Web Information Systems Engineering, pp. 66–75. IEEE Press (2001)

4. Webroot Software, Inc. Webroot survey finds geolocation apps prevalent amongst mobile device users, but 55% concerned about loss of privacy (July 13, 2010), http://pr.webroot.com/threat-research/cons/social-networks-mobile-security-071310.html
5. Marist Institute for Public Opinion (MIPO). Half of Social Networkers Online Concerned about Privacy., http://maristpoll.marist.edu/714-half-of-social-networkers-online-concerned-about-privacy/
6. Dateline NBC. Tracing a stalker, http://www.msnbc.msn.com/id/19253352
7. Voelcker, J.: Stalked by satellite: An alarming rise in gps-enabled harassment. IEEE Spectrum 47(7), 15–16 (2006)
8. Chow, C.-Y., Mokbel, M.F.: Liu,X.:A peer-to-peer spatial cloaking algorithm for anonymous location-based service. In: Proceedings of the 14th Annual ACM International Symposium on Advances in Geographic Information Systems, pp. 171–178. ACM Press (2006)
9. Gruteser, M., Grunwald, D.: Anonymous Usage of Location-Based Services Through Spatial and Temporal Cloaking. In: Proceedings of the 1st International Conference on Mobile Systems, Applications and Services, pp. 31–42. ACM Press, San Francisco (2003)
10. Gedik, B., Ling, L.: Protecting Location Privacy with Personalized k-Anonymity: Architecture and Algorithms. IEEE Transactions on Mobile Computing 7(1), 1–18 (2008)
11. Kalnis, P., et al.: Preventing Location-Based Identity Inference in Anonymous Spatial Queries. IEEE Transactions on Knowl. and Data Eng. 19(12), 1719–1733 (2007)
12. Hu, H., et al.: Privacy-aware location data publishing. ACM Trans. Database Syst. 35(3), 1–42 (2010)
13. Krumm, J.: A survey of computational location privacy. Personal and Ubiquitous Computing 13(6), 391–399 (2009)
14. Sweeney, L.: Achieving k-anonymity privacy protection using generalization and suppression. International Journal of Uncertainty, Fuzziness and Knowledge-Based Systems 10(5), 571–588 (2002)
15. Chow, C.-Y., Mokbel, M.F.: Trajectory Privacy in Location-based Services and Data Publication. SIGKDD Explor. Newsl. 3(1), 19–29 (2011)
16. Beresford, A.R., Stajano, F.: Mix zones: user privacy in location-aware services. In: Proceedings of the Second IEEE Annual Conference on Pervasive Computing and Communications Workshops, pp. 127–131 (2004)
17. Chow, C.-Y., Mokbel, M.F.: Enabling private continuous queries for revealed user locations. In: Papadias, D., Zhang, D., Kollios, G. (eds.) SSTD 2007. LNCS, vol. 4605, pp. 258–275. Springer, Heidelberg (2007)
18. Brinkhoff, T.: A framework for generating network- based moving objects. GeoInformatica 6(2), C153–C180 (2002)
19. Road Networks and Points of Interest, http://www.cs.fsu.edu/-lifeifei/SpatialDataset.html

A General Framework of Nonleakage-Based Authentication Using CSP for the Internet of Things

Licai Liu[1,2,3], Bingxing Fang[1,2,3], and Beiting Yi[1]

[1] Beijing University of Posts and Telecommunications, Beijing, China
[2] Institute of Information Engineering, Chinese Academy of Sciences, Beijing, China
[3] National Engineering Laboratory for Content Security Technologies, Beijing, China
`liulicai@nelmail.iie.ac.cn`

Abstract. Authentication is a slippery and important security property related to verify the identity and authenticity of someone or something, its formal definition is one key aspect of the research into authentication. The existing proposed formal definitions of authentication are not widely agreed upon. Moreover, these definitions cannot reach the requirements of diverse security and privacy in the Internet of Things(IoTs). In this paper, with introducing the notion of non-leakage, we proposed a general framework of authentication property in CSP for the Internet of Things. In the framework, we defined three forms of authentication - entity authentication, action authentication and claim authentication- and three strength levels for each form - weak, non-injective and injective level. We formalized each definition using the process algebra CSP. The framework can easily express different security requirements of the IoTs.

Keywords: Authentication, Trace Model, Communicating Sequential Processes, Internet of Things.

1 Introduction

The Internet of Things (IoTs) is suffering from an increasing number of security threats, particularly those related to authentication, such as unauthorized access and wormholes, with high risks [1]. To provide authentication, many security protocols have been proposed. However, some protocols are complicated and may not satisfy authentication [2]. To solve the problem, an important task is to formally verify protocol. The precondition of formal verification is to formally definite authentication.

Generally, authentication refers to the capability of verify identity and authenticity of someone or something. For example, entity authentication commonly refers to the capability of verifying an entity's claimed identity by another entity in ISO/IEC 9798, while message authentication usually refers to the capability of verifying the authenticity of the content and origin of message. These are informally definitions and thus authentication is considered as a slippery property [3]. Many formal definitions have proposed on different aspects of authentication in the literature, and however they are not widely agreed upon. For example, Burrows et al. [4] proposed a logic (called BAN logic) to describe and deduce the beliefs of trustworthy parties.

W. Han et al. (Eds.): APWeb 2014 Workshops, LNCS 8710, pp. 312–324, 2014.

Gollmann [5] specified entity authentication. Focardi [6] clarified the formal definition of message authentication through non-interference using process algebra CryptoSPA. Lowe [7] identified and formalized four possible definitions of authentication using process algebra CSP. As these definitions of authentication are formalized by different mathematical methods, they are usually hard to compare with each other and are not widely agreed upon.

Those above formal definitions are sometimes difficult to describe diverse security requirements of the IoTs. As we know, the IoTs is a heterogeneous network and has a variety of authentication requirements. For instance, the military sensor network cares the identity of nodes more, while the sensor network is more concerned with the authenticity of the message in environmental monitoring. Further, protecting privacy during authentication is an essential requirement for the IoTs. Moreover, with limited resources in a constrained environment, nodes in the IoTs are usually deficient in the capacity of authentication. Thus, it is a key issue and a significant challenge to propose a general framework to meets the diverse security requirements of the IoTs.

To address the above problem, by introducing the notion of nonleakage, a general framework of authentication property for the IoTs is proposed in this paper. In the framework, we defined three forms of authentication - entity authentication, action authentication and claim authentication- and three strength levels for each form - weak level, non-injective level and injective level. We formalized each definition using CSP. The framework can express different security requirements of the IoTs.

2   Related Work

In this section, we introduce the related work, including the formal definitions and methods of authentication.

2.1   Authentication Property

As authentication reflects many facets of security goal, so, many research works focus on formally defining it in different ways [4-16]. Burrows et.al proposed BAN logic [4] and made a first attempt to formalize authentication by describing and evolving the beliefs of trustworthy parties. However, BAN-type logics are not successful in revealing important flaw [8]. Schneider proposed [9] a message-based authentication in CSP. His idea is to observe when a set of messages T authenticates another set of messages R. Zhou et.al [10] introduced non-repudiation in protocols and aimed to produce evidence about the execution of services, among parties that do not trust each other. Gollmann [5] presented a specification for entity authentication which emphasized the need for the right level of abstraction. Focardi et.al [11][12] used a generalization of non-interference and proposed non-deducibility on compositions (NDC)-based authentication. Its basic idea is: the untrusted users that must not interfere with the other users are characterized by the set of actions that they can execute. However, NDC-based authentication cannot work efficiently as the problem of state explosion.

Woo and Lam [13] specified authentication by introducing the idea of correspon-dence which requires every end-event corresponds to a unique and earlier begin-event with the same name. Gollmann [14] went one step further and investigated to what extent correspondence corresponds to the various flavors of authentication. As one popular formal definition of authentication, Lowe [7] presented agreement property, identified four definitions of authentication (aliveness, weak agreement, non-injective agreement and injective agreement) and formalized them by using CSP. The highest strength level is injective agreement, which requires a complete match of the conver-sation between peer entities. Duo to without strict one-to-one correspondence, agree-ment property cannot reflect the requirements of authentication perfectly. Ahmed and Jensen [15] presented one separation of security and correctness requirements, by introducing the notion of binding sequence as a security primitive. A binding se-quence is the only required security property of an authentication protocol, while the correctness requirements can be derived from the binding sequence.

Although, many protocols are proposed to guarantee the authentication require-ments of the IoTs, such as TAM [17] and FssAgg [18]. Additionally, A few works related to authentication protocol analysis are presented, such as modeling and ana-lyzing the μTESLA protocol using CSP [19], proving the secrecy property of secure WLAN authentication scheme(SWAS) using Extended Protocol Composition Logic [20]. However, the formal definition of authentication in the IoTs is rarely given.

2.2   Formal Method

There are various methods are proposed for formally defining authentication. These formal methods fall into some different categories: belief-based methods [4], process algebra, strand space [21] and induction methods [22] and other methods. Belief-based methods are all based on the entity's knowledge about peer entity and used by belief logic, such as BAN [4]. Process algebra, such as CSP, CCS and Spi-Calculus, is the most widely used method. Agreement property is a trace-based definition using CSP [23]. Focardi and Gorrieri [11] proposed NDC which defined on CCS agents. Abadi and Gordon [24] proposed the Spi-Calculus for formalizing and modeling au-thentication protocols. CSP is a best advantageous formal method among these me-thods, as the following reasons: it has complete semantic and automatically model checker tools such as FDR2, it's widely used in formal verification [25] [26].

3   Preliminaries

In this section, we introduce the Communicating Sequential Process, the concept of trace equivalence and non-leakage.

3.1   The CSP-Based Trace Model

CSP(Communicating Sequential Process) is an abstract formal language proposed for describing behaviors of communication entity in concurrent systems by famous com-puter scientist C.A.R.Hoare [27][28]. CSP has characteristics of process algebra, good semantics and complete calculus ability and is good at describing concurrent events and interactive processes. Lowe first used CSP and model checking to formal analyze security protocols [29]. The approach is that the problems of protocol come down to whether CSP processes meet security specifications and whether protocol meets security properties. Modeling protocol by CSP, then checking the model using FDR2.

In CSP, process is a set of a series action, every action or a series action is an event. Process is defined by communication events it can execute; the communication assumes visible events and actions forms. Σ is the set of all actions, τ is the invisible action. CSP defines complete operators to describe protocols as Table 1.

Table 1. CSP Operators

Operator	Annotation	Operator	Annotation
$Stop$	process does nothing	$Skip$	successful termination
$P;Q$	sequential composition	$\mu p.F(p)$	recursive process
$a \rightarrow b$	event prefix	$a \rightarrow P$	event prefix of process
$?x:A \rightarrow P(x)$	event prefix choice	$c?x:A \rightarrow P(x)$	input prefix choice
$\square x:S$	external choice of set	$P\square Q$	external choice
$P \prod Q$	nondeterministic choice	$P \| Q$	lockstep parallel
$P \,_x\|_y\, Q$	synchronizing parallel	$P\|_Y Q$	interface parallel
$\|\|\| S$	general interleaving	$P \backslash X$	event hiding
$P = F(P)$	recursive definition	$P[R]$	proeess renaming

The *Skip* is a process terminates successfully; *Stop* is the process that does nothing. As (1)(a), P_1 is a process that initially performs event x and y, and then behaves as process *Skip*. P_2 is a recursive process which executes y and x indefinitely as (1)(b).

$$\text{(a): } P_1 = x \rightarrow y \rightarrow Skip; \qquad \text{(b): } P_2 = y \rightarrow x \rightarrow P_2$$
$$\text{(c): } P_3 = P_1 \square P_2; \qquad \text{(d): } P_4 = P_1 \prod P_2 \tag{1}$$

P_3, defined as (1)(c), is a process which offers the environment choice of the first events of P_1 and P_2 and then behaves accordingly. If the first event chose as x, then P_3 behaves like P_1. As (1)(d), P_4 is a process which can behave like either P_1 or P_2

CSP defines several types of channel that parties transmit messages through: *Receive* and *Send* is the standard channel of honest party, *Receive* denotes receiving messages and *Send* denotes sending messages. *Hear* and *say* represent intruder receives and sends messages, respectively. *Leak* means message may be leaked.

3.2   Trace Equivalence

The process P is defined as a set of finite sequences of events that it may possibly perform. This set will always be non-empty as it has an empty trace $\langle\rangle$, and prefix closed. A sequence of events tr is a trace of P if some execution of P performs exactly that sequence of events. This is denoted as $tr \in traces(P)$, where $traces(P)$ is the set of all possible traces of P. If $tr \in traces(P)$, then so is any prefix of tr.

Most of formal definitions of security properties that have been proposed are based on the simple notion of trace equivalence: two processes are equivalent if they show exactly the same execution sequences of events.

Definition 1. Let $P, Q \in \Sigma$, $P \leq_{trace} Q$ iff $tr(P) \subseteq tr(Q)$. So, P and Q are trace equivalent (denoted as $P \approx_{trace} Q$) iff $P \leq_{trace} Q$ and $P \geq_{trace} Q$.

3.3   The Concept of Non-leakage

In order to express the requirement of privacy protection during the authentication procedure of the IoTs, we introduce the notion of non-leakage, similar to [30], in the formal definition of authentication. Let $\vec{P}$ be the set of subsequent processes which may reach from process P in system S.

A process is said to be non-leaking iff for any pair of processes $\vec{P}_i$ and $\vec{P}_j$, while $\vec{P}_i, \vec{P}_j \in \vec{P}$ and $i \neq j$, and observing process O, the two processes resulting from running any sequence from P to $\vec{P}_i$ and $\vec{P}_j$ are indistinguishable for O if $\vec{P}_i$ and $\vec{P}_j$ have been indistinguishable for all output that may (directly or indirectly) interfere with O during the run of P. Or in other words, for any sequence of actions performed by the system, the outcome of O's observations is independent of variations in all subsequent processes except for those related to P, from which (direct or indirect) information flow in the course of P is allowed.

Definition   2. In   system S,   For   all $\vec{P}_i, \vec{P}_j \in \vec{P}$ and $i \neq j$,
$P?x \rightarrow \vec{P}_i!x \rightarrow O?x \rightarrow O!x \rightarrow Skip$, $P?y \rightarrow \vec{P}_j!y \rightarrow O?y \rightarrow O!y \rightarrow Skip$. P is non-leakage in S (denoted as $N_S(P)$) iff $x = y$.

4   Formalizing the General Framework of Authentication

In this section, we propose and formalize a general framework of authentication property in CSP for the IoTs.

4.1   A General Framework of Authentication Property

By analyzing the behavior of users or nodes in authentication protocols, we can summarize some common and interesting characteristic: when an user accesses the system, the system will authenticate the authenticity of its identity; when an user takes action in the system, the system will verify whether its action is allowed to perform or not; when an user presents a claim which states what it has done, sometimes system record is also a claim, the system will authenticate the authenticity of its claim. So, we can give the formal definition of authentication property from these three aspects. In the framework, we define three forms of authentication - entity authentication, action authentication and claim authentication – and three strength levels of authentication for each form – weak level, non- injective and injective level.

Then, we formalize different formal definitions of authentication and consider that the authentication protocol (also called system. Note: A system is considered as "authentication protocol" unless it is expressly stated.) performs a serial of interactive behavior between numerous of agents, which taking roles of system. Now, we first focus on a single interactive behavior (also called session) between two agents. A *Session* with a sequence identifier *sid* refers to a pair of agents, A and B, taking system roles *A-Role* and *B-Role*, respectively, and executing the events *A-event* and *B-event* on data set *ds*, respectively. We consider the problem of whether the session correctly authenticates B to A. For this problem, we define the predicate *1SI-ACCEP* (stands for one session injective acceptance). The basic idea of *1SI-ACCEP* is test whether the system guarantees to an agent A, taking the system role *A-Role* and executing the event *A-event* with sequence identifier *sid*, acceptance with an agent B, taking the system role *B-Role* and executing the event *B-event* with exactly sequence identifier *sid*, on the data set *ds*.

Definition 3. In system S, let $A, B \in Agent$, $A\text{-}Role, B\text{-}Role \in Role$, $A\text{-}event, B\text{-}event \in Event$:

$$1SI\text{-}ACCEP(A, A\text{-}Role, A\text{-}event, sid, ds)(B, B\text{-}Role, B\text{-}event)$$

$$\triangleq B\text{-}event.B\text{-}Role.B\,?\,sid\,?\,A\,?\,ds \rightarrow A\text{-}event.A\text{-}Role.A.sid.B.\overset{.}{ds}$$

Additionally, two weak forms of *1SI-ACCEP*, *1SN-ACCEP* and *1SW-ACCEP*, are defined. The *1SI-ACCEP* requires the one-one relationship between the events of A and B, however, *1SN-ACCEP* (stands for one session non-injective acceptance) doesn't have such requirement and *1SW-ACCEP* (stands for one session weak acceptance) even doesn't guarantee the acceptance on data set.

Definition 4. In system S, let $A, B \in Agent$, $A\text{-}Role, B\text{-}Role \in Role$, $A\text{-}event, B\text{-}event \in Event$, $sid, sid' \in SID$:

$$1SN\text{-}ACCEP(A, A\text{-}event, A\text{-}event, ds)(B, B\text{-}Role, B\text{-}event)$$

$$\triangleq B\text{-}event.B\text{-}Role.B\,?\,sid'\,?\,A\,?\,ds \rightarrow A\text{-}event.A\text{-}Role.A.sid.B.\overset{..}{ds}$$

Definition 5. In system S, let $A, B \in Agent$, $A\text{-}Role, B\text{-}Role \in Role$, $A\text{-}event, B\text{-}event \in Event$, $sid, sid' \in SID$, $ds, ds' \in DataSet$.

$1SW\text{-}ACCEP(A, A\text{-}Role, A\text{-}event)(B, B\text{-}Role, B\text{-}event)$

$\triangleq B\text{-}event.B\text{-}Role.B?sid"?A?ds' \rightarrow A\text{-}event.A\text{-}Role.A.sid.B.ds\,\dot{}$

If $1SI\text{-}ACCEP$ holds, then the events of A and B, $A\text{-}Event$ and $B\text{-}Event$, with the sequence identifier sid have executed correctly and the executing details have been accepted by A. Then, we extend the $1SI\text{-}ACCEP$ to multi-sessions between A and B, with a given set of sequence identifiers SID . Similar to $1SI\text{-}ACCEP$, we define the predicate $MSI\text{-}ACCEP$ (stands for one session injective acceptance).

Definition 6. In system S, let $A, B \in Agent$, $A\text{-}Role, B\text{-}Role \in Role$, $A\text{-}Event, B\text{-}Event \in Event$, $ds, DS \in DataSet$, $ds \in DS$:

$MSI\text{-}ACCEP(A, A\text{-}Role, A\text{-}Event, SID, DS)(B, B\text{-}Role, B\text{-}Event)$

$\triangleq SID!sid_1 \rightarrow SID \setminus sid_1 ?SID \rightarrow$

$\quad 1SI\text{-}ACCEP(A, A\text{-}Role, A\text{-}Event \# sid_1, sid_1, ds)(B, B\text{-}Role, B\text{-}Event \# sid_1) \rightarrow \dot{}$

$\quad MSI\text{-}ACCEP\ (A, A\text{-}Role, A\text{-}Event \setminus sid_1, DS)(B, B\text{-}Role, B\text{-}Event)$

While $A\text{-}event = A\text{-}Event \# sid_1$ represents the event $A\text{-}event$ which has the sequence identifier sid_1 and $A\text{-}event \in A\text{-}Event$. $MSI\text{-}ACCEP$ is defined as a recursive definition. If $MSI\text{-}ACCEP$ holds, then the sessions consist of all events of A and B, $A\text{-}Event$ and $B\text{-}Event$, with the set of sequence identifiers SID have executed correctly and the executing details have been accepted by A. Additionally, two weak forms of $MSI\text{-}ACCEP$, $MSN\text{-}ACCEP$ and $MSW\text{-}ACCEP$, are defined base on $1SN\text{-}ACCEP$ and $1SW\text{-}ACCEP$, respectively.

Definition 7. In system S, let $A, B \in Agent$, $A\text{-}Role, B\text{-}Role \in Role$, $A\text{-}Event, B\text{-}Event \in Event$, $ds, DS \in DataSet$, $ds \in DS$:

$MSN\text{-}ACCEP(A, A\text{-}Role, A\text{-}Event, SID, DS)(B, B\text{-}Role, B\text{-}Event)$

$\triangleq SID!sid_1 \rightarrow SID \setminus sid_1 ?SI \rightarrow$

$\quad 1SN\text{-}ACCEP(A, A\text{-}Role, A\text{-}Event \# sid_1, ds)(B, B\text{-}Role, B\text{-}Event \# sid_1) \rightarrow \dot{}$

$\quad MSN\text{-}ACCEP\ (A, A\text{-}Role, A\text{-}Event, SID \setminus sid_1, DS)(B, B\text{-}Role, B\text{-}Event)$

Definition 8. In system S, let $A, B \in Agent$, $A\text{-}Role, B\text{-}Role \in Role$, $A\text{-}Event, B\text{-}Event \in Event$:

$MSW\text{-}ACCEP(A, A\text{-}Role, A\text{-}Event, SID)(B, B\text{-}Role, B\text{-}Event)$

$\triangleq SID!sid_1 \rightarrow SID \setminus sid_1 ?SID \rightarrow$

$\quad 1SW\text{-}ACCEP(A, A\text{-}Role, A\text{-}Event \# sid_1)(B, B\text{-}Role, B\text{-}Event) \rightarrow \dot{}$

$\quad MSW\text{-}ACCEP\ (A, A\text{-}Role, A\text{-}Event, SID \setminus sid_1)(B, B\text{-}Role, B\text{-}Event)$

4.2   Entity Authentication

Entity authentication commonly refers to the capability of verifying an entity's claimed identity by another entity, however, it has no commonly agreed definition. We consider that entity authentication reflects whether the role of agent B which executing events with agent A is the role agent B claimed.

(a) Injective Entity Authentication

We give the formal definition of injective entity authentication I-$EAuth$. Its basic idea is test whether the system guarantees to the agent executing the events SID with the agent A, taking the system role A-$Role$ and, is exactly the agent B, taking the system role B-$Role$. It also requires there is non-leakage in the events of A and B.

Definition 9. In system S, let $A, B \in Agent$, A-$Role, B$-$Role \in Role$, $Signal.Commit, Signal.Runing \in Event$:

I-$EAuth(A, A$-$Role, SID)(B, B$-$Role)$

$\triangleq N_S(Signal.Runing) \wedge N_S(Signal.Commit) \wedge$

$\quad MSI$-$ACCEP (A, A$-$Role, Signal.Runing, SID,)(B, B$-$Role, Signal.Commit)$

(b) Non-injective Entity Authentication

Additionally, one weak form of I-$EAuth$, N-$EAuth$ is defined. The N-$EAuth$ doesn't require the one-one relationship between the events of A and B.

Definition 10. In system S, let $A, B \in Agent$, A-$Role, B$-$Role \in Role$, $Signal.Commit, Signal.Runing \in Event$:

N-$EAuth(A, A$-$Role, SID)(B, B$-$Role)$

$\triangleq N_S(Signal.Runing) \wedge N_S(Signal.Commit) \wedge$

$\quad MSN$-$ACCEP (A, A$-$Role, Signal.Runing, SID,)(B, B$-$Role, Signal.Commit)$

4.3   Action Authentication

Action authentication refers to verifying the authenticity of the actions of agents and whether these actions are allowed and no harm for other agents. We introduce the notion of non-interference to define the action authentication.

(a)   Injective Action Authentication

We give the formal definition of injective action authentication I-$AAuth$. The basic idea of I-$AAuth$ is test two conditions: one is test whether the system guarantees to an agent A, taking the system role A-$Role$ and executing the event A-$Event$ with sequence identifier SID, acceptance with an agent B, taking role B-$Role$ and executing the event B-$Event$ with exactly sequence identifier SID, on the data set DS. The another one is test whether the events of B interfere the events of A (or it requires trace equivalence). It also requires that there is non-leakage in the events of A and B.

We define traces, which eliminates the interactive events, between A and B, of agent A and the traces , which eliminates the interactive events between A and B, of $(A \| B) \backslash C$ as (2)(a) and (2)(b),respectively, while we consider C as the set of channels that classified processes B use when trying to interfere with other processes.

(a): $PublicTr(A) = tr(A) \backslash \{Signal.Runing, Signal.Commit\}$

(b): $PrivateTr((A \| B) \backslash C) = tr((A \| B) \backslash C) \backslash \{Signal.Runing, Signal.Commit\}$

(c): $tr(A) = PublicTr(A) \|$

$MSI\text{-}ACCEP \ (A, A\text{-}Role, Signal.Runing, SID, DS)(B, B\text{-}Role, Signal.Commit)$

(d): $tr((A \| B) \backslash C) = PrivateTr((A \| B) \backslash C) \|$

$MSI\text{-}ACCEP \ (A, A\text{-}Role, Signal.Runing, SID, DS)(B, B\text{-}Role, Signal.Commit)$

$$(2)$$

Then, the traces of A and $(A \| B) \backslash C$ as (2)(c) and (2)(d), respectively.

Definition 11. In system S, let $A, B \in Agent$, $A\text{-}Role, B\text{-}Role \in Role$, $Signal.Commit, Signal.Runing \in Event$:

$I\text{-}AAuth(A, A\text{-}Role, SID, DS)(B, B\text{-}Role)$

$\triangleq N_S(Signal.Runing) \wedge N_S(Signal.Commit) \wedge tr(A) \approx_{trace} tr((A \| B) \backslash C)$.

(b) Non-injective Action Authentication

Additionally, two weak forms of $I\text{-}AAuth$, $N\text{-}AAuth$ and $W\text{-}AAuth$, are defined. The $N\text{-}AAuth$ doesn't require the one-one relationship between the events of A and B.

We rewrite (2)(c) and (2)(d) as (3)(a) and (3)(b), respectively, and get $N\text{-}AAuth$.

(a): $tr(A) = PublicTr(A) \|$

$MSN\text{-}ACCEP \ (A, A\text{-}Role, Signal.Runing, SID, DS)(B, B\text{-}Role, Signal.Commit)$

(b): $tr((A \| B) \backslash C) = PrivateTr((A \| B) \backslash C) \|$

$MSN\text{-}ACCEP \ (A, A\text{-}Role, Signal.Runing, SID, DS)(B, B\text{-}Role, Signal.Commit)$

$$(3)$$

Definition 12. In system S, let $A, B \in Agent$, $A\text{-}Role, B\text{-}Role \in Role$, $Signal.Commit, Signal.Runing \in Event$:

$N - AAuth(A, A\text{-}Role, SID, DS)(B, B\text{-}Role)$

$\triangleq N_S(Signal.Runing) \wedge N_S(Signal.Commit) \wedge tr(A) \approx_{trace} tr((A \| B) \backslash C)$.

(c) Weak Action Authentication

The $W\text{-}AAuth$ doesn't require the one-one relationship between the events of A and B, and even doesn't guarantee the acceptance on data set.

We rewrite (2)(c) and (2)(d) as (4)(a) and (4)(b), respectively, get $W\text{-}AAuth$.

(a): $tr(A) = PublicTr(A) \,\|$

   $MSW\ ACCEP\ (\Lambda, A\text{-}Role, Signal.Runing, SID)(B, B\text{-}Role, Signal.Commit)$

(b): $tr((A \parallel B) \backslash C) = PrivateTr((A \parallel B) \backslash C) \,\|$

   $MSW\text{-}ACCEP\ (A, A\text{-}Role, Signal.Runing, SID)(B, B\text{-}Role, Signal.Commit)$

(4)

Definition 13. In system S, let $A, B \in Agent$, $A\text{-}Role, B\text{-}Role \in Role$, $Signal.Commit, Signal.Runing \in Event$:

   $W\text{-}AAuth(A, A\text{-}Role, SID, DS)(B, B\text{-}Role)$

   $\triangleq N_S(Signal.Runing) \wedge N_S(Signal.Commit) \wedge tr(A) \approx_{trace} tr((A \parallel B) \backslash C)$.

4.4   Claim Authentication

Claim authentication refers to verifying the authenticity of the claims which states what agent has done before.

(a)   Injective Claim Authentication

We give the formal definition of injective claim authentication $I\text{-}CAuth$. The basic idea of $I\text{-}CAuth$ is test whether the system guarantees to an agent A, taking the system role $A\text{-}Role$ and executing the event $A\text{-}Event$ with sequence identifier SID, acceptance with an agent B, taking the system role $B\text{-}Role$ and executing the event $B\text{-}Event$ with exactly sequence identifier SID, on the data set DS.

Definition 14. In system S, let $A, B \in Agent$, $A\text{-}Role, B\text{-}Role \in Role$, $Signal.Commit, Signal.Runing \in Event$:

   $I\text{-}CAuth(A, A\text{-}Role, SID, DS)(B, B\text{-}Role)$

   $\triangleq N_S(Signal.Runing) \wedge N_S(Signal.Commit) \wedge$

   $MSI\text{-}ACCEP\ (A, A\text{-}Role, Signal.Runing, SID, DS)(B, B\text{-}Role, Signal.Commit)$.

(b)   Non-injective Claim Authentication

Additionally, two weak forms of $I\text{-}CAuth$, $N\text{-}CAuth$ and $W\text{-}CAuth$, are defined. The $N\text{-}CAuth$ doesn't require the one-one relationship between the events of A and B.

Definition 15. In system S, let $A, B \in Agent$, $A\text{-}Role, B\text{-}Role \in Role$, $Signal.Commit, Signal.Runing \in Event$:

   $N\text{-}CAuth(A, A\text{-}Role, SID, DS)(B, B\text{-}Role)$

   $\triangleq N_S(Signal.Runing) \wedge N_S(Signal.Commit) \wedge$

   $MSN\text{-}ACCEP\ (A, A\text{-}Role, Signal.Runing, SID, DS)(B, B\text{-}Role, Signal.Commit)$.

(c) Weak Claim Authentication

The W-$CAuth$ doesn't require the one-one relationship between the events of A and B, and even doesn't guarantee the acceptance on data set.

Definition 16. In system S, let $A, B \in Agent$, $A\text{-}Role, B\text{-}Role \in Role$, $Signal.Commit, Signal.Runing \in Event$:

W-$CAuth(A, A\text{-}Role, SID)(B, B\text{-}Role)$

$\triangleq N_s(Signal.Runing) \wedge N_s(Signal.Commit) \wedge$

$\quad MSW\text{-}ACCEP\ (A, A\text{-}Role, Signal.Runing, SID)(B, B\text{-}Role, Signal.Commit)$

4.5   Check the Authentication of System

If we want to check whether a system meets a specification of authentication property, we can test as (5). Where, $\alpha SPEC$ is the alphabet of specification $SPEC$.

$$SYSTEM \ \ meets \ SPEC \triangleq SPEC \subseteq SYSTEM \setminus (\Sigma - \alpha SPEC) \qquad (5)$$

For example, we can test whether a system meets entity authentication as (6).

$$SYSTEM \ \ meets \ EAuth(Alice, Initiator, SID)(Bob, Responder) \qquad (6)$$

5   Conclusions

As some existing definitions of authentication are formalized by different mathematical methods, they are usually hard to compare with each other and are not widely agreed upon. Moreover, they are sometimes difficult to describe diverse security requirements of the IoTs. To address the problem, by introducing the notion of nonleakage, we proposed a general framework of authentication property for the IoTs in this paper. In the framework, we defined three forms of authentication - entity authentication, action authentication and claim authentication- and three strength levels for each form - weak level, non-injective and injective level. We formalized each definition using the process algebra CSP. The framework can easily express different security requirements of the IoTs.

Acknowledgment. This work was supported by the National Natural Science Foundation of China (61100181, 61100186), the National High Technology Research and Development Program of China (863) (2013AA014002) and Beijing Key Laboratory of IOT Information Security (Institute of Information Engineering, CAS).

References

1. Li, S., Xu, L., Zhao, S.: The internet of things: a survey. Information Systems Frontiers, 1–17 (2014)
2. Ahmed, N., Jensen, C.D.: Definition of entity authentication. In: Proc. of 2010 2nd International Workshop on Security and Communication Networks (IWSCN), pp. 1–7 (2010)

3. Focardi, R., Gorrieri, R., Martinelli, F.: A comparison of three authentication properties. Theoretical Computer Science 291(3), 285–327 (2003)
4. Burrows, M., Abadi, M., Needham, R.M.: A Logic of Authentication. Proc. of the Royal Society of London. Series A. Mathematical and Physical Sciences 426(1871), 233–271 (1989)
5. Gollmann, D.: What do we mean by entity authentication? In: Proc. of 1996 IEEE Symposium on Security and Privacy, pp. 46–54 (1996)
6. Focardi, R., Gorrieri, R., Martinelli, F.: Message Authentication through Non Interference. In: Rus, T. (ed.) AMAST 2000. LNCS, vol. 1816, pp. 258–272. Springer, Heidelberg (2000)
7. Lowe, G.: A hierarchy of authentication specifications. In: Proc. of the 10th IEEE Workshop on Computer Security Foundations, Rockport, MA, USA, pp. 31–43 (1997)
8. Kurkowski, M., Srebrny, M.: A Quantifier-free First-order Knowledge Logic of Authentication. Fundamenta Informaticae 72(1), 263–282 (2006)
9. Schneider, S.: Security properties and CSP. In: Proc. of 1996 IEEE Symposium on Security and Privacy, pp. 174–187 (1996)
10. Zhou, J.Y., Gollmann, D.: A fair non-repudiation protocol. In: Proc. of 1996 IEEE Symposium on Security and Privacy, pp. 55–61 (1996)
11. Focardi, R., Gorrieri, R.: An Information Flow Security Property for CCS. In: Proc. of the Second North American Process Algebra Workshop (NAPAW 1993), Cornell, Ithaca, pp. 1–11 (1993)
12. Focardi, R., Gorrieri, R., Martinelli, F.: Classification of security properties - (Part II: Network security). In: Focardi, R., Gorrieri, R. (eds.) FOSAD 2001. LNCS, vol. 2946, pp. 139–185. Springer, Heidelberg (2004)
13. Woo, T.Y.C., Lam, S.S.: A semantic model for authentication protocols. In: Proc. of 1993 IEEE Symposium on Security and Privacy, Oakland, CA, USA, pp. 178–194 (1993)
14. Gollmann, D.: "Authentication by correspondence. IEEE Journal on Selected Areas in Communications 21(1), 88–95 (2003)
15. Ahmed, N., Jensen, C.D.: Demarcation of Security in Authentication Protocols. In: Proc. of 2011 First SysSec Workshop (SysSec), pp. 43–50 (2011)
16. Yunchuan, G., Bingxing, F., Lihua, Y., Yuan, Z.: A Security Model for Confidentiality and Integrity in Mobile Computing. Chinese Journal of Computers 36(7), 1424–1433 (2013)
17. Younis, M., Farrag, O., Althouse, B.: TAM: A Tiered Authentication of Multicast Protocol for Ad-Hoc Networks. IEEE Transactions on Network and Service Management 9(1), 100–113 (2012)
18. Ma, D., Tsudik, G.: Extended Abstract: Forward-Secure Sequential Aggregate Authentication. In: Proc. of 2007 IEEE Symposium on Security and Privacy (SP 2007), pp. 86–91 (2007)
19. Wang, M., Zhu, H., Zhao, Y., Liu, S.: Modeling and Analyzing the (mu)TESLA Protocol Using CSP. In: Proc. of 2011 Fifth International Symposium on Theoretical Aspects of Software Engineering (TASE), pp. 247–250 (2011)
20. Singh, R., Sharma, T.P.: Proof of the Secrecy Property of Secure WLAN Authentication Scheme (SWAS) Using Extended Protocol Composition Logic. Journal of Safety Engineering 2(A), 7–13 (2013)
21. Fábrega, F.J.T., Jonathan, C.H., Joshua, D.G.: Strand spaces: proving security protocols correct. Journal of Computer Security 7(2), 191–230 (1999)
22. Paulson, L.C.: Proving properties of security protocols by induction. In: Proc. of the 10th Computer Security Foundations Workshop, pp. 70–83 (1997)

23. Evans, N., Schneider, S.: Analysing Time Dependent Security Properties in CSP Using PVS. In: Cuppens, F., Deswarte, Y., Gollmann, D., Waidner, M. (eds.) ESORICS 2000. LNCS, vol. 1895, pp. 222–237. Springer, Heidelberg (2000)
24. Abadi, M., Gordon, A.D.: A Calculus for Cryptographic Protocols: The Spi Calculus. Information and Computation 148(1), 1–70 (1999)
25. Mazur, T., Lowe, G.: CSP-based counter abstraction for systems with node identifiers. Science of Computer Programming 81, 3–52 (2014)
26. Dinh, T., Ryan, M.: Verifying Security Property of Peer-to-Peer Systems Using CSP. In: Gritzalis, D., Preneel, B., Theoharidou, M. (eds.) ESORICS 2010. LNCS, vol. 6345, pp. 319–339. Springer, Heidelberg (2010)
27. Roscoe, A.W.: "On the expressiveness of CSP," Technical report Oxford University (2011), http://www.cs.ox.ac.uk/files/1383/expressive.pdf (accessed May, 2014)
28. Roscoe, A.W.: The Theory and Practice of Concurrency. Prentice-Hall, Upper Saddle River (2010)
29. Lowe, G.: Breaking and fixing the Needham-Schroeder Public-Key Protocol using FDR. In: Margaria, T., Steffen, B. (eds.) TACAS 1996. LNCS, vol. 1055, pp. 147–166. Springer, Heidelberg (1996)
30. von Oheimb, D.: Information Flow Control Revisited: Noninfluence = Noninterference + Nonleakage. In: Samarati, P., Ryan, P.Y.A., Gollmann, D., Molva, R. (eds.) ESORICS 2004. LNCS, vol. 3193, pp. 225–243. Springer, Heidelberg (2004)

A Security Routing Mechanism against Sybil Attacks in Mobile Social Networks

Yan Sun[1,2] and Lihua Yin[1,3]

[1] Institute of Information Engineering, Chinese Academy of Sciences, Beijing, China
[2] Beijing University of Posts and Telecommunications, Beijing, China
[3] Beijing Key Laboratory of IOT Information Security, Beijing, China
{sunyan,yinlihua}@nelmail.iie.ac.cn

Abstract. Mobile social networks (MSNs) are a kind of delay tolerant network that consists of lots of mobile nodes with social characteristics. Recently, many social-aware algorithms have been proposed to address routing problems in MSNs. However, it also brings more security and privacy concerns. In this paper, we discuss a specific type of Sybil attack in MSNs, which few researches has been focused on. We proposed a security mechanism to detect Sybil nodes and eliminate them to ensure the routing security while routing forwarding. It needs to cooperate by two sides, which is, respectively measuring distance in client side and eliminating Sybil nodes in server side. We also demonstrate the solution is correct and analyze its energy costs.

Keywords: Sybil Attacks, Mobile Social Networks, Routing Security.

1 Introduction

In Mobile Social Networks (MSNs), the mobile devices carrying by users are abstracted as mobile nodes with social characteristics. The topological connections between nodes not only represent both the physical device links, but also describe their social relationships. Therefore, the node in MSNs reflects the social ties of device holders while moving. Recently, many social-aware algorithms have been proposed to address routing problems in MSNs [1, 2, 3]. These algorithms design efficient routing strategies by making accurate analysis of social network properties. These social-aware algorithms have improved routing performance. But at the same time, these social properties in routing also bring a variety of security and privacy problem.

Sybil attacks are to disrupt routing protocols by forged identities or location, especially in the multicast routing and geographical routing. It also occurs in MSNs since MSNs usually use multicast routing and geographical routing. In a mobile social network, a user wants to forward its message to an objective user by its neighbor friend nodes. If there is a malicious node as its neighbor, its object is to cheat its neighbor node for routing selection by creating virtual nodes that are called Sybil nodes. A Sybil node has a forged identity and location and also reports its virtual location information to servers looking like a normal node. It is easy for the malicious node to

W. Han et al. (Eds.): APWeb 2014 Workshops, LNCS 8710, pp. 325–332, 2014.

forge reasonably virtual locations to disrupt routing if the malicious node knows the location information of its neighbors.

Based on the attack analysis for Sybil attacks in mobile social networks, our challenge is how to detect these virtual nodes and eliminate them to ensure the routing security while routing forwarding. In this paper, we proposed a security mechanism to detect these Sybil nodes and eliminate them while selecting routing forwarding nodes. It needs to cooperate by two sides, which is, respectively measuring distance in client side and eliminating Sybil nodes in server side. We also demonstrate the solution is correct and analyze its energy costs.

2   Related Work

With the development of Internet, the online social network (Online Social Network, OSN) [1] has a great success. In recent years, mobile terminal technology and wireless access technology are widely used. It makes it possible to enjoy social network service in the process of moving whenever and wherever, which is called mobile social networks (Mobile Social Networks MSNs) [2, 3]. MSNs are a type of important application in the social network services, for example, intelligent transportation and so on. Mobile social networking [2] is social networking where individuals with similar interests converse and connect with one another through their mobile phone and/or tablet. It combines the social network with mobile communication network, having both mobility and sociality. MSNs are a special type of Delay Tolerant Networks (Delay Tolerant Network, DTN), in which intelligent terminals are networked based on self-organized network, and mobile terminals do not need the third party server. It exchanges and shares information with Bluetooth, Wi-Fi and other wireless access technology, when users opportunistic contacts in their respective communication scopes.

In MSNs, routing is the basis of networking and information transmission. Its routing strategy is Store-Carry-and-Forward (SCF). At present, there are some routing algorithms using nodes' social attributes (such as social roles, personal preferences, and social purposes and so on) to design routing algorithms, so as to improve the routing performance. For an instance, the time critical content delivery algorithm [4], message duplication reduction algorithm [5], friendship based routing algorithm [6, 7] and WSW algorithm [8], etc. However, it also provides more available resources for attackers to launch an attack.

Sybil attack [10] is a malicious attack that malicious devices or nodes occur with multiple identities illegally. The malicious devices or nodes are usually called Sybil devices or nodes. Sybil attack was originally developed by Douceur in P2P. Then, Karlof[11] and Newsome[12] pointed out that Sybil attacks also occur in sensor networks to disrupt its routing mechanism. Newsome [12] give a scientific classification on Sybil attack and analyzed its threats on many functions in sensor networks, for example, routing, resource allocation and illegal behavior detection etc. In a word, Sybil attacks threaten multicast routing and geographical routing. For mobile social networks, it uses store-carry-forward routing mechanism. It always employs multicast routing and geographical routing mechanism. Therefore, there are the same security threats in mobile social networks.

3   Sybil Attack Model

In the mobile social network scenario as shown in Figure 1, Alice want to send a message to the destination node Eva. Alice broadcast a forwarding request message to its neighbor nodes within its communication range. When a neighbor node receives the forwarding request message, it checks the social relationship with the destination node. If the relationship list of the neighbor node contains the destination node, the neighbor node sends a response to Alice. On the contrary, the neighbor node does not give any response if the relationship list of the neighbor node does not contain the destination node. If one of the neighbor nodes is a malicious node, it forages multiple social identities with position information and relationship information. It is easy to make it by long-term collection of encounter frequency, meeting time information and other information between nodes in mobile process. These virtual nodes are called Sybil nodes and want to cheat Alice when selecting routing relaying nodes. A Sybil node has a forged identity and location and also reports its virtual location information to servers looking like a normal node. It is easy for the malicious node to forge reasonably virtual locations to disrupt routing if the malicious node knows the location information of its neighbors.

Based on the attack analysis for Sybil attacks in mobile social networks, our challenge is how to detect these virtual nodes and eliminate them to ensure the routing security while routing forwarding.

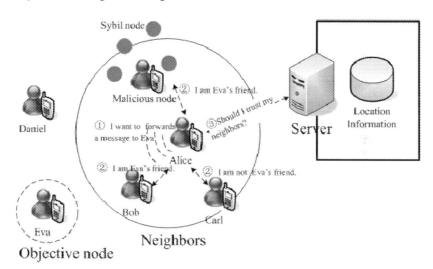

Fig. 1. A scene for Sybil attacks in mobile social networks

4   Proposed Defense Mechanism

4.1   Problem Statements

In mobile social networks, the node report location information to the location servers periodically. The location information is shared via location servers to determine the next routing relaying node. Therefore, the location sharing or query creates a condition for Sybil nodes to forge reasonable location information.

If the position information of Sybil node cannot be forged reasonably according to the existing information, then we can easily detect the Sybil node using inconsistency of location information. Therefore, our problem is transferred to detect the unreasonable location information. To solve the problem, this requires two premises.

Premise 1. The forged position information of Sybil nodes is not reasonable.

The attack goal of a malicious node is to cheat its neighbor node by creating virtual nodes that are called Sybil nodes. A Sybil node has a forged identity and location and reports its virtual location information to servers. It is easy for the malicious node to forge reasonable virtual locations if the malicious node knows the location information of its neighbors. To solve it, we must answer the issue that is how the malicious node could not forge a reasonable virtual location. We give the theorem 1.

Theorem 1. A malicious node cannot forge a Sybil node that has a reasonable location if it does not know the locations information of its neighbor nodes, but knows the distance information to its neighbor nodes.

Proof: we assume that the location of a malicious node is O, and it obtains locations of its three neighbor nodes, A, B, C, satisfying $OA>OB>OC$. Then, it can infer that the three neighbor nodes are in the concentric circles with the center O. The malicious node has to report these distances to A, B, C to server to forge a virtual location O'. It is difficult for it to generate three distance values satisfying a certain constraint relationship if it doesn't know the precious location. When there is a certain density of nodes in concentric circles, the malicious node can generate several groups of distance values, some of which may be inconsistent to the constraint condition. It will be confused to determine choosing which group of distance values. Therefore, it is almost impossible to forge a reasonable virtual location.

Premise 2. Find the conditions of unreasonable position information. According to the irrationality, we detect Sybil nodes and filter them.

Thus, our problem is changed to how to eliminate the virtual locations of Sybil nodes with a high probability by the inconsistent property. To solve it, we define these location coordinates in a two-dimensional coordinate system. In the system, the real Euclidean distance between nodes exists. Algorithm 1 is proposed in the server side. It use the distance set between neighbor nodes as input. The output is a set of candidate routing nodes. If there are Sybil nodes taking part in the computation, there will be a higher dimensional location to be output. The faith is that there is no such location in a two-dimensional plane, so that it could be inferred that the node is a forged node, Sybil node. Therefore, it must be removed.

4.2   Our Proposed Solution

Based on the two premises above, in order to prevent malicious nodes spoofing routing, we propose a security routing mechanism against Sybil attacks based on mobile client- server:

- In the mobile client side, nodes report their position information periodically to location servers. The location information contains two parts. One is its position and the other is distance information measured to their neighbor nodes by RSSI. The shared information is not the location information, but the distance information to their neighbor nodes.
- In the location server side, location servers not only collect the position information of nodes, but also distance information among neighboring nodes. The location servers detect whether the nodes' locations are reasonable according to the actual position and the computed distance by distance information between neighbor nodes. If the result is inconsistency, then the node is a Sybil node, which is filtered in routing relaying selection.

Therefore, our solution is divided into two parts: distance measurement in client side and Sybil nodes detection and filtering in server-side. There are some notations are used in the following, and we give their illustration in Table 1.

Table 1. Notations and its description

Notations	Description
v	A source node
$Nbr(v)$	The neighbor node set of a node, v
N	The total of neighbor nodes
P_k	The location of a node, k, in the coordinate system
d_{ij}	The measured distance between two nodes, i, j
C_{ij}	The calculated distance between two nodes, i, j
S	The set of measured ditances

Distance Measurement

In client side, distance measurement is computing the distance set between initiator node v and its neighbor nodes using encrypted communication to ensure the process security. A node v measures the distance to its neighbor node when each encounter period comes. We adopt distance measurement based on RSSI. Its specific process is as follows:

Setp1: A node v launches a distance measurement. It sends a full power radio signal to its neighbor nodes $u \in Nbr(v)$.

Step 2: Node v sends a distance measurement packet to its neighbor node u with a random RSSI strength p_{tx}.

Step 3: The node u receive records the received power value p_{uv}.

Step 4: The node u send the power value p_{uv} to the node v.

Step 5: The node v computes the distance to its neighbor node d_{ij} using free-space wave propagation and Gaussian fitting formula.

Sybil Node Detection and Filtering
In the server side, we design an algorithm to detect and filter Sybil node. The initiator node v selects randomly two neighbor nodes, i, j, to define a plane, so that they establish a local coordinate system L.   We define a set $G(V, E)$, where V is the set consisting of initiator node v and its neighbor nodes, E is the set of distances between any two nodes in V. Initially, G is null. In the local coordinate system L, the relative positions of nodes v, i, j, k, are unique. We employ the measured distance d_{kv}, d_{ki}, d_{kj}. The location of a neighbor node k is determined with nodes v, i, j, using trilateration. The specific process is as follows:

Step 1: A node v randomly selects two neighbor nodes i and j to determine a plane. The local coordinate system L is established by their three edges.

Step 2: initialize the undirected graph $G(V, E)$, where V is a set of node v and its neighbor nodes, and E is a set of distance measurement between two nodes.

Step 3: Compute the position of node k, p_k, for each neighbor node $k \in Nbr(v)$. It is determined with nodes v, i, j, using trilateration.

Step 4: For two nodes, $i, j \in V$, find their measured distance d_{ij}. Using both of location, compute the calculated distance $c_{ij} = |p_i - p_j|$.

Step 5: Judge if the two distances is consistent: if $|d_{ij} - c_{ij}| \leq \varepsilon$, then the edge $e(i, j)$ is added to E.

Step 6: $G(V, E)$ is a subset C in which the distances of nodes are consistent. For each node in V, step 1 to 5 is repeated. Then, find the edges with largest number of connection in E, which are defined the maximum distance consistent subset C.

The maximum distance consistent subset C is a route candidate result. We give the algorithm as Algorithm 1.

Algorithm 1. *Eliminate_Sybil_Node(v)*

Input: The initiator node v which is waiting for selecting honest routing forward nodes ;
Output: the set of honest routing forward nodes neighboring v.

1: Define Plane P: $(v, i, j) \rightarrow P$, where i, j are randomly selected and satisfy $i \in Nbr(v), j \in Nbr(v)$ and $i \neq j$;

2: Define local coordinate system L:$(e(v,i), e(v,j), e(i,j)) \rightarrow L$;

3: **Initialize** $G(V, E)$:
$$V = v \cup Nbr(v)$$
$$E = \Phi;$$

4: **For** each neighbor node $k \in Nbr(v)$ **do**
p_k= Trilateration(d_{kv}, d_{ki}, d_{kj});
// The k's location p_k in L is computed using trilateration with the measured distance to nodes, d_{kv}, d_{ki}, d_{kj}.

   End For

5: **For** any two nodes $i, j \in V$ **do**
$c_{ij} = |p_i - p_j|$; // find the measured distance between them, d_{ij}, and obtain the computed distance, c_{ij}.

6:        **If** $|d_{ij} - c_{ij}| \leq \varepsilon$ **Then**
$E = E \cup e(i, j)$; // The edge $e(i, j)$ could be added to E satisfying $|d_{ij} - c_{ij}| \leq \varepsilon$

$c_{ij} = c_{ij} + 1$;

End For

$C = \{ e(i, j) \mid v \in e(i, j) \text{ and } c_{ij} \text{ is the maximal in the edge containing the node } v.\}$

7: **Return** C;

End

---

5   Algorithm Analysis

5.1   Algorithm Proof

Theorem 2. If the randomly selected vertices are real nodes, then there is no Sybil node in the produced set of routing forward nodes using Algorithm 1.

Proof: We assume the node v's location is p_1. It selects two neighbor node p_2, p_3 to define a plane P. They are all real nodes. After measuring the distance, E contains the edges between honest neighbor nodes. On the other hand, the forged locations of Sybil nodes have not the property of consistency in P. Such locations will be in another plane. Therefore the edge between honest node and Sybil node will not to be created, and finally Sybil node is removed.

Theorem 3. If there is at least one Sybil node in the randomly selected vertices, then the amount of the produced set of routing forward nodes is less than the one of the produced set by all real nodes.

Proof: We assume the node v's location is p_1. It selects two neighbor node p_2, p_3 to define a plane P', in which p_2 is a Sybil node. Some honest nodes will be removed because there is no consistent information between honest nodes and p_2. Therefore, the amount of the produced set of routing forward nodes in P' is less than the one of the produced set by all real nodes in P.

5.2   Algorithm Cost

For mobile terminals, they have limited energy and computation resource. In our proposed solution, the energy cost is mainly the communication cost while measuring distance in client side. We assume that the initiator node v has n-1neighbour nodes, and then it will need to collect n-1 distance information to these neighbors. One node needs at least 2n+1 packets switching. There are n nodes, so the whole process needs (2n+1)n packets switching.

The computation cost is produced to remove Sybil nodes in server side. For the algorithm 1, it is consumed by computing trilateration and comparing the measured distance with calculated distance.

6   Conclusions

In this paper, we discuss a specific type of Sybil attack in MSNs. For resisting this, we proposed a security mechanism to detect these Sybil nodes and eliminate them while routing forwarding to ensure the routing security. It needs to cooperate by the mobile terminals and servers, that is, distance measurement in client side and Sybil node elimination in server side. We also demonstrate the solution is correct and analysis its energy cost. It not only takes advantages of the social-based routing performance, but also resists Sybil attacks.

Acknowledgment. This work was supported by the National Natural Science Foundation of China (61100181), the National High Technology Research and Development Program of China (863) (2013AA014002) and Beijing Key Laboratory of IOT Information Security(Institute of Information Engineering, CAS).

References

1. Online social network,
   `http://en.wikipedia.org/wiki/Social_networking_service`
2. Mobile social network,
   `http://en.wikipedia.org/wiki/Mobile_social_network`
3. Vastardis, N., Yang, K.: Mobile Social Networks: Architectures, Social Properties, and Key Research Challenges. IEEE Communications Surveys & Tutorials 15(3), 1255–1371 (2012)
4. Nazir, F., Ma, J., Seneviratne, A.: Time Critical Content Delivery Using Predictable Patterns in Mobile Social Networks. In: Proc. of International Conference on Computational Science and Engineering, pp. 1066–1073 (2009)
5. Kawarabayashi, K., Nazir, F., Prendinger, H.: Message duplication reduction in dense mobile social networks. In: Proc of the 19th International Conference on Computer Communications and Networks, pp. 1–6 (2010)
6. Bulut, E., Szymanski, B.K.: Friendship based routing in delay tolerant mobile social networks. In: Proc. of IEEE Conference on Communications Society, pp. 1–5 (2010)
7. Bulut, E., Szymanski, B.K.: Exploiting Friendship relations for efficient routing in mobile social networks. IEEE Trans. on Parallel and Distributed Systems 23(12), 2254–2265 (2012)
8. Khosravi, A.: Pan Jian-ping. Exploring personal interest in intermittently connected wireless mobile social networks. In: Proc. of IEEE Consumer Communications and Networking Conference, pp. 3–508 (2011)
9. Kayastha, N., Niyato, D.: Applications, architectures, and protocol design issues for mobile social networks: a survery. Proceedings of the IEEE 99(12), 2130–2158 (2011)
10. Douceur, J.R.: The Sybil attack. In: Druschel, P., Kaashoek, M.F., Rowstron, A. (eds.) IPTPS 2002. LNCS, vol. 2429, pp. 251–260. Springer, Heidelberg (2002)
11. Karlof, C., Wagner, D.: Secure routing in wireless sensor networks: attacks and countermeasures. Ad Hoc Networks 1(2-3), 293–315 (2003)
12. Newsome, J., Shi, E., Song, D., Perrig, A.: The Sybil Attack in Sensor Networks: Analysis & Defenses. In: IPSN 2004, April 26-27, pp. 259–268. ACM Press, Berkeley (2004)

Attribute-Role-Based Hybrid Access Control in the Internet of Things

Kaiwen Sun[1,2] and Lihua Yin[2,*]

[1] Beijing University of Posts and Telecommunications, Beijing 100876, China
[2] Institute of Information Engineering, CAS, Beijing 100093, China
sunkaiwen@nelmail.iie.ac.cn

Abstract. Internet of Things has been penetrating into many aspects of human lives as the Informationization develops rapidly in the world. And yet traditional access control models, such as RBAC, have some shortage on the environment of large-scale dynamic users due to the real time and dynamic characteristics of Internet of Things, resulting in various problems especially on the disclosure of private information. We propose an access control model based on attribute and role to solve the scenarios of large scale dynamics users. The model put forward a policy language of attribute rules and a method to solve the policy conflict and redundancy. We also illustrate the feasibility of the model with an example of Wechat. The results indicate our model could simplify the complexity of traditional ABAC in the aspect of permissions assignment and policy management.

Keywords: Internet of Things, Access Control, RBAC, ABAC.

1 Introduction

As an advanced information technology, the Internet of things (IOT) has become an indispensable part of human lives. Therefore, the development of IOT has great significance on promoting the development of economics and technology.

As an extension of the Internet, IOT identifies the real world through various sensors and RFIDs etc., then transfers information over the network, which enables the communication between people and objects, and even between objects and objects. However, with the rapid development of IOT, more and more security issues appears, such as the disclosure of taxis information in taxi systems and the disclosure of location privacy in Wechat. The access control models which are widely used in the Internet cannot be used in IOT due to the characteristics of the real time and dynamic. Therefore a fine-grained access control model needs to be researched to assign permissions automatically.

* This work was supported by the National Natural Science Fundation of China (61100181), the National High Technology Research and Development Program of China (863) (2013AA014002) and Beijing Key Laboratory of IOT Information Security(Institute of Information Engineering, CAS).

W. Han et al. (Eds.): APWeb 2014 Workshops, LNCS 8710, pp. 333–343, 2014.

Currently, access control is mainly used in the Internet[1]. But in IOT, the users, time and location are changing with the constant changes of automotive systems, mobile terminals and other equipments. It is so dynamic that we cant assign permissions in advance and numerous users also increase the workload of assignment in advance. Therefore, role-based access control (RBAC) etc. widely used in the Internet cant be applied to IOT directly. Although the attribute-based access control (ABAC) can solve the dynamic problem, but the process of authorization is complex, inflexible and thus not applicable to real-time scenarios. Meanwhile, the number of rules will rapidly increase with the number of users and attributes. Therefore, a new access control model – attribute-based and role-based hybrid access control (ARBHAC) is proposed in this paper. The model has the advantage of both ABAC and RBAC, which solves the problem that RBAC cannot satisfy large-scale dynamic users, and simplifies the complexity of traditional ABAC in the aspect of permissions assignment and policy management.

The rest of the paper is organized as follows. Section 2 describes the relevant research of access control. Section 3 proposes the ARBHAC scheme in detail. Section 4 analyzes the scheme and illustrates some instances. Section 5 is the conclusions.

2   Related Work

2.1   The Role-Based Access Control (RBAC)

The RBAC determines roles associated with permissions in advance, the user is related to certain role thus obtains the corresponding permissions. In the access, the users permission depends on roles related to the user and permissions associated to the role.

As a fundamental research, RBAC has been continuously developed, which includes RBAC96[2], ARBAC97 (Administrative RBAC97)[3], ARBAC99[4], ARBAC02[5] and NIST RBAC (National Institute of Standards and Technology RBAC)[6] and a series of more sophisticated RBAC models. These RBAC models overcame the potential safety issues of the discretionary access control (DAC) and the limitations of the mandatory access control (MAC). However, these access control models fail to consider the impacts of time and location. Bertino et al.[7] proposed the GEO-RBAC model, which allows the user to activate the corresponding role in a specific location based on the location of the user. However, this model fails to consider the impacts of the spatial limitation of the role hierarchy, separation of duties and user-role assignment. To address them, Chen and Crampton[8] develop the graph-based representation for the spatio-temporal RBAC. The temporal and spatial domains are separated, such that the spatio-temporal constraints is extremely complex. Later, Bertino et al.[9] proposed a generalized spatio-temporal role-based access control (GSTRBAC) model, which abstract location and time into one entity, thus reduces the number of entities to manage and also prevents the creation of new roles or permissions when spatio-temporal constraints associated with them change. This model decreases the spatio-temporal impact on prerequisite constraints.

The RBAC solves the problem of permission issues while the time and location change. However, the scenario that a large number of user's access is still unresolved. Therefore the RBAC could combine with other access control models to resolve the dynamic characteristics of the users.

2.2   The Attribute-Based Access Control (ABAC)

In ABAC, all entities are described using the word 'attribute'. Although the attribute is a variable, policy is relatively fixed. This makes ABAC can be applied to the scenario where users are dynamic change. In order to achieve fine-grained and scalable access control, ABAC describes complex authorization process and constraints through defining the relationship between attributes of time, location, behavior, resource types and other information associating with access.

The policy of access control is the core of ABAC. Policies are generated by the Boolean expressions of attributes. The initial ABAC policies are mostly based on the X.509 Attribute Certificate. But the authentication method of the attribute certificate is not flexible enough. It even does not work in the environment where attributes change constantly. So Al-Kahtani and Sandhu[10] introduced attribute to the process of user-role assignment to solve the problem of large scale dynamic users, which can overcame the limitation of RBAC. Zhu et al.[11] proposed a general attribute based Role-based Access Control (GAR-BAC). This model distinguishes users based on different attributes expressions, then the users is associated with certain roles. In this model, the number of GARBAC rules increases with the number of users and attributes linearly, thus outperforms the performance of ABAC of Yuan and Tong[12]. However, they failed to consider the impact of policy conflict and redundancy.

Although ABAC can be applied in the scenarios of scale dynamic users, the complexity of the policy and authorization process, and the expansion of the rules make it difficult to be used. Therefore combining ABAC with RBAC may have a better performance.

3   Architecture of Model

3.1   ARBHAC Model

The following model is based on the traditional RBAC and ABAC, which is shown in Fig 1.

In the process of attributes-based assigning roles, the policy enforcement point (PEP) receives natural access request (NAR). The NAR is the most primitive information, which includes all kinds of users attributes, such as age, location, identity etc., and the way to access the resource. Then PEP extracts the required attribute values and formats it as one unified form. The attribute-based access request (AAR) is constructed according to these attribute values extracted. AAR

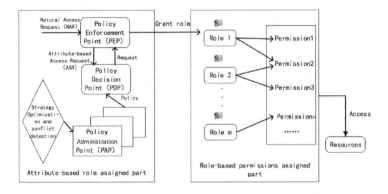

Fig. 1. The overall structure of the model

can be represented by a four-tuple (s, o, a, e), where s represents the subject attribute, o represents the object attribute, a represents the behavior attribute, e represents the environmental attribute. PEP transfers AAR to the policy decision point (PDP). The PDP makes judgment of AAR according to the policy obtained from the policy administration point (PAP) and transfers the judgment result to PEP which performs the access decision. This PAP stores the default access policies, which need to be optimized and conflicts detected when the policies are updated. Role-based permissions assignment is relatively simple. Define multiple roles based on real needs, and put appropriate permissions corresponding to these roles. Therefore, if a user assigning to a certain role means obtaining the corresponding permissions.

3.2   Correspondence between Attribute Expressions and Roles

Attribute expressions and roles are one-to-one or many-to-one relationships in this model. If there have been a one-to-many relationship, eliminate it by the following method.

There are three kinds of relationships among the permissions corresponding to roles—conflict, containment and irrelevance, so:

- The conflict relationship can be solved by designing the policy to deny access;
- The containment relationship can be solved by setting senior roles according to the priority;
- The irrelevance relationship can be solved by combining two roles into one and then assign it.

Now even if multiple attributes of a user conform to a number of rules, only one role can be obtained. This makes conflict detection from dynamic to static, reduces the complexity of policies greatly, enhances the readability of the policies as well as makes it convenient for modifying roles and permissions, thus optimizes the model.

Example 1. Network Studio formulates:

$age > 11 \rightarrow Juvenile$;

$age > 18 \rightarrow Adult$ *and* $Juvenile(Juvenile$ *and* $Adult$ *are roles*);

Then a 20-year-old man clearly conforms to the second rule, therefore two roles are obtained. But the two roles have a containment relationship obviously Juvenile contains Adult. The policy is modified and priority is defined:

$age > 11 \rightarrow Juvenile$;

$age > 18 \rightarrow Adult(Adult$ *is senior than Juvenile*);

Then only one role—Adult is obtained by the 20-year-old user.

3.3   Policy Optimization and Conflicts Detection

Policy Grammar

Context-free grammars are used to define our policy language[13]. This language is simple to understand and easy to extend, as shown in Fig 2.

```
1) The terminal symbols: {and, or, xor, not, <, =, >, ≤,≠, ≥, in,
   "subjected to" ,  "revoked if not" , 0, 1, 2, 3, 4, 5, 6, 7, 8, 9}
2) The non-terminal symbols: {Attribute_Expression, Attribute_Pair,
Relation_Operator, Operator.Attribute, Roles, Constraints, Constraint, Num, Digit,
Conditions, Set, Range, Role, Attribute_Value}
3) The start symbols: Rule
4) The production rules:
   Rule :=Attribute_Expression subjected to[Constraints]revoked if not[Conditions]
       ->Roles;
 Attribute_Expression:=Expression;
 Conditions := Expression;
 Expression := Attribute_Pair | Expression Operator Expression | (Expression Operator
       Expression);
 Attribute_Pair := Attribute Relation_Operator Attribute_Value | Attribute [NOT] IN Set
       | Attribute [NOT] IN Range;
 Roles := [not] Roles | Roles Operator Roles | (Roles Operator Roles) | Role;
 Constraints := Constraint | Constraint Operator Constraint | (Constraint Operator
       Constraint)
 Operator := and | or | xor;
 Relation_Operator := < | = | > | ≤ | ≠ | ≥;
 Attribute := {specified by organization};
 Attribute_Value := {specified by organization};
 Set := {specified by organization};
 Range := (Num...Num);
 Num := Num Digit;
 Digit := 0 | 1 | 2 | 3 | 4 | 5 | 6 | 7 | 8 | 9;
 Role := {specified by organization};
 Constraints := {specified by organization};
```

Fig. 2. Policy Language

The Specific Method of Policy Optimization and Conflict Resolution

As defined in RFC3198 by IETF, policy conflict refers to two or more policies are satisfied simultaneously, but the actions cannot be performed simultaneously. Therefore, the entity cannot determine which action should be performed.

Example 2 Define:

Rule1: $sex = man \rightarrow playing\ football$;
Rule2: $sex = woman \rightarrow playing\ badminton$;
Rule3: $age > 18 \rightarrow swimming$;
Obviously, if the priority is not defined, Rule3 conflicts with the other two; if the priority is defined (Rule1 and Rule2 have higher priority than Rule3), Rule3 will be redundant. Therefore, it is indispensable to optimize the policy and avoid conflicts.

Definition 1

1. *R represents an attribute expression, and S represents the set of attribute keywords;*
2. *$R[i]$ represents the attribute value (domain) of attribute keyword i;*
3. *$R_1 = R_2$ represents $R_1[i] = R_2[i]$ for $\forall i \in S$;*
4. *$R_1 \subset R_2$ represents $R_1[i] \subset R_2[i]$ or $R_1[i] = R_2[i]$ for $\forall i \in S$ and $\exists i \in S, R_1[i] \subset R_2[i]$;*
5. *$R_1 \infty R_2$ represents $R_1[i] \cap R_2[i] \neq \emptyset$ for $\forall i \in S$, but the condition $R_1[i] \subset R_2[i]$ or $R_1[i] = R_2[i]$ and the condition $R_2[i] \subset R_1[i]$ or $R_1[i] = R_2[i]$ cannot be true simultaneously;*
6. *$R_1 \not\infty R_2$ represents that $\exists i \in S$ makes $R_1[i] \cap R_2[i] = \emptyset$.*

According to the definition[14] above, we can divide the policy optimization and conflict problems into four categories.

(1) Exception Conflict
   Case a: If $R_1 \cap R_2$ or $R_1 = R_2, R_1 \rightarrow Role_1, R_2 \rightarrow Role_2$, and $Role_1 \neq Role_2$, the exception conflict is generated.

   The so-called exception conflict is that $Rule_2$ includes $Rule_1$ (including the case of $R_1 = R_2$), then when a user accesses resources, the attribute values provided by the user must also conform to R_1 if they conform to R_2, because R_1 is an exception of R_2. Therefore it is defined as exception conflict.

Example 3 An Entertainment Store is currently open to the European Union, the USA, Japan and China (without regard to other countries due to religious issues). Since China does not allow young children use the Internet alone, the store makes the following rules:

if:
Rule1: $age \geq 18$ *and country in* $\{EU,\ America,\ Japan, China\} \rightarrow Adult$;
Rule2: $age \geq 16$ *and country in* $\{EU,\ America,\ Japan,\ China\} \rightarrow Adolescent$;
   Obviously, "$age \geq 18$ *and country in* $\{EU,\ America,\ Japan, China\}$" is an exception of "$age \geq 16$ *and country in* $\{EU,\ America,\ Japan,\ China\}$".

(2) Policy Redundancy
   Case b: If $R_1 \subset R_2$ or $R_1 = R_2, R_1 \rightarrow Role_1, R_2 \rightarrow Role_2$, and $Role_1 = Role_2$, the policy redundancy is generated.

Table 1. The role assignment of the entertainment store

Age	Country	Role	Authority(level of movies)
≥ 3	the European Union, the USA, Japan	Child	1 level
≥ 11	the European Union, the USA, Japan	Juvenile	1,2 level
≥ 16	the European Union, the USA, Japan, China	Adolescent	1,2,3 level
≥ 18	the European Union, the USA, Japan, China	Adult	1,2,3,4 level

The so-called redundancy is that when the rules are satisfied, we have the same effect performing any one. Above example, if:

Rule1: *age $\geq$ 16 and country in {EU, America, Japan} $\rightarrow$ Adolescent*;
Rule2: *age $\geq$ 16 and country in {EU, America, Japan, China} $\rightarrow$ Adolescent*;

Obviously, $Rule_1$ is useless. Because "*age $\geq$ 16 and country in {EU, America, Japan}*" $\subset$ "*age $\geq$ 16and country in {EU, America, Japan, China}*". Deleting Rule1 has no effect. As for the special case of $R_1 = R_2$, they are the exactly the same rules.

(3) Policy Hidden

This situation is generated under the premise of the predetermined priority.
Case c: If $R_1 \subset R_2$ or $R_1 = R_2, R_1 \rightarrow Role_1, R_2 \rightarrow Role_2, Role_1 \neq Role_2$, and the priority of R_2 is higher than R_1, policy hidden is generated.

In this case, the attribute values must conform to $Rule_1$ if they conform to $Rule_2$ and $Rule_2$ (having higher priority) executes first, so $Rule_1$ can never be executed like being hidden. Above example, if:

Rule1: *age $\geq$ 18 and country in {EU, America, Japan, China} $\rightarrow$ Adult*;
Rule2: *age $\geq$ 16 and country in {EU, America, Japan, China} $\rightarrow$ Adolescent*; (Rule2 has higher priority)

Now for a 16-year-old boy, the role—Adolescent is obtained because $Rule_2$ is executed first.

(4) Association Conflict

Case d: If $R_1 \infty R_2, R_1 \rightarrow Role_1, R_2 \rightarrow Role_2$, and $Role_1 \neq Role_2$, the association conflict is generated.

This situation is generated when the users range defined by R_1 and R_2 has intersection, but due to the different actions in two rules, the users in the intersection don't know how to perform actions. Above example, if:

Rule1: *age $\geq$ 16 and country in {EU, America, Japan, China} $\rightarrow$ Adolescent*;
Rule2: *age $\geq$ 11 and country in {EU, America, Japan} $\rightarrow$ Juvenile*;

Now users in the intersection "*age $\geq$ 11 and country in {EU, America, Japan}*" do not know how to perform actions.

According to the four categories above, the following theorem is obtained:

Theorem 1. *Any optimization (only for redundancy) and conflict of policy must conform to one of the four cases a-d above.*

Proof. for $\forall R_1, R2, R_1 \rightarrow Role_1, R_2 \rightarrow Role_2$, R_1 and R_2 conforms to $R_1 \not\propto R_2$ or one of the following four relationships:$R_1 = R_2, R_1 \subset R_2, R_2 \subset R_1$ and $R_1 \infty R_2$.

1. If R_1 and R_2 conforms to $R_1 \not\propto R_2$, then $\exists i \in S$ makes $R_1[i] \cap R_2[i] = \emptyset$. We can divide all users into two disjoint parts according to i. So conflict or redundancy does not exist in this case.
2. If R_1 and R_2 conforms to one of the following four relationships: 1)$R_1 = R_2$; 2)$R_1 \subset R_2$; 3)$R_2 \subset R_1$; 4)$R_1 \infty R_2$, then:
   - If there doesn't exist priority:
     for 1) 2) 3), exception conflict will be generated if $Role_1 \neq Role_2$. Define priority can solve it;
   - If there exists priority:
     - for 1), policy hidden will be generated if $Role_1 \neq Role_2$, and policy redundancy will be generated if $Role_1 = Role_2$. Delete one rule can solve it;
     - for 2) 3), policy hidden will be generated if $Role_1 \neq Role_2$ and policy redundancy will be generated if $Role_1 = Role_2$. For the first case, the priority can be changed to solve it; For the second, we can remove the rule which is included by the other;
   - For 4)
     if $Role_1 \neq Role_2$, association conflict must be generated whether the priority is defined. There are two solutions. One is to define the priority of the role, and the priority of the policy is determined by the priority of its assigned role. However, usually we cannot determine the priority of the role, therefore a new rule $R_3[i] = R_1[i] \cap R_2[i]$ is generated for $\forall i \in S$, and then exclude R_3 from R_1 or R_2 optionally. We assume that R_1 excludes R_3, then R_1 and R_2, R_3 are irrelevant, so there will be no redundancy or conflict, and we just need to execute (1), (2), (4) for R_2 and R_3.

Therefore, the cases a-d contain all the possible conditions of redundancy and conflict.

4   Model Analysis

4.1   Model Analysis

ABAC is applied in order to solve the problem of large-scale dynamic users. Because all the existing access control models are for known-user access except ABAC, and only ABAC can realize the unknown-user access. The ARBHAC model have a great advantage than traditional ABAC on managing permissions. The relevant argumentation are given on modifying permissions, adding permissions and deleting permissions in the following.

Proof. For traditional ABAC model, assume that:

$R_1 \rightarrow authority_1,\ authority_2,\ ......,\ authority_i,\ ......,\ authority_n;$

$R_2 \rightarrow ......,\ ......,\ authority_i,\ ......,\ ......;$

......;

$R_m \rightarrow ......,\ ......,\ authority_i,\ ......,\ ......;$

......;

$R_k \rightarrow ......,\ ......,\ authority_j,\ ......,\ ......;$

- Modifying permissions

  If $authority_i$ need to be modified we have to traverse all the k rules to find the m rules containing $authority_i$ and then modify it. But if:

  $R_1 \rightarrow Role_1;\ R_2 \rightarrow R_2;\ ......;\ R_m \rightarrow Role_m;\ ......;\ R_k \rightarrow Role_k;$

  The number of roles is normally far less than the number of rules, caused by different rules performing the same operations. Therefore, the number of roles we need to traverse and modify is far less than the number of rules. The worst case is that each rule performs each operations, resulting in each rule corresponds to one role. Then the number of roles we need to traverse and modify is the same as the rules.

- Adding permissions

  When we need to add permissions to all the rules which conform to a certain expression, we have to traverse all the k rules to find the l rules which should add permissions to. In this case, even if the role is introduced, the time complexity of traversal will not be reduced. But when the number of roles is less than the number of rules we could reduce the time on writing data.

- Deleting permissions

  It can be divided into two categories:

  - Delete permissions on the entire system and the complexity will be the same as that of modifying permissions;
  - Delete permissions for the users who have particular conditions, and the complexity will be the same as that of adding permissions.

In general, introducing roles could make it more flexible and convenient in managing permissions and faster in executing permissions of modifying, adding and deleting permission. In the worst case, the performance of ARBHAC is the same as ABAC.

However, conflict may occur when policy is modified. This needs to execute the algorithm of policy optimization and conflicts detection, which is time-consuming.

4.2 An Example

The people nearby can be searched by using the function of "neighborhood people" or "neighborhood groups" on the popular social software Wechat, which also brought a huge challenge to security. Unfortunately, the security issue is not solved properly in Wechat. A woman lived in Shenyang was murdered because of leaking information of photos and location in Wechat on May 17, 2013.

The process of access is similar with the traditional Qzone, which only has two kinds of settings: Open or Privacy. If we choose "Open", it will be open to everyone who wants to click and search; if we choose "Privacy", the users will must answer certain questions before entering the page, which is equivalent to a file encryption. It would be unsafe if its non-encrypted for strangers could enter freely. However if it is encrypted, it could be cumbersome to tell others password. Therefore an access control model which can realize self-recognition and automatically implement is needed for large-scale strangers, and ARBHAC is such a model.

All the users can be divided into five categories: phone contacts (phone numbers saved), Wechat friends, QQ friends, people whose address of school or work is the same in Wechat, and strangers. Now four roles can be obtained: friends, known, colleague, unknown, and the priority are defined as follows:

Table 2. The role assignment of the entertainment store

Priority	Role	Reason	Authority
1	friends	We think the people whose number in your phone are your friends	Read, TimeSeen, PlaceSeen, Forwarding
2	known	We think you just know them because many Wechat friends, QQ friends are never seen in real life	Read, TimeSeen, PlaceSeen(12hour later), Forwarding
3	colleague	We think you might be schoolfellow or colleagues but dont know each other when you have the same school or unit	Read, TimeSeen, PlaceSeen (12hour later)
4	unknown	No intersection, and we think you are strangers to each other	TimeSeen

*(**Note:** Read represents photos, logs, etc. are visible; TimeSeen represents the publishing time is visible, PlaceSeen represents the publishing place is visible, Forwarding represents forwarding comments is allowed)*

In accordance with the policy language above, policy is defined as follows:

Rule1: $phone - number\ in\ Phone - Contacts \rightarrow friends$;

Rule2: $Wechat - number\ in\ Wechat - friends \rightarrow known$;

Rule3: $QQ - number\ in\ QQ - friends \rightarrow known$;

Rule4: $place = my - place \rightarrow colleague$;

Rule5: $Identity = interviewer \rightarrow unknown$;

According to the definition of Section 3, the following conflicts are generated:

– Association conflict is generated within $Rule_1$, $Rule_2$, $Rule_3$ and $Rule_4$;
– Exception conflict is generated between $Rule_5$ and $Rule_1 - Rule_4$;

Define priority of the policy: priority is decreased from $Rule_1$ to $Rule_5$, therefore all conflicts would be resolved.

We apply ARBHAC in Wechat by this way. If Person A searches Person B in "neighborhood people", and clicks photos to visit, then the policy above will be executed. Person A would get only one role, and he is impossible to know Person Bs current location if he doesnt know Person B.

5   Conclusion

In this paper, we propose an ARBHAC model and give the method of building the attribute rules, policy optimization and conflict resolution. The model solves the scenarios of large scale dynamics users, and has greater advantage in policy management and permissions assignment compared with RBAC. Yet its not optimal for only avoiding policy conflict and redundancy. It also requires administrator to manage the roles and authorities permissions when they are changed. Therefore we need to do many researches to make ARBHAC model more applicable.

References

1. Fang, B.X., Guo, Y.C., Zhou, Y.: Information Content Security on the Internet: the Control Model and Its Evaluation. J. Science in China Series F: Information Sciences 53, 30–49 (2010)
2. Sandhu, R., Coyne, E., Feinstein, H., et al.: Role-Based Access Control Models. J. IEEE Computer 29, 38–47 (1996)
3. Sandhu, R., Bhamidipati, V., Munawer, Q.: The ARBAC97 Model for Role-Based Administration of Roles. J. ACM Transactions on Information and System Security 2, 105–135 (1999)
4. Sandhu, R., Munawer, Q.: The ARBAC99 Model for Administration of Roles. In: Proceedings of the15th Annual Computer Security Applications Conference (ACSAC 1999), pp. 229–238. IEEE Computer Society, USA (1996)
5. Oh, S., Sandhu, R., Zhang, X.W.: An Effective Role Administration Model Using Organization Structure. J. ACM Transactions on Information and System Security 9, 113–137 (2006)
6. Ferraiolo, D.F., Sandhu, R., Gavrila, S., et al.: Proposed NIST Standard for Role-Based Access Control. J. ACM Transactions on Information and System Security 4, 224–274 (2001)
7. Bertino, E., Catania, B., Damiani, M.L., Perlasca, P.: GEO-RBAC: A Spatially Aware RBAC. In: Proc. 10th ACM Symp., pp. C29–C37. SACMAT (2005)
8. Chen, L., Crampton, J.: On Spatio-Temporal Constraints and Inheritance in Role-Based Access Control. In: Proc. ACM Symp., pp. C205–C216. ASIACCS (2008)
9. Abdunabi, R., Al-Lail, M., Ray, I., et al.: Specification, Validation, and Enforcement of a Generalized Spatio-Temporal Role-Based Access Control Model. J. IEEE Systems Journal 7, 501–515 (2013)
10. Al-Kahtani, M.A., Sandhu, R.: A Model for Attribute-Based User-Role Assignment. In: Proceedings of the 18th Annual Computer Security Applications Conference, pp. 353–362. IEEE Computer Society, Washington (2008)
11. Zhu, Y., Li, J., Zhang, Q.: General Attribute-Based RBAC Model for Web Services. J. Wuhan University Journal of Natural Sciences 13, 81–86 (2008)
12. Yuan, E., Tong, J.: Attributed-Based Access Control (ABAC) for Web Services. In: Proceedings of the IEEE International Conference on Web Services, pp. 561–569. IEEE Computer Society, Washington (2005)
13. Hong, F., Yao, S., Duan, S.: Attribute-Based Model of Permissions-Role Assignment. J. Computer Applications 24, 153–155 (2004) (in Chinese)
14. Hu, Y.: Graph-Based Network Security Strategy Research of Conflict. Master's Degree Thesis of Nanhua University (2007) (in Chinese)

A Dynamic Feature-Based Security Detector in Wireless Sensor Network Transceiver

Xin Tong, Jikang Xia, Lan Chen, and Ying Li

Institute of Microelectronics of Chinese Academy of Sciences Beijing, China
{tongxin,xiajikang,chenlan,liying1}@ime.ac.cn

Abstract. As various sensing nodes find increasing adoption in Wireless sensor network (WSN) where different attacks have the potential to be catastrophic, security strategy against hardware intrusion become first-class transceiver design constraint. When an operation is running to send or get information in the transceiver, node's behavior can be extracted and evaluated by security criterions to help judging the existence of attacks. Thus, this paper presents a novel feature-based intrusion detection method on mostly hardware level of a WSN transceiver, which targets 3 types of typical attacks difficult to be detected in conventional software method. Simulation of FPGA implementation proves how effective the detector acts by using AMBA Bus Function model.

1   Introduction

Wireless Sensor Networks (WSNs) have been used for a variety of applications over the last three decades. While initially developed for security purposes, the security problem of WSN itself still suffers severe challenges, such as Denial of Service (DoS) attacks, Sybil attacks, traffic analysis attacks and attacks against privacy [1]. Besides, physical attacks or the hardware security threats in WSN sensing nodes are more considerable and risky, which results from the property of open medium, limited power supply, and unattended operations. Therefore, security strategy against hardware intrusion should become first-class design constraint.

However, the present WSN security solutions– cryptographic technology and intursion detection technics– mainly focus on network layer or above network layer [2][3][4][5], ignoring the effect of hardware security, especially the intursion detection technics. Applying such intrusion detection in physical layer, more evidences can be extracted and evaluated by security criterions to help judging the existance of attacks, which largely improves the efficiency of the whole intursion detection system. In light of this, a novel feature-based intrusion detection method is proposed in this paper, as well as its hardware implementation in a WSN transceiver. By extracting, detecting and combining the features of physical behaviors, the detector confirms existence of 3 types of attacks difficult to be detected in conventional method.

The rest of the paper is organized as follows. Section 2 introduces the related works and provides a more extensive discussion of existing researches in the intrusion

W. Han et al. (Eds.): APWeb 2014 Workshops, LNCS 8710, pp. 344–352, 2014.

detection and hardware security field. Section 3 explains the WSN baseband transceiver and framework. Section 4 describes the proposed feature-based security detector method. Section 5 shows the feature-based security detector design in transceiver and analyses the simulation result. Finally section 6 concludes our work and discusses future direction.

2   Related Work

It is generally accepted three kinds of intrusion detection techniques: signature-based detection, specification-based detection, and anomaly detection [4]. C. E. Loo et al. [6] first put forward 12 traffic features in WSNs for ID. And then Mitchell and Chen [7] proposed and analyzed a behavior-rule specification-based technique for intrusion detection in a medical cyber physical system. But they all aim to detect system anomalies without hardware inside involved.

In the WSN hardware field on the other hand, the security study mainly includes three aspects. The first ones are Advanced Encryption Standard (AES), key distribution, management and authentication mechanism [8] [9]. Second, how to improve the channel transfer reliability to assure the communication security, such as decreasing the channel bit error rate [10], ameliorating the antenna system [11] and so on. The last sort of security techniques refers to the passive radio frequency (RF) devices. Although these are inspiring to some extent, researches about intrusion detection are rarely discussed.

So the state-of-the-art security method requires considering both the WSN hardware and intrusion detection since quite of the physical and MAC attacks involves the signal level behavior of transceiver in sensing node. It should benefit the efficiency, precision and cost to monitoring the secure situation in single node and the whole system. That is the origin of our work. Lacking mature research achievement though, we draw lessons from the existing intrusion detection schemes and proposed a feature-based intrusion detection method adapted in the physical layer of a WSN baseband transceiver.

3   Wireless Sensor Network Baseband Transceiver

Wireless Sensor Network usually has a structure of application layer, network layer, link layer and physical layer from up to down. IEEE 802.15.4 standard defines the protocol and interconnection of devices via radio communication in it.

A typical baseband system consists of three fundamental components as shown in Fig.1: an embedded microprocessor, a digital hardware transceiver for signal processing and a RF analog frontend. The embedded microprocessor (CPU) is mainly responsible for network management, data stream control and user interaction with software applications between network entities. The digital baseband takes charge of reliable data transmission and reception with RF interconnection. The RF analog frontend up-converts the modulated signal into 2.45 GHz band and down-converts the radio signal into baseband.

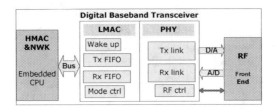

Fig. 1. An example of IEEE 802.15.4 systems

The refined digital transceiver in Fig. 2 consists of two parts: low MAC layer (LMAC) and Physical layer (PHY). LMAC is mainly responsible for buffering the data by FIFOs (First-In First-Out), connecting the upper-layer protocols processors by AMBA (Advanced Microcontroller Bus Architecture) bus, and implementing low-energy strategy by FSM (Finite State Machine). In the PHY transmitter (Tx), the bit stream from MAC layer is wrapped as a packet in fixed formats. All the binary data are encoded by bit-to-chip spreading, O-QPSK modulation and half-sine pulse shaping. Then the resultant packet is transmitted by radio. Through the noisy channel, the receiving RF signal is down-converted to baseband. In the PHY receiver (Rx), after channel listener's detecting and AGC's adjusting, the received signal is down-sampled with a low-pass filter and then synchronized on phase and symbol by preamble and SFD (start-of-frame delimiter). After demodulation, de-spreading and frame parsing, the data is recovered and stored in LMAC RX FIFO.

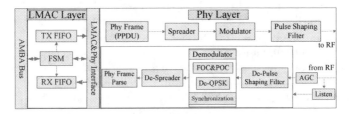

Fig. 2. The baseband transceiver architecture

4   Feature-Based Security Method

To realize the intrusion detection function in WSN physical layer, a Security-Detector module (SDIP) using the feature-based intrusion detection method to detect misbehavior is added in the transceiver above.

In [6], the authors present four standards for a good behavioral feature used for anomaly detection: (1) the feature must discriminate between normal (benign) and anomalous (attack) activity; (2) the feature must be stable with benign values observed now representative of benign values observed later; (3) the feature must be one that can be modeled; and (4) the feature must be efficiently calculable and comparable. They also point that the features based on the rate are not suitable for intrusion detection feature because they relate with network traffic and vary for too many benign reasons. Unlike network layer, when it comes to the PHY and MAC layer, the rate is nothing to do with the concerns above and then becomes a good feature to be selected.

Fig.3 shows the proposed feature based intrusion detection scheme. The scheme is composed of three main phases: the preprocessing, the feature-based intrusion detection, and the post processing. In this paper, we focus on the feature extraction and hardware feature configuration principle in phase 1. After that a brief explanation about other phases is followed.

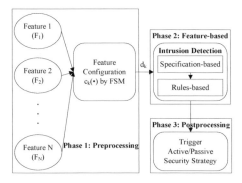

Fig. 3. Feature-based intrusion detection scheme

Phase 1:

1. In view of transceiver, physical behaviors can be extracted as features of four categories:

    (a) Features of source, such as node power or location.

    (b) Features of channel, like channel condition free or busy,and frequency.

    (c) Features of information, meaning the PHY and MAC frame head or content.

    (d) Features of communicating e.g. the rate of sending or receiving.

How to choose and define features is the key of security detector. Due to different frames from different nodes through the transceiver, and asynchronous signals from multiple sources inside the transceiver, which have diversity both in space and time, the security detector must pay great attention to it and take measures to assure its credibility. These are some principles we considered:

    (a) Concerning system, remember the detector consumes extra resource of baseband, optimize clock tree and lower the influence on other modules on chip as much as possible.

    (b) Concerning time, accumulate time point into time slice and normalize the data microscopically, while use time beacon to synchronize data in a specific period of time macroscopically.

    (c) Concerning storage, parse frames based on frame head, only retaining data frames and ACK frames, as well as keep address field down for later use.

    (d) Concerning processing ability, behavior features are extracted by hardware security detector in transceiver, and complex computing can be done by CPU as an assist.

(e) Concerning calculation, apply appropriate manners to extract different features. For example, compare a time series of physical signals with dynamic threshold, while check the incessancy of sequence number when the received frames are suspected missing.

2. Accordingly, define the physical feature set $PF = \{F_1, F_2, ..., F_N\}$ with N feature elements for intrusion detection. For every node in the network, The configuration rule for features can be formulated as follows:

(a) Select k F_i for $i \in \{1, ..., N\}$ from PF, form the subset $F_N^k \subseteq PF$.

(b) Define: combination function $c_k (\bullet)$.

(c) Denote: the binary decision $d_k = c_k(F_N^k) \in \{0, 1\}$

$$d_k = \begin{cases} 1, & \text{normal} \\ 0, & \text{abnormal or under attack} \end{cases}$$

3. Similarly, for m nodes in the network, define $D_m = C_m(d_k^m)$, where D_m is the binary decision of the security condition, while $C_m(\bullet)$ is the intrusion detection rule given each node's condition d_k^m.

Phase 2:

$C_m(\bullet)$ is setup by two steps in Phase 2:

1. Specifications-based: The actions of sensor nodes are either pre-scheduled inside chips or triggered to respond outside events in the predefined way. Such relatively predictable working manner makes it easy to build accurate node specification. As a result, security detector use specification-based intrusion detection rule first.

2. Rule-based: Also called rule-based anomaly detection, which not only detects incorrect behaviors (which violate specifications), but also detects abnormal behaviors (which do not violate specifications) based on the defined rules. It is served as a complement to (a) in the security detector.

Phase 3:

When there is any threat detected, the security detector will trigger either active or passive security strategy in the node. This part is beyond the range of discussion in this paper.

In the transceiver of WSN baseband, features like number of PHY frames, receiving rates, etc. are practical. When nice and available features settled, the detector should set the initial threshold of detection rules according to some experienced values, which will change dynamically by data mining or machine learning. After that, every state of FSM performs the combining function in feature-based intrusion detection scheme. By jumping to different states, there comes final diagnosis of security condition from FSM output.

In this way, every node with this module does preliminary security detection process, which holds good for WSNs because distributed processing is more desirable than centralized processing.

5   A Feature-Based Security Detector in Transceiver

5.1   A Feature-Based Security Detector Design

This module SDIP implements the former feature-based detection scheme, whose framework is shown in Fig.4, and its interaction with other parts in transceiver is predicted in Fig.5. The detector gets features through FIFOs and AMBA bus. After short buffering, the data enter the main part of this module SD_top, which performs the whole intrusion detection method. The result of SDIP will write to a random access memory (RAM) and respond to the baseband and upper CPU.

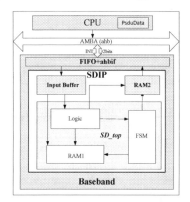

Fig. 4. Feature-based Security Detector Framework

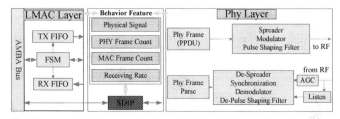

Fig. 5. The baseband transceiver with feature-based security detector

The SD_top is made up with 3 sub-modules: Logic, FSM and RAM.

(a) Logic is responsible for preprocessing, who transforms data in baseband to available features.
(b) FSM accomplishes the defined c_k ($\bullet$) in the above feature-based intrusion detection method.
(c) RAM1 stores the feature data preprocessed and detection rules, which is updated with time and rules gradually.

Based on the transceiver, we extract features from valid physical signal, physical frames, MAC frames and receiving rate in Logic. The accordingly components of main module SD_top are shown in Fig.6. The sub-modules written by verilog in Logic are:

(a) ValSig.v gets the feature of valid signal lasting time;
(b) PhyFrmCnt.v gets the feature of counts on valid PHY frames and receiver opening number of times;

(c) MacFrmCnt.v gets the feature of counts on valid MAC frames;

(d) RxRate.v gets the feature of receiving rate and incessancy of MAC frames.

(e) ValWarn.v, PhyWarn.v, MacWarn.v and RxWarn.v sets the alarming thresholds for each corresponding features. As far as we concerned, the values are based on some experience values, but they will be dynamic self-adopting later.

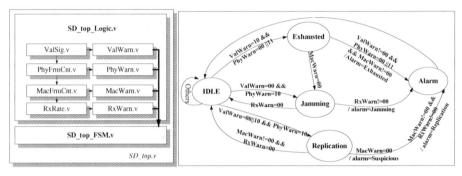

Fig. 6. Components of SD_top   **Fig. 7.** State Transforming of FSM

Using these features, we also designed a FSM whose state switching diagram is Fig.7. The FSM utilizes specification-based detect scheme. When it finished, one whole period ends and resets all the signals, its result delivers to CPU and triggers post-processing phase.

5.2   Simulation Result and Analysis

Implemented in Virtex 6 FPGA, the SDIP consumes hardware resource in Table 1. During a simulation period of 27µs, the Logic in SDIP calculates whether the Valid-Signal, PhyLayerFrames, MacLayerFrames and ReceivingRate suspicious or not. If any suspicious conditions are found, it produces corresponding Warns and enables the FSM part. The simulation result is shown in Fig.8.

Table 1. Key Hardware Consumption of SDIP

Device	Number
Adder/ Subtractors	11
Flip-Flops	86
Comparators	5
Multiplexers	81

In Fig.8, after the input signal 'rxon' is valid, SDIP begins to work. Every time 'fifo_write_lmac' is valid, it is reckoned that there is a PHY layer frame. If 'Int1' turns into 1 at the negative edge of 'fifo_write_lmac', it adds the MAC layer frame count. At the end of this counting period, 'fsm_en' is valid, and it produces 'alarm' and 'int' to request the AMBA bus.

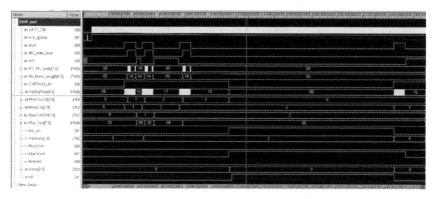

Fig. 8. Simulation Result of SDIP

Then, when the receiver opens again, enabling signal of SDIP takes effect with it, and all the registers go back to zero preparing for a new round.

When SDIP detects the valid signal lasts too long, but the PHY frames count or receiver opening frequency is not beyond the suspicious threshold, it makes the 'Val-Warn' 'High', 'PhyWarn' 'Normal'. In the meanwhile, the valid MAC frames are too less than the accumulated PHY frames, the MacWarn is in force and other warning signals are normal. That makes the FSM jumps into the alarm state and tells post-processing the sensing layer suspects of 'Exhausted' attacks. More relationship between behavior features and attacks we used can be found in Table 2.

Table 2. Relationship between Behaviors and Attacks

Behavior features	Exhausted	Jamming	Replication
Phy Valid Signal	10	00‖10	00‖10
PHY frame	00‖11	11	10
MAC frame	11	X	10
Rx Rate	X	11	X

Normal=00, High=10, Low=11

For better comprehension, Table 3 shows some characters of literature [7] and our work, which both concentrate on behavior rule based intrusion detection. The difference is [7] mainly focuses on medical cyber physical system, and its intrusion detection uses system level human behaviors as rules to against unspecified attacks. Our work, on the other hand, aims at specified attacks in terms of sensor node behavior.

Table 3. Comparison with literature

Literature	Application	Attacks	Detection Base	Level
Mitchell and Chen [7]	Medical devices	Unspecified	Behavior rule, human behavior	System
Our Work	Terminal device chips	Specified	Behavior feature, sensor behavior	Node

6   Conclusion and Future Work

In this paper, we present a feature-based intrusion detection method, as well as one practical hardware implementation. The hardware security detector extract and combine features of physical bahaviors from a WSN transceiver. After performing the intrusion detection scheme, the detector will confirm existence of 3 typical attacks. It overcomes disadvantages of conventional network software intrusion detection techniques which are communication overhead consumptive, routing protocol limited and incomprehensive. What's better, the distributed detection parameter comes from every node inside and does not depend on any routing topology or protocol and so can be adapted to multiple application scenes.

This work is exploratory in this realm; more works on feature set expansion, dynamic detection parameters and so are expected to present in the future. We hope that this method will widen the mind of suitable security solution for WSN applications.

References

1. Butun, I., Morgera, S., Sankar, R.: A Survey of Intrusion Detection Systems in Wireless Sensor Networks (2013)
2. Bao, F., Chen, R., Chang, M.J., et al.: Hierarchical trust management for wireless sensor networks and its applications to trust-based routing and intrusion detection. IEEE Transactions on Network and Service Management 9(2), 169–183 (2012)
3. He, D., Bu, J., Zhu, S., et al.: Distributed access control with privacy support in wireless sensor networks. IEEE Transactions on Wireless Communications 10(10), 3472–3481 (2011)
4. Farooqi, A.H., Khan, F.A., Wang, J., et al.: A novel intrusion detection framework for wireless sensor networks. Personal and Ubiquitous Computing 17(5), 907–919 (2013)
5. Sooyeon, S., Taekyoung, K., Gil-Yong, J., et al.: An Experimental Study of Hierarchical Intrusion Detection for Wireless Industrial Sensor Networks. IEEE Transactions on Industrial Informatics 6(4), 744–757 (2010)
6. Loo, C.E., et al.: Intrusion Detection for Routing Attacks in Sensor Networks. International Journal of Distributed Sensor Networks 2, 313–332 (2006)
7. Mitchell, R., Chen, R.: Behavior rule based intrusion detection for supporting secure medical cyber physical systems. In: 2012 21st International Conference on Computer Communications and Networks (ICCCN), pp. 1–7. IEEE (2012)
8. Mukherjee, A., Fakoorian, S.A.A., Huang, J., Swindlehurst, A.L.: Principles of physical layer security in multiuser wireless networks: A survey. Computing Research Repository (CoRR) abs/1011.3754 (2010), http://arxiv.org/abs/1011.3754
9. Meguerdichian, S., Potkonjak, M.: Security primitives and protocols for ultra low power sensor systems. In: 2011 IEEE Sensors, pp. 1225–1227. IEEE (2011)
10. Kimura, T., Aoki, Y., Yoshida, N., et al.: A Wireless Dual-Link System for Sensor Network Applications. In: IEEE International Solid-State Circuits Conference, ISSCC 2008. Digest of Technical Papers, pp. 534–633. IEEE (2008)
11. Shi, H.Z., Tennant, A.: Secure physical-layer communication based on directly modulated antenna arrays. In: Antennas and Propagation Conference (LAPC), pp. 1–4. IEEE, Loughborough (2012)

A Highly Adaptable Reputation System
Based on Dirichlet Distribution

Xia Xiao, Lan Chen, and Ying Li

Common Technology Research Department
Institute of Microelectronics, Chinese Academy of Sciences
Beijing, China
{xiaoxia,chenlan,liying1}@ime.ac.cn

Abstract. Conventional reputation system follows 2-dimensional Beta-distribution to do one rule observation. However, complex environments with high BER and propagation delay, strong interference or moving protocol will cut down the accuracy of detection. Apart from that, the response times of reputation value are various from different circumstances, which demand an adaptable mechanic. Therefore in this paper, we propose a highly adaptable reputation system with the ability of results category, fluctuation toleration, response time adjustment, and attack prevention. The results show our system has great performance on fuzzy state detection, effectively filters the burst fluctuation and rapidly responds to the long term variation with low power cost. Practically, the performance under wormhole attacks such as jam-and-reply attacks is promising.

1 Introduction

1.1 Background

Reputation management has been widely studied to improve the security, reliability and robustness of wireless sensor networks. The reputation value can be extensively adopted to evaluate critical processes such as routing, localization, synchronization, data aggregation, topology control and security [1] [2] [3] [4]. Conventional reputation system [5] introduces binomial distribution to match the interaction possibility of target nodes, which defines the interaction result as cooperative or non-cooperative. This approximation under the theoretical derivation is reasonable because the result can only be either positive of negative [6], and there remains no intermediate selection. Unfortunately, during actual experiments, the simple result varies in complicated surrounding conditions. It's hard for a monitor node to category a controversial behavior as definitely cooperative or non-cooperative. We define the controversial observation results as "fuzzy observations". The phenomenon appears because the monitor take the judgment based on the received messages, which might be different from what is real happened.

In addition, potential attacks such as bad-mouth attacks and ballot-stuffing attacks highly threatened the reputation system [7]. Many efforts are done to prevent such attacks [8] [9], but an effective solution remains undiscovered. Moreover, the

W. Han et al. (Eds.): APWeb 2014 Workshops, LNCS 8710, pp. 353–364, 2014.

monitoring power consumption is considerable which need to be reduced in practical use, and a fluctuation tolerance system is required to meet the complex circumstances of moving networks.

To solve these problem, we introduce a reputation system based on dirichlet distribution, containing several effective mechanics to achieve a dynamic adjusted, fluctuation tolerated, attack prevented, low power consumed trust management system. In section 2, we summarize several works related to our work. Section 3 provides theoretical derivation of our system. Section 4 discusses the experimental evaluation of our results. Section 5 summarizes the system and discusses further directions.

1.2   Related Work

The proposal of a complete Beta-reputation system is proposed in [10], which based on the mathematical method [11]. The author established an integrated system including reputation generation, bootstrapping, aging and second-hand information. However, the author failed to solve the latent ballot-stuffing attacks, and the threat that compromising of a high reputation node was remained. In addition, the limitation which only cooperative second-hand information is permitted weakened the versatility of the system.

Chen [12] introduced 3-dimensional evident space to expand the conventional reputation evident state space of 2-dimensional. The new state "unknown", which deterrent the attackers pretended as newcomers, is quite suitable to networks in complex environment, and the author successfully mapped the 3-states evident space to 2-states trust space. Otherwise, newcomer nodes will not be able to engage in interactions with existing nodes and the network will not be able to grow. However, obeying 2-deminsional beta distribution limits the expansion of "fuzzy-states", and hinders further applications based on it.

Hierarchical based reputation method [13] is proposed to establish a robust system where the cluster head monitors all of the member and the member in turn monitor the cluster head and inform the higher level nodes through extra route if misbehavior happens. However, the method failed to quickly locate misbehaving high reputation member nodes, and extra energy is consumed by the control messages.

Qin [14] proposed an adaptive aging factor (history factor) mechanic to adjust the weight of history observations, to adapt to the changing reputation. In their theory, if a non-cooperation message is detected, the history factor sharply drop from 0.9 to 0.1, which leads to a losing weight of history observations and gaining weight of current results. In this way, reputation is made "hard to earn but easy to lose."

2   Dirichlet Distribution System

2.1   Multinomial Distribution

The traditional distribution of observation results, such as Beta, follow the binominal distribution when assuming the result can only be either cooperative or non-cooperative. Unfortunately, such all-or-nothing classification is limited concerning the complexity environment of WSN when deployed. To break the limitation of binomial distribution, we recommend multinomial distribution as the measurement of node behavior, where controversial results can be precisely defined. The multinomial distribution can be defined [15] as:

$$Multi(\vec{p} \mid \vec{n}, N) = \binom{N}{\vec{n}} \prod_{k=1}^{K} p_k^{n_k} \tag{1}$$

Where K refers to the number of results of a single monitor, $p_k^{n_k}$ refers to the possibility of a certain result, $\vec{n}$ refers to observations state space and N represents the total number of sampling. Given K=5, we can easily define five possible monitoring results state space as very good, possibly good, fair, possibly bad, and very bad.

2.2   Dirichlet Distribution

As dirichlet distribution is both the prior and posterior distribution of Bayesian method [19] when the observation results conformed to multinomial distribution mentioned above, we use multi-elements dirichlet distribution to predict the possible behavior of target nodes. Compared to Beta distribution system, where the state space is binary, the dirichlet distribution provides K-dimensional state space, which is useful to solve fuzzy observation problem. The dirichlet distribution [20] is defined as:

$$Dir(\vec{p} \mid \vec{\alpha}) = \frac{\Gamma(\sum_{k=1}^{K} \alpha_k)}{\prod_{k=1}^{K} \Gamma(\alpha_k)} \prod_{k=1}^{K} p_k^{\alpha_k - 1} \tag{2}$$

Where $\vec{\alpha}$ refers to state space, $\vec{p}$ refers to possibility space, and K refers to the total dimension of observation space. For mutually exclusive observation event, there exists limitation that:

$$p_k \geq 0, and \sum_{k=1}^{K} p_k = 1 \tag{3}$$

And the Bayesian method can be showed as:

$$Dir(\vec{p} \mid \vec{\alpha}) + \sum_{i}^{M} Multi(\vec{p} \mid \vec{n}_i) = Dir(\vec{p} \mid \vec{\alpha} + \sum_{i}^{M} \vec{n}_i) \tag{4}$$

2.3   The Expression of On-demand Reputation

Dirichlet distribution provides multi-dimensional space for observation result, and the reputation value can be specified by the mathematical expectation of it. If we take priori base (the ratio when no results are received) into account, for K dimensional vector $\vec{p}$, the expectation can be defined as:

$$R(\vec{p}) = E(\vec{p}) = \left\{ \frac{\alpha_1 + Cb_k}{C + \sum_{k=1}^{K} \alpha_k}, \frac{\alpha_2 + Cb_k}{C + \sum_{k=1}^{K} \alpha_k}, \cdots, \frac{\alpha_K + Cb_k}{C + \sum_{k=1}^{K} \alpha_k} \right\} \tag{5}$$

Where C and b_k are bootstrapping factors which will be discussed in part 3.4.

During practical implementation, the limitation of storage restricts the types of observation as well as the dimension K, there is no need to save all of the $E(\vec{p})$ value. Mostly, only part of the reputation value is in common use. For example, if no attack or malfunction occurred within a certain amount of time, then average security level is applied and all the monitor results above average are acceptable, we have:

$$R_{ave} = E(\alpha < \alpha_T) = \frac{\sum_{k=1}^{T}(\alpha_k + Cb_k)}{C + \sum_{k=1}^{K} \alpha_k} \tag{6}$$

$R_{ave}(\vec{p})$ expresses reputation under average situations when T=average, and can be commonly adopted as regular reputation value. The average situation is an approximation to Beta distribution when all of the fuzzy observations are categorized as cooperative or non-cooperative. While this kind approach is easy to implement, it always produce inaccuracy judgments in complex circumstances, where some underpinning features such as high latency or high BER can be exploited to launch attacks. Thus we adopt value of K as 3 or more, which provides at least one fuzzy state for further decision making.

Here, we call this adapting reputation mechanic on-demand reputation, which means that each node decides which environment they are in and therefore which level of reputation is accepted. With on-demand reputation, specific evident which is more convincing than in Beta reputation system are provided for nodes to take precise judgment in different circumstances. In Beta reputation, if higher security level is required, the reputation threshold is increased to wipe off non-cooperative nodes. However this method is inaccurate because malicious nodes will take advantage of network feature (ex. High jamming rate, BER, propagation delay) to pretend as normal nodes. The promising solution is to establish more severe rule, which unfortunately introduces high false positive rate. On-demand system solves the problem by enacting a strict rule, a flexible rule or more rules simultaneously, and adopt the abeyance record of each one of them when needed.

2.4   Critical Factors

1. Bootstrapping

According to [21], C is a priori constant which equals to the cardinality of the state space over which a uniform distribution is assumed, and b_k is the base rate which equals to $1/k$ if all the alternatives are impartially weighted. These two factors determine:

   a.   The original reputation without any observation at the beginning of network which limits the original threshold or when.

   b.   How the starting several observations influence the initial stage value at the beginning of a newcomer.

To ensure the equity of each alternative, we assume that for K-dimensional reputation state space, $C = k$ and $Cb_k = 1$, thus for every new node, the reputation equals to:

$$R(\vec{p}) = E(\vec{p}) = \left\{ \frac{1}{K}, \frac{1}{K} \dots \dots \right\} \tag{8}$$

And for the first observation, the reputation is calculated as:

$$R_i = \frac{\alpha_i + 1}{K + \sum_{k=1}^{K} \alpha_k} \tag{9}$$

Because of the priori factor, the starting k observations cannot strongly affect the reputation value, which contributes to the steadiness and fault tolerance in network starting process. However, in different circumstances, the requirement of bootstrapping stage is extraordinarily different, so the ratio value should be carefully designed to be suitable to the environment and application of the network.

2. Aging

The aging factor determines the respond speed toward situation changes in networks. As mentioned above, most of the dynamic aging factor mechanic devotes to accelerate the respond speed of fluctuation which is suitable to strict networks. While, fluctuations in networks are inevitable due to the wireless characteristic of WSN, so almost completely abandon history reputation value is unwise. Here we proposed a dynamic aging factor mechanic which reflect the changing of the system. If the fluctuation is severe and for long-term, the aging factor rapidly decreases and historical value is weakened. On the contrary, the aging factor growth and new observation result is weakened.

$$\alpha_{real} = \mu \alpha_{history} + (1 - \mu)\alpha_{new} \tag{10}$$

3. Second-hand information

Observations of only one node is insufficient and the process is slow, so we need extra help from other neighbors. Unfortunately, the neighbors might be potential attackers which provide misleading information, such as bad-mouth attacks and ballot-stuffing attacks [19]. To prevent those attacks, we just use second-hand information as fluctuation reference but not take them into account, and the second-hand information just be transmitted once in a certain period, thus the energy of transmission and calculation is reserved. The mechanic of second-hand information is described in section 3.4.

2.5   Fluctuation Prediction and Toleration Mechanics

2.5.1   Time-Chips

Refreshing the reputation value after each interaction is harsh and misguided. Instead of evaluating every observation after each interaction, we define an average observation result within a certain time period called time chips, which is defined as:

$$aveob_j = \frac{1}{n} \times \sum_{i=1}^{T} ob_i \tag{11}$$

The average results successfully conquer the weakness that only sparse exceptional observations (such as one or two) will lead to the severe change of reputation especially when the aging factor is low concerning the now existed theory. The time chips mechanic is shown in Fig.1, and the average observation value is shown in Fig.2.

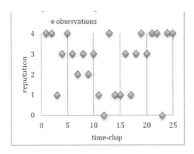

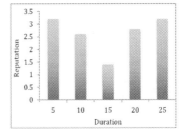

Fig. 1. Time-chip mechanic   **Fig. 2.** Average reputation of time-chip

2.5.2   Fluctuation Prediction

We define the fluctuation as the variation of observation results between different time chips, which is:

$$F = \frac{1}{k} \times |aveob_j - aveob_{j-1}| \qquad (12)$$

Where k is the dimension of observation. The fluctuation factor F represents the extent of fluctuation of target node due to the complexity of interference, the moving topology, or a certain exceptional behavior, which derived from the state changing of networks leading to the reputation changing.

2.5.3   Dynamic history Factor

With the variation of F, which reflects the level of fluctuation, the aging factor ought to follow the network status. If F is great which means the network is suffering a severe surging, the aging factor diminishes to put more weight on the current observations and decrease the weight of history observations because the current consequences rather than history behavior most likely represent the actual changing of target node. Inversely, aging factor increases when F is low. The dynamic aging factor is defined as:

$$\mu = 1 - F \qquad (13)$$

2.5.4   Dynamic Time-Chips

Time-chips are the primary mechanic of fluctuation toleration. When the network is stable, few burst turbulences ought to be minimized, which means that the duration of a single time-chip extends in the stable circumstances to exclude the sudden fluctuations and keep the reputation steady. On the contrary, when the network is unstable and frequently changing, duration of every time-chip shorten to configure a dynamic reputation value which swiftly keep track of the actual value of target node. The duration of time chip is defined as:

$$Chip_i = 2^k Chip_{i-1} \qquad (14)$$

Considering the sudden turbulence of network, we don't simply define the chip duration as the direct reflection of factor F, rather, the double and halve method is more steady and suitable. Fig.3 shows that the higher the fluctuation, the shorter the time-chip, the more stable the network, the longer the duration.

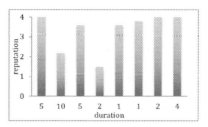

Fig. 3. Dynamic time-chip mechanic

2.5.5   On-demand Second-Hand Information

The exchange of second-hand information is the weak point of potential attacks. As a consequence, several restrictions are established.

Firstly, the second-hand information can be used only as reference but not in the calculate process because the dynamic aging factor mechanic mentioned in the next section will guarantee the speed of convergence of reputation value, so that extra risks introduced by second-hand information are diminished.

Secondly, the second hand information is transmitted on-demand rather than spontaneously, and the second-hand information just be transmitted once in a certain period, thus the energy of transmission and calculation are reserved. In that the reputation value is relatively stable if the fluctuation factor F remains low, there is no need to require the second-hand information during that period. Rather, when F is high indicating the fiercely fluctuation of network, the second-hand information ought to be required more frequently as the growing of factor F.

In the end of each period, if the deviation of average value of second hand information and local reputation value, another requirement will be generated at the end of the smallest time chip to indicate that the local reputation value is inaccurate. To prevent the potential attacks, the iteration above is forced to suspend in times of R.

Thirdly, attackers will provide confliction evidences comparing to those normal one. However, the confliction cannot totally ensure the reputation attackers but just to provide probably evidences. For example, if an attacker selectively discards the information of a certain node but performed well to others, the reputation of the attacker form that certain node will be distinguished to others. But the certain node is a victim rather than an attacker. Thus, if the confliction occurred, a warning of potential attacker will be generated and reported for further decision. During that time, the information provided by special node will not be concerned.

These mechanics restrains the number of actions of attacks' behavior and the extent of destruction of potential attacks.

3   Experimental Evaluation

3.1   Reputation Evaluation

To verify the advantage of dirichlet distribution system against Beta distribution system, we define a relatively harsh environment to evaluate the performance of both system. Here, we deployed 100 nodes, assumed the aging factor $\mu=0.9$, channel fading rate is 10%, excellent working rate = 50%, positive fuzzy state rate = 20%, negative fuzzy state rate = 20%, and multinomial distribution probability vector $\vec{p} = \{0.5, 0.2, 0.2, 0.1\}$.

For dirichlet distribution system, the observations can be categorized as excellent, good, average, and bad. The experimental results are shown as:

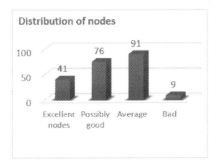

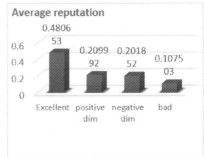

Fig. 4. Average reputation of 100 node   **Fig. 5.** Distribution of 100 nodes

The dirichlet distribution consequence showed an extension of 2-dimensional observation which precisely matched the origin possible vector $\vec{p}$ in multinomial distribution behaviors. Fig.5 shows an accurately observation classification which every node can be clearly categorize as one of the pre-defined states. By using the on-demand reputation, different number of nodes can be trusted under different security or reliability level. Here we have 41 trusted nodes if the level is set to excellent, and 91 trusted nodes if the level is above average. Thus, the goal of reputation adaptation is achieved. In variation of circumstances, the network has the ability to choose the most suitable level to make it function well.

Scenario1: The reaction possibility vector of target changed from p={0,0.05,0.05,0.9} to p={0.5,0.3,0.1,0.1}. This scene simulate the occasion that target node is compromised or malfunction. The result is shown in Fig.6 and Fig.7.

Our design quickly responded to the immediate action vector change of target node. Compared to the traditional dynamic aging factor mechanic, our design is more stable and steady. Compared to the fixed aging factor, our design performed a more accurately approximate to the real vector.

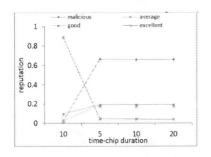

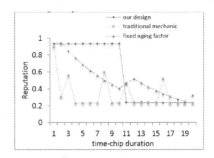

Fig. 6. Reputation distribution   **Fig. 7.** Reputation comparison

Scenario2: The action vector changed from p={0,0.05,0.05,0.9} to p={0.5,0.3,0.1,0.1} for period of five, and then reversed back to p={0,0.05,0.05,0.9} for period of five. The loop was examined twice, and finally remained p= {0, 0.05, 0.05, 0.9}. This model is set to evaluate the sudden fluctuation of target node. The result is shown in Fig.8 and Fig.9:

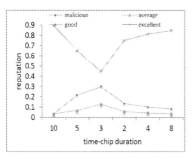

Fig. 8. Reputation distribution

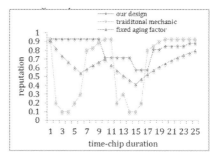

Fig. 9. Reputation comparison

Our design performed outstanding in the fluctuation scene. The transition window decreased when encountered with the deviation of observed value to the calculated value, and increased when such deviation is diminished. Our design successfully eliminated the sudden burst of misbehavior of target node, and the reputation value came to 0.5 at the end of second loop, and quickly reversed back to the normal value. The reputation is relatively stable compared to traditional dynamic mechanic and accurate compared to the fixed aging factor.

Scenario3: The action vector fluctuated between p={0,0.2,0.2,0.6}, p={0.5,0.3,0.1,0.1} and p={0.25,0.25,0.25,0.25}. For each of the vector, the duration is five. The model is set to simulate the sustaining changing network. The result shown in Fig.10 and Fig.11:

The transition window decreased to the minimal value (=2) because of the continuous network fluctuation. The aging factor approached to zero when the transition window approached to minimal value, which means the current observations weight more value. The mechanic successfully calculated the accurate value approximated to the real value. Compared to the traditional mechanic, our design performed stable and accurate. The minimal transition window ensured that the calculated value kept stable rather than sharply wave up and down as traditional mechanic did.

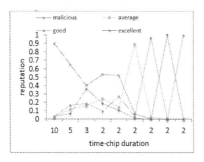

Fig. 10. Reputation distribution

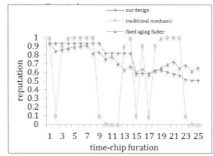

Fig. 11. Reputation comparison

3.2   Wormhole Attacks Evaluation

This simulation focused on jam-and-reply attack [20]. Node A is a wormhole attacker, which jammed all messages transmitted from node S, and replied another certain message containing malicious attempt to R. Node R expected data from S but always failed because of the existing of attacker A. So the reputation value of S is low.

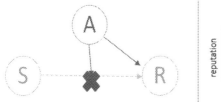

Fig. 12. Wormhole attack   **Fig. 13.** Reputation of malicious node

The traditional beta reputation system failed to detect this kind of attack because a message followed by a jamming can be either good or bad. The misuse detection categorizes it as qualified, which overlooks this attack. And anomaly detection classifies it as abnormal, leading to high a false positive rate if the jamming occurs frequently especially in harsh environments. The dirichlet reputation system put the characteristic into vague state, high ratio of which triggered a warning of a potential attacker and the suspicious node A was eliminated from interaction. The result is shown in Fig.13:

This is a representative experiment of many wormhole attacks, which conceal the characteristic by acting vague behaviors, and the dirichlet distribution system is a promising solution to detect suspect wormhole attacks.

3.3   Potential Attacks Evaluation

In this simulation, we considered the occasion about bad-mouth attack. Bad-mouth attack is fatal to the network when non-cooperate second-hand information is permitted. To prevent the fatal attacks, we introduced an iteration parameter to stop the excessive iteration caused by abnormal second-hand information value. In this scenario, node R evaluated the reputation transition of node S, which encountered with a short-term channel fading and performed limited times of malfunction. Nodes A is a bad-mouth attacker which always returns false reputation values, and node B is a normal node which returns actual values. The maximum iteration factor R equals to 3.

The second-hand information from node A and B is distinct. If the deviation is severe enough, the situation is recorded as a confliction. If such confliction occurs frequently, a report of potential attacker will be generated. Because the second-hand information will not be used directly, the devastative consequences will not happen. The existence of attackers will only partly increase the limited power consumption rather than inaccurate elimination.

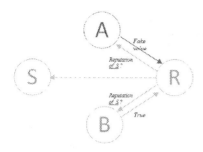

Fig. 14. Bad-mouth attack

Table 1. Detection result of bad mouth attack

Value of R	3	4	5
Conflictions detected	10	1	18
Reputation of node S	0.88	0.92	0.93

Table 1 shows the result of our detection method. The bad mouth attack doesn't influence the reputation value of node S, because the confliction values are discarded. And the suspicious conflictions are reported.

4 Conclusion

We introduce a vague states detective, fluctuation tolerated, attack prevented trust management system to wireless sensor network in complex environments. Compared to restricted 2-dimensional result space of conventional beta reputation, the extra states provided by dirichlet distribution can accurately category the observations by level through the enacted rules, and all misuse detection, anomaly detection and even more specific rules can be adopted simultaneously to satisfy diversity complex applications. The result of any rule can be rapidly revealed if needed and the hierarchical classification method conforms to already existing protocol. The IEEE802.15.4 and Zigbee protocol can take advantage the different level of results provided by dirichlet distribution to implement security level and security routing.

By introducing fluctuation prediction and dynamic factor mechanics, during the transition state, the weighting of history value and current value, and the average reputation are dynamic adjusted. Comparing to traditional system, our proposal filter burst fluctuation and keep the reputation relatively steady.

Besides, the dirichlet distribution system successfully distinguishes jam-and-reply attack and provides possible tool to detect deeply hidden wormhole attacks, which remains one of the most intractable problem in intrusion detection system of WSN. Moreover, the damage of bad-mouth attack and ballot-stuffing attack are diminished, and potential attackers can be reported. Our system solves those attacks even if both cooperative and non-cooperative second-hand information are permitted.

References

1. Wang, X., Ding, L., Bi, D.: Reputation-Enabled Self-Modification for Target Sensing in Wireless Sensor Networks. IEEE Transactions on Instrumentation and Measurement 59(1), 171–179 (2010)

2. Bao, F., Chen, I., Chang, M., Cho, J.: Hierarchical trust management for wireless sensor networks and its applications to trust-based routing and intrusion detection. IEEE Transactions on Network and Service Management 9(2), 169–183 (2012)
3. Maarouf, I., Baroudi, U., Naseer, A.R.: Efficient monitoring approach for reputation system-based trust-aware routing in wireless sensor networks. IET Communications 3(5), 846–858 (2008)
4. Harbin, J., Mitchell, P.: Reputation routing to avoid sybil attacks in wireless sensor networks using distributed beamforming. In: ISWCS 2011 (2011)
5. Momani, M., Aboura, K., Challa, S.: RBATMWSN: Recursive Bayesian Approach to Trust Management in Wireless Sensor Networks. In: Intelligent Sensors, Sensor Networks and Information (ISSNIP 2007), Melbourne (2007)
6. Resnick, P., Kuwabara, K., Zeckhauser, R., Friedman, E.: Reputation systems. Communications of the ACM, 45–48 (2000)
7. Yu, H., Shen, Z., Miao, C., Leung, C., Niyato, D.: A Survey of Trust and Reputation Management Systems in Wireless Communications. Proceedings of the IEEE 98(10), 1755–1772 (2010)
8. Hongjun, D., Zhiping, J., Xiaona, D.: A Entropy-based Trust modeling and Evaluation for Wireless Sensor Networks. In: International Conference on Embedded Software and Systems (ICESS 2008), Sichuan (2008)
9. He, Q., Wu, D., Pradeep, K.: SORI:a secure objective reputation-based incentive scheme for ad-hoc networks. In: Wireless Communications and Networking Conference (WCNC 2004) (2004)
10. Ganeriwal, S., Balzano, L.K., Srivastava, M.B.: Reputation-based framework for high integrity sensor networks. ACM Transactions on Sensor Networks (TOSN) 4(3), 1–37 (2008)
11. Jøsang, A., Ismail, R.: The Beta Reputation System. In: 15th Bled Electronic Commerce Conference, pp. 1–14 (2002)
12. Ren, Y., Zadorozhny, V.I., Oleshchuk, V.A., Li, F.Y.: A Novel Approach to Trust Management in Unattended Wireless Sensor Networks. IEEE Transations on Mobile Computing (2013)
13. Chen, H., Wu, H., Gao, C.: Agent-based Trust Model in Wireless Seneor Networks. In: Software Engineering, Artificial Intelligence, Networking, and Parallel/Distributed Computing, Qingdao (2007)
14. Qin, T., Yu, H., Leung, C., Shen, Z., Miao, C.: Towards a trust aware cognitive radio architecture. ACM SIGMOBILE Mobile Computing and Communications Review 12(2), 86–95 (2009)
15. Rasmusson, L., Jansson, S.: Simulated social conreol for secure Internet commerce. In: Proceedings of the 1996 Workshop on New Security Paradigms, pp. 18–25 (1996)
16. Leonard, T.: A Bayesian Approach to Some Mulinomial Estimation and Pretesting Problems. Journal of the American Statistical Association 72(360a), 869–874 (1977)
17. Wang, X., Ma, X., Grimson, E.: Unsupervised Activity Perception by Hierarchical Bayesian Models. In: Computer Vision and Pattern Recognition, CVPR 2007, Minneapolis (2007)
18. Jøsang, A., Haller, J.: Dirichlet Reputation Systems. In: The Proceedings of the 2nd International Conference on Availability, Reliability and Security (ARES 2007), Vienna (2007)
19. Buchegger, S., Boudec, J.-Y.L.: A Robust Reputation System for P2P and Mobile Ad-hoc Networks. In: PP2PEcon 2004, Cambridge, MA (2004)
20. Zhou, R., Hwang, K.: PowerTrust: A Robust and Scalable Reputation System for Trusted Peer-to-Peer Computing. IEEE Transactions on Parallel and Distributed Systems 18(4), 460–473 (2007)
21. Hu, Y.-C., Perrig, A., Johnson, D.: Wormhole attacks in wireless networks. IEEE Journal on Selected Areas in Communications 24(2), 370–380 (2006)

Research on Covert Channels Based on Multiple Networks[*]

Dongyan Zhang[1], Pingxin Du[1], Zhiwen Yang[1], and Lan Dong[2]

[1] School of Computer Science and Technology, University of Science and Technology Beijing,
No.30 Xueyuan Road, Haidian District, Beijing, China
[2] School of Computer and Information Technology, Beijing Jiaotong University Beijing,
No.3 Shangyuancun, Haidian District, Beijing, China
zhangdy@ustb.edu.cn, {1259920914,huanshui_2005}@qq.com,
donglan@bjtu.edu.cn

Abstract. Covert channel is a kind of method, used to transmit message secretly usually. Till now, most covert channel uses one network with protocol header or filter bits, time interval of packets etc. With the development of networks, using multiple networks transmit data getting more common. In this paper, a new kind of covert channel is proposed using multi-networks to transmit ciphertext. One is used to transmit key, and the others are used to transmit ciphertext. In the experiment, the key is transmitted with 3G network, while the cipher using WiFi network in the covert channel. For making the covert channel more robust, we use one time padding to encrypt our message. We also measure the accuracy with different transmission speed. The result shows that the covert channel we propose has high accuracy with different transmission speed.

Keywords: Covert Channel, Multi-network, One-time pad, WiFi, 3G.

1   Introduction

Covert channels implement unmonitored communication using existing resources onpublic network. A traditional covert channel is usually built based on a lower-layer protocol in the network model formulated by the International Standardization Organization (ISO), for example, identifier field in IP protocol, ISN field in Transmission Control Protocol (TCP), and filler field in TCP or IP protocol. In recent years, with the construction of network infrastructure and the popularity of the broadband network, there have been covert channels based on time, on Hypertext Transfer Protocol (HTTP), and on public facility resource modifications.

With the development of network, it becomes more and more common to use multiple networks to transmit data. A multi-homing system uses multiple access networks concurrently, which can get more effective bandwidth and connectivity, improve the capability of error recovery, and reduce network latency. Now the most commonly used multiplex transmission methods are designed for wireless devices including 3G,

---

[*] The Fundamental Research Funds for the Central Universities Contract No. 2012JBM030.

W. Han et al. (Eds.): APWeb 2014 Workshops, LNCS 8710, pp. 365–375, 2014.

4G and Wi-Fi devices. In relatively small signal coverage, Wi-Fi can provide good communication quality at low deployment cost and is most commonly used. However, 3G, 4G and cellular networks support wide signal coverage with high cost of deployment and maintenance.

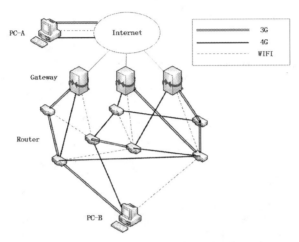

Fig. 1. The Network Topology of Mobile Network

During Wi-Fi,3G or 4G network transmission between wireless mobile devices, the link layer is usually treated by TCP. Studies show that currently more than 95% Internet traffic is transmitted using TCP, including a variety of applications, such as email, web, and social network applications. Due to the huge network traffic, it becomes increasingly difficult to detect data transmission in covert channels. Even if we can detect these channels using advanced methods, the message has already expired after data transmission and encryption analysis. Detection of these channels may be useful only for digital forensics.

For the covert channels with only one network, it may fragile to the attackers'wreckage easily. Once the attacker detects the covert channel, he can crack it easily or just change some of the packets which make it unreadable to the receiver.

As we know, one-time pad was proved to be the most security encryption. If we want to use one-time pad in covert channel with only one network, it will be inconvenience. We should mark the key and ciphertext in different way andseparate key and ciphertext at the receiver correctly.By using two kinds of network, it is convenience to transmit key and ciphertextseparately. I t also increase the difficulty for attacker to destroy the whole channel.

This paper proposed a method of using multiple networks to build covert channels transmitting data. Using multiple networks, you can integrate the covert channel technology and encryption technology;separate the key from ciphertext to improve the security of covert channels. If an attacker detects the covert channel, the only way he gets the original text is capture the key and ciphertext and crack them.

The remaining part of this paper is organized as follows. In Section 2,we describe the latest research of studies about multiple-network application security. The details about specific method implementation are in Section 3. In Section 4, we give the details of the experiment. We then summarized this paper and proposes future work prospect in Section 5.

2   Related Work

There is a significant body of related work that has appeared. There are some studies about covert channels and multiple-network security problems. Studies about covert channels mainly focus on detection and defense of covert cannels.[1][2][3][4]

For instance, in [5] the authors proposed a method of designing covert channels based on time, which cannot be detected in polynomial time. This method simulated a covert channel as a differential communication channel, and set up encoding and decoding equations based on the communication model. Lin and Ding[6] explored how to build a robust and efficient covert channel on the operating system. They proposed three models of covert channel communication protocols, including the basic protocol, dual-frequency transmission protocol, and adaptive protocol. In [7], a new type of covert channel was proposed, which minimize the application footprint and can be used on the Android operating system while bypassing mainstream detection technologies such as TaintDroid. In [8] the authors studied a mixed covert channel attack mode on the cloud and used two or more types of covert channels to transmit data. Wu and others[9] also proposed a new covert channel attack mode and implemented a robust communication protocol that can attack multiple types of virtual machines running an X86 operating system. In [17], a new method to quantify the information leakage for a fully probabilistic system was proposed, which can quantify the implicit leakage of information.

Studies about multiple-network security problems mainly include Hashimetc[10]'s applying biological rules of human immune system and epidemiology to detect and control attack on multiple networks, including detection of Denial of Service (DoS), Distributed Denial of Service (DDoS), and worm attacks and control of their spread on multiple networks. Zhang and Hayashi[11] proposed an efficient Session Initiation Protocol (SIP) security solution, which implemented SIP security and supported mobility on heterogeneous mobile access networks. Gondi and Agoulmine[12] used the Geographic Information System (GIS) model to simulate mobile terminal position prediction and access network selection, which implements seamless access and selection on multiple heterogeneous mobile access networks. Hsiao and others[13] propose a management architecture for heterogeneous wireless networks, which is adjustable based on users' mobile statuses and ensures continuity of security policy implementation.

To the author's knowledge, the method proposed in this paper is the first one that integrates covert channel with multiple-network transmission. Covert channel built in this paper can use multi-networks to transmit data. One network is used to transmit encrypted data, and the othersare used to transmit encryption key. Building covert channels using multi-networks can improve channel robustness and remarkably increase the cost of viewing data transmitted through covert channels.

3   Method of Building Covert Channels Using Multiple Networks

This section describes the method of building covert channels and the application method of multiple networks.

3.1   Building Covert Channels Based on a Protocol Header

Covert channels can be built based on the network, operating system, or hardware.[8] Traditionally, covert channels are built based on the network. These covert channels usually hide data in the reserved field of a protocol header. Currently, protocols used for building covert channels include IP, TCP, UDP, and HTTP, and some streaming media protocols.

In this paper, we use HTTP to build covert channels.

3.1.1   Covert Channel Based on HTTP Header

Reasons for using HTTP to build covert channels are as follows:

1. HTTP can be used to transit any types of data, which increases the difficulty of detection and analysis due to data complexity and further disguises data transmission process.
2. HTTP uses the public channel so that the access control system cannot separate covert data packets from other packets. Enabling or disabling the public channel will affect functions of the entire system.
3. Using HTTP to transmit data can attain higher bandwidth than using any other protocols.[14][15]

Fields of HTTP general header include Cache-Control, Connection, Date, Pragma, Transfer-Encoding, Upgrade, and Via. Considering the length of fields, the "connection" field is used to build covert channels and embed data. The sender first converts data into binary strings, then utilizes case sensitivity to modify each byte of the connection field before sending data. Bit 0 is in lower case and bit 1 is in upper case. The remaining bits are flag bits. Data is embedded only when all flag bits are in upper case.

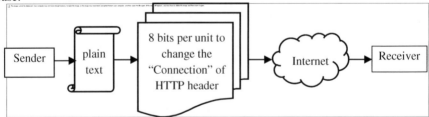

Fig. 2. Process of Sending Data through Covert Channels based on an HTTP Header

3.2   Covert Channels Built with Multiple Networks

This paper builds covert channels using two types of networks (Wi-Fi and 3G networks) to transmit data, which further disguises data transmission process.

3.2.1   Basic Process

The sender first uses the data encryption standard (DES) algorithm to encrypt data and transmits the key to the receiver through the 3G network. Then the sender uses the Wi-Fi network to transmit encrypted data through covert channels as described in the preceding section. After receiving the key and cipher text, the receiver uses the key to decrypt data, read data, and send feedback messages.

The key used in the DES algorithm is short (56 bits + 8 verification bits). To ensure data security, the sender must use a different key each time when transmitting data. By doing this, we can have a one-time pad effectiveness approximately which increases the difficulty of data cracking.

Figure 3 shows the entire process:

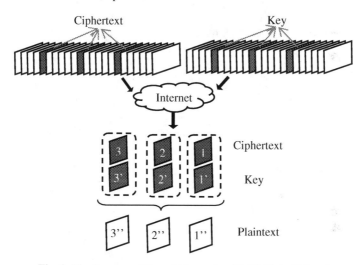

Fig. 3. The Process of Covert Channels with Multiple Networks

3.2.2   Matching of the Covert Channel

HTTP is used by the covert channel in this paper to transmit data. Besides the packets with ciphertext or key, the covert channel also transmits plenty of normal packets. So we set that the packets was embedded with message only when the flag bits in "Connection" all 1s. In this way, we can distinguish the packets with ciphertext from normal HTTP packets.

The covert channel in this paper uses multi-network, and one-time pad encryption, so it's important to ensure the matching of key and ciphertext at the receiver. Only the receiver receives key and ciphertext successfully, and the key and the ciphertext is matching does decrypt the correct plain text.

To ensure matching between key and ciphertext, we use "user-agent" as a mark. At the sender side, we mark the key and ciphertext with changing the case of the last 8 letters of "user-agent". Then the range of mark is between 0 and 255, the detail is the same as the encoding of ciphertext. The receiver will receive the key and ciphertext at the same time normally. At the time of encoding, we make the sum of serial number of key and ciphertext equals to 255 which preventing the attacker's wreckage, and make them difficult to analysis the relationship between the two packets.

3.2.3   Self-adaption of Covert Channel

The covert channel in this paper uses multiple networks to transmit data, so we need to consider the network congestion. In this paper, we design a self-adaptation mechanism which inspire from [16].

To prevent the receiver packet matchless due to a bad network, we modify the TCP/IP stack of receiver and add a detector in the transport layer to detect network state. The detector is used to monitor the state of two networks. We also establish two buffers at the sender system to storage data which will be transmitted by different networks when the network state goes bad.

When the network state goes bad, the detector at the transport layer of the receiver will send a signal to the sender's encrypt system. When the encrypt system receives the signal at the sender side, the sender will use the two buffers storage data respectively. When the buffer is full, the sender stops to handle data. When the network recoveries, the sender side will first transmit the data in the buffers and then handle the remaining data. The process is as follow in Fig.4.

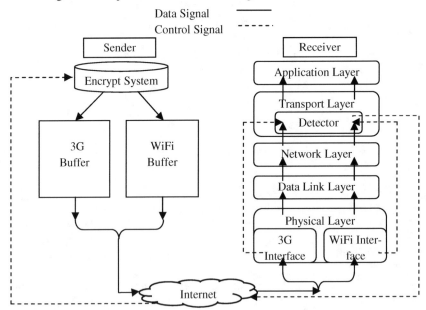

Fig. 4. The Process of Self-Adaptation

3.2.4   Security Analysis of the Covert Channel

Two kinds of networks, 3G network and WiFi network is used by the covert channel in this paper in which WiFi network used to transmit the ciphertext, while 3G network used to transmit key. Most 3G network use CDMA technology while WiFi use IEEE 802.11 standard. So if somebody wants to crack the channel, he must mast the two kinds of technology, then he may get the chance to crack the covert channel. In this way, by using two kinds of networks, we can enhance the difficulty of detection, so as to the difficulty of cracking the ciphertext.

When encrypt the plain text, in our covert channel, we use one-time pad technology. The key will be changed once it used to encryption, and the key is generated randomly. We combine one-time pad with multiple networks, as one-time pad proved is proved to be the most security encryption method, the transmission of message in our covert channel is safety enough.

4   Experiments

We use the 3G and Wi-Fi networks in experiments to verify proposal feasibility.

4.1   Introduction of the System

The experiment program described in this paper includes two parts: the covert channel main body program and the 3G network connection program.

4.1.1   Covert Channel Main Body Program

The covert channel main body program is in the Client/Server (C/S) architecture based on HTTP. The sender (client) first encrypts data using a key, calculates required transmission times (one byte each time), and then modifies the HTTP header to send transmission requests to the receiver (server). Meanwhile, the sender connects to the receiver terminal using 3G network and sends the key. The receiver responds to each HTTP request.

After receiving the data and the key, the receiver decrypts data and replies to the sender using the same method.

4.1.2   3G Network Connection Program

In this paper, the program transmitted through the 3G network is also based on Client/Server (C/S) architecture which can connect different computer and realize simple message transmission. The connection of 3G network through 3G wireless network cards and the 3G network used in this experiment belongs to China Unicom.

4.2   Experiment Details

4.2.1   Accuracy of Covert Channel

In this experiment, the indicator we measured is accuracy of message transmission under different transmission speed (only for WiFi network). We defined this indicator as follow:

$$A=S'/S \qquad (1)$$

And the Transmission speed is:

$$V =S'/t \qquad (2)$$

in which S represents the total bits transmitted, S' represents accurate bits transmitted, t represents the transmission time.

One of the tasks of covert channel is transmitted message covertly, so we use covert channel to transmit a little message normally. In our experiment, we measure the accuracy with different speed of covert channels for 60 times. When the speed is low, it means that the communication in the network is active.

The results are as follow:

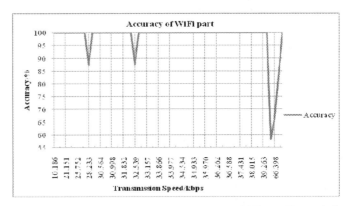

Fig. 5. Accuracy of the Covert Channel(WiFi part) under Different Transmission Speed

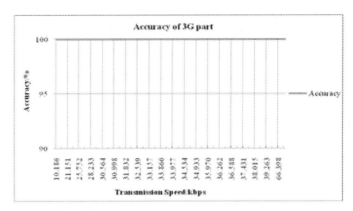

Fig. 6. Accuracy of the Covert Channel(3G part) under Different Transmission Speed

4.2.2   Transmit Speed of Covert Channel

In this experiment, we transmit a file which size is 2896Bytes for 30times, under different network state.

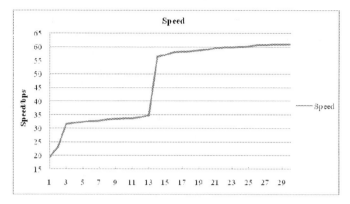

Fig. 7. Transmission Speed of the Covert Channel under Different Network States

4.3   Result Analysis

From the first two figures above, we can see that in most of times the covert channel can have an accuracy of 100% in WiFi network. The worst accuracy drop to 58.3%, it may be caused by the glitch of the network. While during the experiment, the 3G network has an accuracy of 100%, which means a very good accuracy and robustness.

From the third figure, we know that under different network states, the transmit speed of the covert channel is vary from less than 20bps to more than 60bps. So we can conclude that, the speed of this covert channel depends on network states partly.

From the results of the preceding experiments we can conclude that covert information transmission can be implemented using covert channels built on two networks, and that using a channel to transmit a different key each time when transmitting data can ensure the on-time pad approximately and improve data security.

5   Conclusion and Discussion

This paper first introduces latest research of studies on multi-channel transmission through heterogeneous networks and covert channels, and then proposes thoughts about covert channels for wireless transmission using heterogeneous networks. In section 4 we implement experiments to verify these thoughts and build covert channels and utilizing 3G and Wi-Fi applications. After analyzing experiment results, we can conclude that this method is effective in implementing covert data transmission secretly.

When using this covert channel, we can generates lots of simple web pages using HTTP protocol randomly on both sender and receiver sides. In this way, the sender and the receiver are requesting for web page data apparently, in fact they are using HTTP covert channels to transmit data. This further disguises data transmission process and improves channel's robustness. Considering the high cost of 3G networks which will increase the cost when transmit amount of data. In the future, we can increase more networks, such as Bluetooth network and 4G networks, to self-adaption or user based chosen. When transmit little data with short distance, we can use Bluetooth network, on the other hand, when the transmitting distance is long and the data need to be transmitted in a short time, we can choose 4G networks. We also can use more networks as covert channels, as the number of networks increasing, the time of covert channels to live will increasing too.in the future we will choose other networks such as 3G or 4G networks to replace the Bluetooth network and utilize more such networks to implement data transmission. As the number of utilized networks increases, the service life of covert channels increases.

References

1. Jadhav, M.V.: &Kattimani, S. L.: Effective detection mechanism for TCP based hybrid covert channels in secure communication. In: International Conference Emerging Trends in Electrical and Computer Technology (ICETECT), pp. 1123–1128. IEEE Press, India (2011)
2. Liu, X., Xue, H., Dai, Y.: A Self Adaptive Jamming Strategy to Restrict Covert Timing Channel. In: 2nd International Symposium Intelligence Information Processing and Trusted Computing (IPTC), pp. 1–4. IEEE Press, Bangkok (2011)
3. Sun, Y., Zhang, X.: A kind of covert channel analysis method based on trusted pipeline. In: 2011 International Conference on Electrical and Control Engineering (ICECE), pp. 5660–5663. IEEE Press, Yi Chang (2011)
4. Rezaei, F., Hempel, M., Peng, D., Sharif, H.: Disrupting and Preventing Late-Packet Covert Communication Using Sequence Number Tracking. In: Military Communications Conference, MILCOM 2013, pp. 599–604. IEEE Press, San Diego (2013)
5. Ahmadzadeh, S.A., Agnew, G.: Turbo covert channel: An iterative framework for covert communication over data networks. In: Proceedings of the IEEE INFOCOM, pp. 2031–2039. IEEE Press, Turin (2013)
6. Lin, Y., Ding, L., Wu, J., Xie, Y., Wang, Y.: Robust and Efficient Covert Channel Communications in Operating Systems: Design, Implementation and Evaluation. In: 7th International Conference on Software Security and Reliability-Companion (SERE-C), pp. 45–52. IEEE Press, Gaithersburg (2013)
7. Lalande, J.F.: Hiding Privacy Leaks in Android Applications Using Low-Attention Raising Covert Channels. In: 2013 Eighth International Conference on Availability, Reliability and Security (ARES), pp. 701–710. IEEE Press, Regensburg (2013)
8. Okhravi, H., Bak, S., King, S.T.: Design, implementation and evaluation of covert channel attacks. In: Technologies for Homeland Security (HST), pp. 481–487. IEEE Press (2010)
9. Wu, Z., Xu, Z., Wang, H.: Whispers in the hyper-space: High-speed covert channel attacks in the cloud. In: The 21st USENIX Security Symposium (Security 2012) (2012)
10. Hashim, F., Munasinghe, K.S., Jamalipour, A.: Biologically inspired anomaly detection and security control frameworks for complex heterogeneous networks. IEEE Transactions on Network and Service Management 7(4), 268–281 (2010)
11. Zhang, L., Miyajima, H., Hayashi, H.: An effective SIP security solution for heterogeneous mobile networks. In: IEEE International Conference on Communications, ICC 2009, pp. 1–5. IEEE Press, Dresden (2009)
12. Gondi, V.K., Agoulmine, N.: Low Latency Handover and Roaming Using Security Context Transfer for Heterogeneous Wireless and Cellular Networks. In: 2010 IEEE Asia-Pacific Services Computing Conference (APSCC), pp. 548–554. IEEE Press, Hang Zhou (2010)

13. Hsiao, W.H., Su, H.K., Chen, K.J., Zheng, R.H., Chen, J.S.: Mobility-aware security management for heterogeneous wireless networks. In: 2010 International Computer Symposium (ICS), pp. 818–823. IEEE Press, Tai Bei (2010)
14. Liang, Q., Bin, L., Mingzeng, H.: Research of Network Covert Channel Based on HTTP Protocol. Computer Engineering 31(15), 224–225 (2005)
15. Qixiang, W., Zumeng, L., Hua, M.: Research on Covert Channel Based on HTTP Protocol. Information Security and Communications Privacy 1, 73–74 (2009)
16. Shengyang, C., Yuan, Z., Muntean, G.-M.: An energy-aware multipath-TCP-based content delivery scheme in heterogeneous wireless networks. In: Wireless Communications and Networking Conference (WCNC), pp. 1291–1296. IEEE, Shang Hai (2013)
17. Yunchuan, G., Lihua, Y., Yuan, Z., Binxing, F.: Quantifying Information Leakage for Fully Probabilistic Systems. In: 2010 IEEE 10th International Conference on Computer and Information Technology (CIT), pp. 589–595. IEEE, Bradford (2010)

Detecting Insider Threat Based on Document Access Behavior Analysis

Rui Zhang[1,2,3,*], Xiaojun Chen[2,3], Jinqiao Shi[2,3], Fei Xu[2,3], and Yiguo Pu[2,3]

[1] Beijing Jiaotong University, Beijing, China
[2] Institute of Information Engineering, Chinese Academy of Sciences, Beijing, China
[3] National Engineering Laboratory for Information Security Technologies, Beijing, China
zhangrui@nelmail.iie.ac.cn,
{chenxiaojun,shijinqiao,xufei,puyiguo}@iie.ac.cn

Abstract. In recent years, the major source of information leakage is due to insiders. In order to detect information leakage by some internal insiders, anomaly detection using individual and community behavior models have been developed. The basic assumption of anomaly detection is each user has his/her own profile of activities and anomaly detection algorithm attempts to identify any deviation from the basic profile by each user. Both models neglected the possibility of change of individual user profile, e.g. change of individual interests. We propose here an anomaly detection model of insider threat using file content. The proposed model uses the document segmentation and Naive Bayes algorithm to classify the contents of files in an organization. We then set up the correlation matrices between users and their interests, and also the user community and their interests. We then propose a comprehensive model to detect the insider threat, which takes into consideration of the deviations of individual users' current behaviors, their historical behaviors and their associated community behaviors simultaneously. According to the experimental test results, the proposed model can successfully detect the anomaly access to files in the internal systems.

Keywords: insider threats, anomaly detection, text classification, individual behavior, community behavior.

1 Introduction

The diversity of threats against information systems increased significantly recently. Besides the external attackers, more attacks are found within the systems, i.e. from insiders.

A report by Clearswift, a Britain security company, indicates that in data security, 58% of the threats are from the inner-enterprises, which include existing staffs, departed staff and partners. The causes are either due to misuse or malicious attacks.

As the insiders who launched the attack would have some permissions to access the systems, being compared with the external attacks, they have the following characteristics.

---

* Corresponding author.

W. Han et al. (Eds.): APWeb 2014 Workshops, LNCS 8710, pp. 376–387, 2014.

- High risk: For an information system that is exposed to outsider, the external attacks may occurred more frequently than insider attacks, the severity of the latter is usually much greater.
- Latency: With a proper user identities and permissions to access the information systems, an insider can operate and control the system under the cover of the normal functions, which makes it more difficult to be discovered.
- Impersonation: An insider may make use of permissions of their trusted colleagues so as to perform malicious activities using others' user identities.

The insider attackers could make full use of these convenient situations to access files which they are not permitted. In June 2013, Snowden, the former staff of the CIA, stole the data from his employer, the National Security Agency (NSA), which revealed the existence of mass surveillance system "PRISM".

In the "PRISM" scandal, the insider Snowden accessed and stole internal classified documents from NSA. In a computer system, files are the collective of the resources and the carriers of the information. It is therefore necessary to protect the files from the abnormal accesses, which should be part of the security control of computer systems within an organization.

Within an organization, user downloading or uploading files to the Internet or sending files via email are potential paths of information leakage. If users abnormal activities can be detected by an automated system, , it will greatly reduce the risk of information leakage. The problem becomes how to prevent malicious activities on internal files.

Sandhu and his colleagues [1] proposed RBAC. In RBAC, the permission is correlated with a user's role. Users are assigned specific roles so as to acquire appropriate permissions. Users' all permissions are granted by the union of all roles' permissions. The major disadvantages of RBAC are its tedious operation and poor flexibility.

Maloof and Stephens [2] hold the opinion that users are only allowed to access the relevant information. Based on their theory, they built the Elicit System to monitor users' activities, including browsing, inquiring, printing, and downloading which puts emphasis on the outsider characteristics of accessing activities. When two files have different outsider characteristics with similar content, it is difficult to find all insider threats by this method. Chen and his colleagues [3] designed another system called CADS, which extracts information from the users' activities and calculates the distance among the users' activities by the K-nearest neighbor method. The CADS is improved to the MetaCADS which includes the users' accessing data activities. This method focused in the anomaly in the users' neighbors and ignores the abnormal activities of the users.

Faced with the above problems, this paper presents a model for the anomaly detection of insider threats based on file content. It detects the abnormal access to files in the systems from two different perspectives: individual user and user groups. The main contributions are:

- A method that combines the deviant of individual behavior and the deviant of community behavior with respect to file access activities.

- A model for the anomaly detection of insider threats based on the file content. The model also analyze the anomaly from an individual user and the relations among user groups.
- Design practical tests to crosscheck the model. According to the testing results, the model could pinpoint effectively the malicious access to files in the systems.

In the second section of the paper, we will introduce the relevant research. In the third section, we will present the principles and algorithms for the implementation of the model. We then present some test results that verify the validity of the model in anomaly detection of insiders.

2   Related Works

Many researches have done in the detection of the insider threats, e.g. the CMO model by Wood [4]. CMO models the attack from the users' prospective. —It proposed 3 factors, namely, ability, motivation and opportunity, are needed in order to launch launching. Ray and his colleagues [5] proposed to use attack trees to analyze insider attack. In their model, the users could login the systems, only after submitting their intents. The systems then monitor the users' behavior according to their intents. Zheng and his colleagues [6] adopted the biological monitor method, which monitor the identity theft by adopting the user's mouse activities. All these methods perform anomaly detections using the users' biometrics. They have not taken into consideration the file content in the computer system.

In file protection, Zhang and his colleagues [7] used vectors to do feature quantifications of permission activities of the information systems so as to determine potential insider threats. These detecting methods focused to model the permissions of the users, then studied and analyzed the relative characteristic information so to identify anomaly. They all put emphasis on the outsider features of accessing activities. When two files with different outsider characteristics but contain similar content, it is difficult to identify all potential insider threats.

From file accessing behavior point of view, Chen and his colleagues [3] built the bipartite graphs between users and objects, and identified anomaly through the deviated accessing relationship among users, their groups and objects.

Our research focuses on protecting files from insider attacks and proposes to use users' outsider features of activities and insider features of accessing files.

3   The Combination Model of Individual and Community Behavior

The thought of my anomaly detection in the first section is a comprehensive one, which detect the degree of deviation from the individual and its community access behavior model to the user's current access behavior simultaneously. In order to achieve the goal, the model for the anomaly detection of insider threats on the files contents has three main parts: 1) Data preprocessing; 2) Analysis of behavior model; 3) Anomaly detection. The relationship of the components is shown in Fig. 1.

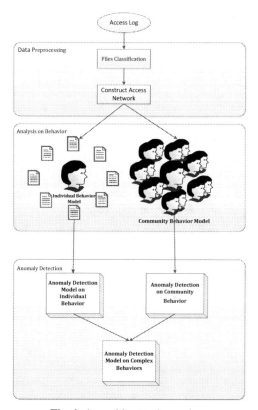

Fig. 1. An architectural overview

The input of the whole system is the access logs of users. The access logs are the record information of users' access to files, which include the information of users and files (using u and f to refer to them respectively). And we take U and F to indicate the set of all users and files respectively. To make good use of the access logs for the users' interests and behavior models, this paper adopts the idea of a text classification which classify the files according to their contents and each category is known as a topic using $T=\{t_1,t_2,\ldots, t_k\}$ to refer to.

The data preprocessing are mainly to classify the set of files F and to calculate the access relationship graph of <users (u), files (f) and topic (t)>.

The analysis of behavior model analyzes the access behavior of users from two respects. The mode of individual behavior is used to analyze the access logs of specific users and set up the individual behavior model of their own. According to the access topics, the mode of community behavior is used to classify the users into different categories. So one user may belong to several communities and the users of different communities may be overlapped as well. We need to set up the influence factors of correlational tables for communities. The tables indicate the influence factors of different communities, which represent the level of interest level of the members of community A to the topics of community B, and the influence factors represent the correlation of different communities.

The last component is anomaly detection. For a document access log <u_i, f_j>, my paper will use a comprehensive detection which detect the degree of deviation from the historical individual and its community access behavior models to the user's current access behavior simultaneously.

The paper will illustrate each part of them in the following.

3.1   Data Preprocessing

Access logs describe the relationship between users and files. And we use $U=\{u_1,u_2,...,u_n\}$ to indicate the sets of users. Assuming that the number of sets in the insider system is limited and we use n to indicate the number of sets. $F=\{f_1, f_2, ..., f_m\}$ represents the set of files and m means the number of different files. As each file has an own topic, T is used to point the number of their topics. And the whole access relationship is shown in the following Fig. 2.

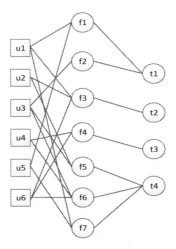

Fig. 2. Relationship between user, file and topic

According from the figure, we can draw the access list in which $N(u_i, f_j)$ stands for the numbers of the user i accessing to the file j:

$$F(u_i) = \sum_{j=1}^{n} N(u_i, f_j) \tag{1}$$

Table 1. The number of users's access to files

	f_1	f_2	f_3	f_4	f_5	f_6	f_7	F
u_1	15	0	6	0	0	0	9	30
u_2	0	0	23	0	5	0	0	28
u_3	0	1	0	0	17	16	0	34
u_4	0	0	0	25	0	4	0	29
u_5	9	0	0	0	0	0	2	11
u_6	0	0	14	8	0	0	0	22

From the Fig. 2, we can see that the set of accessing files F={f_1, f_2, f_3, f_4, f_5, f_6, f_7}. In the first step of files preprocessing, we need to categorize the set of accessing files.

Files Classification

In the insider network, we are assuming that all users access the same certain type of files. And there is a certain sample training set for each type. According to the accessing contents of a certain user, we can use the Naive Bayes algorithm to classify the accessing contents into different types.

Assuming that there are for categories T={t_1, t_2, t_3, t_4}, we will assign the files of F to the appropriate category.

Document Segmentation

In order to get the features from the texts, we should split up the text files, and express them into vector space model (VSM) [8]. The vector space model is used to represent the text files: W={w_1, w_2, w_3, ..., w_i, ..., w_n}. All the files are represented in the same vector space. Here we take the TF-IDF(Term Frequency-Inverse Document Frequency) [9] formula to normalize the results:

$$\text{tfidf}(t, d, D) = \frac{n_{i,j}}{\sum_k n_{k,j}} * \log \frac{|D|}{1 + |\{j : t_i \in d_j\}|} \tag{2}$$

where $n_{i,j}$ represent the occurrence number of t_i in d_j, and $\sum_k n_{k,j}$ represent the sum of the occurrence number of all words. In $\log \frac{|D|}{1+|\{j:t_i \in d_j\}|}$, $|D|$ represent the sum of all file, and $|\{j : t_i \in d_j\}|$ represnt the file number contained t_i, adding 1 in case of denominator is 0. We can use this formula to calculate the vector of each file.

Naive Bayes Algorithm

Naive Bayes algorithm is a simple and effective classification algorithm [10]. We take it as a feature whether a certain word would appear in the files so as to get a vector table of multiple features. Then every file can be described by the vectors of the same dimension.

Assuming that each vector is independent, we can use the Naive Bayes algorithm to classify the files:

$$p(c_i|w) = \frac{p(w|c_i)p(c_i)}{p(w)} \tag{3}$$

w represent the vector space set of words, and considering the independence of words, $p(w|c_i)$ can be represented as $p(w_1 * w_2 * ... * w_n|c_i)$, resulting $p(w_0|c_i)p(w_1|c_i)p(w_2|c_i) ... p(w_n|c_i)$.

According to the $p(w_j|c_i)$, we can calculate the probability which any word vector is belonged to the various categories.

3.2   Analysis on Behavior Models

Individual Behavior Model

In this section, we will set up an access behavior model for individual users according to the historical access records of specific users.

As is shown in Table 2, we could get the probability graph of access from the access logs. And each file has a kind of topic, so it is easy to calculate that $t_1=\{f_1, f_3\}$, $t_2=\{f_2, f_5\}$, $t_3=\{f_4, f_7\}$, $t_4=\{f_6\}$ based on the result of previous section. Then we could get the matrix UT of the corresponding relation between users and topics, among which ut_{ik} indicates the affinity of u_i and topic t_k:

$$ut_{ik} = p(u_i, t_k) = \frac{\sum N_{u_i f m}}{\sum N_{u_i}} \ (f_m \in t_k) \tag{4}$$

$p(u_i, t_k)$ indicates the probability of u_i's access to the document category t_k; $\sum N_{u_i f m}$ means the number of u_i's access to the document category t_k; and $\sum N_{u_i}$ represents the sum of user u_i's access to the documents.

The results are shown in Table 3. For instance, in the 30 access logs of user u_1, 70 percent of access is concentrated in the topic t_1 and the affinity of u_1 and t_1 is 0.7; 30 percent of access is concentrated in the topic t_3 and the affinity of u_1 and t_3 is 0.3.

Table 2. Frequency of users's access to files

	f_1	f_2	f_3	f_4	f_5	f_6	f_7
u_1	0.50	0	0.20	0	0	0	0.30
u_2	0	0	0.82	0	0.18	0	0
u_3	0	0.03	0	0	0.50	0.47	0
u_4	0	0	0	0.86	0	0.14	0
u_5	0.82	0	0	0	0	0	0.18
u_6	0	0	0.64	0.36	0	0	0
	t_1	t_2	t_1	t_3	t_2	t_4	t_3

Table 3. Frequency of users's access to topics

	t_1	t_2	t_3	t_4
u_1	0.70	0	0.30	0
u_2	0.82	0.18	0	0
u_3	0	0.53	0	0.47
u_4	0	0	0.86	0.14
u_5	0.82	0	0	0.18
u_6	0.64	0	0.36	0

Community Behavior Model

In the insider system, users are not isolated. And a community will be set up by them, which is defined as a set of community. In the community, all the users have the same interests. Well, the simplest generation way of community is to define each type of files as a kind of interests. For instance, the sports files can be seen as an interest of sports, and all the sports lovers can be seen as a community.

The formal definition of the community is as follow:

$$C(t_k) = \{u_j | u_j \geq t_k\} \tag{5}$$

"$u_j \geq t_k$" means u_j is interested in t_k. We could judge whether a certain user is interested in one topic or not from the relational table of users and topics. For that we can set a threshold y. If $ut_{ik} \geq \alpha$, we will consider that u_j is interested in t_k; if not, we will think that u_j is not interested in t_k.

In order to describe the behavior model of the community, the paper will take use of the matrix CT of the corresponding relation between users and topics. In the matrix CT, ct_{ik} indicates the interest level of the community ci in the topic tk. According to the definition of the community, we know that the community c_k is a group of users who have the same interest in tk. So ct_{ik} indicates the interest level of the community in the other topic t_k except the topic t_i. Consequently the matrix CT is also used to describe the internal relations of topics, that is, how much the community that is interested in the topic 1 is interested in another topic.

The specific calculation of the matrix CT is as below:

$$ct_{ij=}p\big(C(t_i), t_j\big) = \frac{\sum p(u_m, t_j)}{Count(C(t_i))} \big(P(u_m, T_j) > á\big) \tag{6}$$

$ct_{ij=}P\big(C(t_i), t_j\big)$ indicates the probability of $C(t_i)$'s interest in the topic t_j; $\sum P(u_m, t_j)$ indicates the sum probability of all the communities' (including u_m) access to the T_j; $Count\big(C(t)\big)$ indicates the number of the comminities which including u_m.

3.3   Anomaly Detection

After having the two above matrix UT and CT, the problem on anomaly detection is whether an access record Rec=<u_i, f_j> is anomaly or not, or how large the confidence of the anomaly is? In the section, I will describe the anomaly detections on individual behavior, community behavior and the both behavior successively.

Anomaly Detection Model on Individual Behavior
The anomaly detection model on individual behavior is to judge the anomaly by detecting the deviation between the current access behavior and the user's historical access behavior.

In the section 3.2.1, we have got the matrix of users and topics. Then the most common method of anomaly detection is to judge whether the current access object is in the historical access areas or not. And specifically, the anomaly detection model of individual behavior can be described by the following formula:

$$A_{single}(Rec) = 1 - ut_{ik} \tag{7}$$

Rec=<u_i, f_j>, and f_j is belonging to the topic t_k. $A_{single}(Rec)$ puts an anomaly score of the access record Rec of which interval is [0,1]. And if the number is bigger, the possibility of anomaly will be higher.

Anomaly Detection on Community Behavior
We can detect the anomaly by judging whether the current access behavior is diverged from its belonging community or not. This method is based on that one user' behavior should be in line with its belonging community's.

In the section 3.2.2, we have defined the form of community $C(t_k)$ and the matrix of the corresponding relation between users and topics. To evaluate whether Rec=$<u_i,f_j>$ is anomaly or not is parallel to evaluate whether the community's (including u_i) access to the topic (including f_i) is anomaly or not. The matrix CT of the corresponding relation between users and topics has just defined the model of the community's access to the topic. And specifically, the anomaly detection on community behavior can be described in the following formula:

$$A_{community}(Rec) = 1 - 1/m \sum_{i=0}^{i=m} ct_{ik} \qquad (8)$$

Rec=$<u_i,f_j>$. Assuming that ui is belonged to the community $\{c_1, c_2, ..., c_m\}$ and f_j is belonged to the topic t_k. $A_{community}(Rec)$ puts an anomaly score of the access record Rec of which interval is [0,1]. And if the number is bigger, the possibility of anomaly will be higher.

Anomaly Detection Model on the Combination of Individual and Community Behaviors

From the above mentioned, the anomaly detection model on individual or community behavior has its own disadvantages. If the interest of individual user is changed, the former detection model cannot solve it. Because the former model only detects anomaly from the historical records of individual access interets. For a user u_i who has not accessed to the t_k in its records, if there is a $<u_i, t_k>$, it will be taken as an anomaly. In fact, a user may be interested in t_k because of a recommendation of his friend, so the anomaly detection model on individual behavior will become invalid. At the same time, the latter model also has a problem of neglecting individuality. It does not accord to facts that the latter model only considers the deviation of community behavior without considering the historical access behavior of individual users. For these reason, my paper proposes the anomaly detection on the combination of individual and community behaviors. Specifically, the formula of it is as follows:

$$A(Rec) = â * A_{single} + (1 - â)A_{community} \qquad (9)$$

â is used as a weight to adjust the individual historical behavior and community behavior in the anomaly detection, and it can be adjusts by the users themselves.

4   Experiments

In this paper, we take the corpus which provided by Sogou laboratory as the primary data in the experiments [11]. There are 10 categories files of automobile, finance, IT, health, sports, tourism, education, recruitment, cultural and military in the corpus. Each category has 1990 files, so the total number of files is 19900. We chose 1200 files of each category as the training set, and finally we received 12000 files as the training set. Then we randomly selected 2500 files from the rest 7900 files as the test set and assign it to 40 users. In this paper, we control the total number of each person's access to the range of 150 and 350 and assuming that each person will be mainly access to 3 to 5 categories of files. After randomly selecting a number of categories, we allow users to access these categories with the probability of 90%, and to access others with the other 10%. According to this method, the paper finally generated 9886 simulated access records.

4.1   Text Categorization

According to the generated simulate data sets and the Sogou corpus, we assign the files of the data sets to the 10 categories. Here we can use T1-T10 to represent the 10 categories. The final result is shown in Table 4.

Table 4. The Number of Files and Users in Each Topic

	T_1	T_2	T_3	T_4	T_5	T_6	T_7	T_8	T_9	T_{10}
The number of files	230	342	211	267	196	256	212	332	189	265
The number of users	27	33	31	29	29	26	31	29	30	31

Table 4 shows the number of files and users in each topic when the classification is completed.

4.2   The Experiments of Individual Behavior Detection

Section 3.2.1 and Section 3.3.1 have described the algorithm of generating the individual behavior patterns. Using this algorithm and the result of classification can calculate the probability of an exception occurs when each user access to different categories. As shown in Table 5, we have calculated the access frequency table of u_1.

Table 5. U1's Access Number and Frequency of Each Category

	T_1	T_2	T_3	T_4	T_5	T_6	T_7	T_8	T_9	T_{10}
Access number	123	0	2	1	43	1	0	1	2	176
Access frequency	0.352	0	0.006	0.003	0.123	0.003	0	0.003	0.006	0.504

The Table 5 lists the u_1's access number and access frequency of each category. As can be seen from the table, the main accesses of u_1 are T_1, T_5 and T_{10}. Because we generate the data based on a 90% probability of access records to distribute them into several categories. So we take this threshold value of 0.1 as the threshold value α in the community model. The access behavior is represented in Fig. 3.

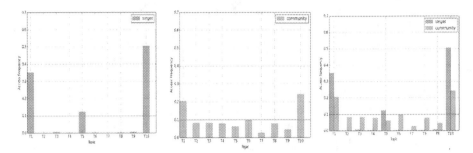

Fig. 3. The access of u_1 in individual model distribution

Fig. 4. The access of u_1 in the community model distribution

Fig. 5. The access of u_1 in the individual model and the community model distribution

4.3 The Experiments of Community Behavior Detection

Section 3.2.2 and Section 3.3.2 have introduced the community behavior model. And we can apply the algorithm to the data in Table 4. As is shown in Fig. 4, we finally get the result of u_1's community model (when $\alpha = 0.1$).

In Fig. 4, we put together the calculation results of individual model and the community model, so it is clearly to see the difference of the two models.

From the Fig. 5, we can see that the distribution of the community model is more moderate than that of the individual model. But the former model is parallel to the latter one.

4.4 The Experiments of Individual-Community Behavior Detection

According to the test results, there will be some deviation between the individual behavior model and the community behavior model. Section 3.3.3 describes the emendation of the deviation. Therefore, we will get different results for the different β values. When the β value of these three models take 0.2, 0.5 and 0.8 respectively, u_1's access frequency distribution is shown in Fig. 6.

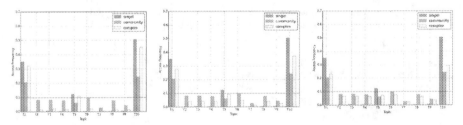

Fig. 6. The access of u_1 in three models distribution (when $\beta=0.2$)

Fig. 7. The access of u_1 in three models distribution (when $\beta=0.5$)

Fig. 8. The access of u_1 in three models distribution (when $\beta=0.8$)

As can be seen from the figure, different values of β will have different effects on the final results. In a real system, the user can adjust the size of the β value to get best results.

5   Conclusions

Through these studies, we can judge the user's abnormal access behavior from the user's individual access habits and their community access habits. By this we can solve the problems of the transfer of individual interests in the anomaly detection of individual behavior and the neglecting of individuality in the anomaly detection of community behavior simultaneously. According to the results of our experiments, the model can play some part in the protection of files in the internal systems.

However, this method is relatively dependent on the existing corpus, so in practice we need to invest a lot of experience to create the own corpus of internal information system.

In the next study, we will do the following things: 1) to make the text classification not dependent on the corpus but cluster automatically; 2) continue to improve the method of obtaining the model values of α and β, so that the detection model will be more accurate.

Acknowledgment. This work is supported by the Strategic Priority Research Program of the Chinese Academy of Sciences under Grant No. XDA06030200.

References

1. Sandhu, R.S., Coynek, E.J., Feinsteink, H.L., et al.: Role-Based Access Control Models yz. IEEE Computer 29(2), 38–47 (1996)
2. Maloof, M.A., Stephens, G.D.: ELICIT: A system for detecting insiders who violate need-to-know. In: Kruegel, C., Lippmann, R., Clark, A. (eds.) RAID 2007. LNCS, vol. 4637, pp. 146–166. Springer, Heidelberg (2007)
3. Chen, Y., Nyemba, S., Malin, B.: Detecting anomalous insiders in collaborative information systems. IEEE Transactions on Dependable and Secure Computing 9(3), 332–344 (2012)
4. Wood, B.: An insider threat model for adversary simulation. SRI International, Research on Mitigating the Insider Threat to Information Systems 2, 1–3 (2000)
5. Ray, I., Poolsapassit, N.: Using attack trees to identify malicious attacks from authorized insiders. In: de Capitani di Vimercati, S., Syverson, P., Gollmann, D. (eds.) ESORICS 2005. LNCS, vol. 3679, pp. 231–246. Springer, Heidelberg (2005)
6. Zheng, N., Paloski, A., Wang, H.: An efficient user verification system via mouse movements. In: Proceedings of the 18th ACM Conference on Computer and Communications Security, pp. 139–150. ACM (2011)
7. Hongbin, Z., Qingqi, P., Chao, W., Meihua, W.: Sensing insider threat based on access vectors. Journal of Xidian University (2014)
8. Salton, G., Wong, A., Yang, C.S.: A vector space model for automatic indexing. Communications of the ACM 18(11), 613–620 (1975)
9. Manning, C.D., Raghavan, P., Schutze, H.: Scoring, term weighting, and the vector space model. Introduction to Information Retrieval 100 (2008)
10. Harrington, P.: Machine Learning in Action. Manning Publications Co. (2012)
11. The corpus of test categorization of Sogou. [OL] (May 20, (2014), http://www.sogou.com/labs/dl/c.html

Enhanced Sample Selection for SVM on Face Recognition

Xiaofei Zhou[1], Wenhan Jiang[2], and Jianlong Tan[1]

[1] Institute of Information Engineering, Chinese Academy of Sciences, Beijing, 100095, China
[2] First Research Institute of Ministry of Public Security, Beijing, 100048, China
zhouxiaofei@iie.ac.cn

Abstract. For SVMs, large training samples will lead to high computing complexity of convex quadratic programming, even difficulty in running. Sample selection as a preprocessor of classification can greatly reduce the computational cost of training and test. In this paper, we present an enhanced sample selection frame based on convex structure for SVM. By learning the approximate errors of chosen set, we realize automatic control of sample scale for SVMs. Experimental results on face recognition show that our sample selection methods can adaptively select fewer high quality samples while maintaining the classification accuracy of SVM

Keywords: Sample selection, SVMs, Classification, Convex Hull, Subspace, Face Recognition.

1 Introduction

Support Vector Machine (SVM) is a powerful classification methodology in data mining and artificial intelligence fields [1-5]. However, large scale of training data will meet the difficulties of huge memory requirement and long running time. Especially in training phase the computing complexity of convex quadratic programming problem is quadratic proportional to the number of training samples. For test phase, the inner production computing times in decision function also equal to the number of training samples. Thus samples reduction or selection is effective strategy for decreasing total complexity of SVMs. Currently many emerged sample selection methods have sped up SVMs, such as random method[6], sampling near the separation margins[7], confidence measure-based method[8], Hausdorff distance-based method[8], k-means cluster based method [9], homogeneous cluster method[10], kernelized ionic interaction model [11], fuzzy C-means cluster sampling[12], active learning sampling[13-15], subclass convex hull method[16] and subspace sample selection [17] etc. Among these sample selection methods, the mentioned in [7-15] are inter-class methods, and methods in [6][16][17] are intra-class methods. Inter-class sample selection methods consider the information from different category, and try to select the boundary samples of each binary SVM. Intra-class methods choose samples according to the information of the homogeneous, without any relationship to other classes [16]. Inter-class method and intra-class methods all aim to select boundary samples, however, the former focuses on selecting boundary samples near to decision boundary, and the latter mainly considers the representation of essential distribution in each class [16],[17].

W. Han et al. (Eds.): APWeb 2014 Workshops, LNCS 8710, pp. 388–401, 2014.

As the optimal problem of finding maximal margin in SVM is equivalent to find-ing a pair of nearest points in two class convex hulls [18][19]. That implies SVM separate two convex hulls with maximal margin, and the convex hull of each class directly influence the decision distinguished to other classes. However, for high di-mensional data, to find convex hull of samples is not easy work, and also a NP hard problem. Some intra-class sampling measures directly select the boundary samples to approximate the convex hull, for example convex distance sampling [16] and sub-space sampling [17], which not only have greatly reduced the scale of training set, but also for multi-class problem the times of sample selection are also less than that of inter-class method[16],[17]. But for practical application, how to decide the size of chosen set or how to degree the estimated convex hull which is suitable for classifier, is important problem. In this paper, we present an enhanced intra-class sampling frame by approximation error bound and classification feedback to control the sam-pling for estimated convex hull. Experiments on two face databases validate the effec-tiveness of the proposed strategies.

The remaining of this paper is organized as follows. In section 2, we briefly intro-duce two intra-class convex hull sampling measures[16][17]. In section 3, we present our motivation to enhance the convex hull samplings. In section 4, we give our en-hanced algorithms, enhanced convex distance sampling (ECS) and enhanced subspace sampling (ESS). In section 5 we conduct some experiments on MIT-CBCL face data-base [20] and Head Pose database [21]. At last, we give the conclusions and acknowl-edgements in the last section.

2   Relative Works

In order to effectively reduce the scale of training data and maximally preserve origi-nal convex hull structure, selecting boundary samples to support the contour of class convex hull is a good strategy[16][17]. The two methods separately mentioned in [16][17] according to different sampling measure to iteratively select samples of con-vex hull. One is to select the furthest sample to the convex hull of chosen set [16], and the other utilizes subspace distance to pick out the linear irrelative boundary samples [17]. Both of them can rapidly approximate original convex hull of each class, and some experiments on larger face images data also show good performances [16][17]. In this section, we introduce the two sampling measures and prove that the samples selected by the two measures are the boundary samples of convex hull.

2.1   Convex Distance Sampling (CS)

For convex distance sampling, in a class data set, one uses convex distance to itera-tively pick out the furthest support sample to the convex hull of chosen samples, and the subclass convex hull spanned by chosen samples is the approximation of original convex hull. Eq. (1) is the distance between a candidate sample and chosen convex hull in each iterative process, where $co(\mathbf{S})$ denote convex hull spanned by chosen sample set $\mathbf{S} = \{\mathbf{z}_1, \mathbf{z}_2, ..., \mathbf{z}_l\}$, $\mathbf{x}$ is a candidate sample, and $\mathbf{z}_i$ ($i = 1, 2, \cdots, l$) are cho-sen samples.

$$d^2(\mathbf{x}, co(\mathbf{S})) = \min_{\alpha} \left\| \mathbf{x} - \sum_{i=1}^{l} \alpha_i \mathbf{z}_i \right\|^2, \text{ s.t. } \sum_{i=1}^{l} \alpha_i = 1, \ \alpha_i \geq 0, \ i=1,\cdots,l \tag{1}$$

In each iterative sampling, the candidate sample with maximal $d^2(\mathbf{x}, co(\mathbf{S}))$ will be selected. In theory, this sampling measure can give the best estimate for convex hull with enough samples, because when the maximal distance between candidate samples to chosen convex hull is zero, the convex hull spanned by the chosen set is just the original convex hull.

2.2 Subspace Distance Sampling (SS)

Convex distance is a convex quadratic programming which is sensitive to samples size [16]. Compared with convex distance sampling, subspace distance as sampling measure can perform more quickly to select boundary samples of convex hull in homogeneous training data. Eq. (2) is the subspace distance function for a candidate sample, which can be directly solved with much faster speed than convex hull sampling process.

$$d^2(\mathbf{x}, H(\mathbf{S})) = \min_{\alpha} \left\| \mathbf{x} - \sum_{i=1}^{l} \alpha_i \mathbf{z}_i \right\|^2 \tag{2}$$

Subspace sampling method uses subspace spanned by chosen set to approximate the distribution of samples. Those chosen samples are not only the boundary samples, but also linear independence. That leads to a rapid dimensionality approximation of original convex hull. It means that when the maximal subspace distance between sample and chosen set is zero, the subspace spanned by chosen set is just the space of all samples, and the dimensionality of subclass convex hull spanned by chosen set is equivalent to the dimensionality of original convex hull. As the distance $d^2(\mathbf{x}, H(\mathbf{S}))$ for selected sample can not be zero, it is obviously the chosen samples are linear independence.

In Figure 1, we simply show the sampling process of the two methods [16][17].

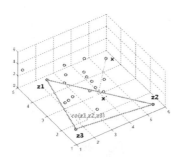 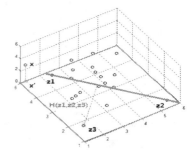

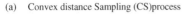

(a)   Convex distance Sampling (CS)process   (b)   Subspace distance Sampling(SS)process

Fig. 1. Two convex structure sampling process

In (a) (b), z1 and z2 are the first two chosen samples, and z3 is the next selected sample with furthest distance (convex distance in (a) and subspace distance in (b)). In (a), the convex hull of chosen sample z1, z2, z3 is co(z1,z2,z3), the next selected sample **x** is the furthest sample to co(z1,z2,z3), and in (b) the H(z1,z2,z3) is the subspace of chosen set, the next selected sample **x** is the sample with furthest subspace distance to H(z1,z2,z3). The project point **x'** is the nearest point to candidate sample **x** in chosen convex hull (in (a)) or subspace (in (b)).

2.3   Proof

In the following, we briefly prove that the selected samples by above two sampling measures are the boundary samples of class convex hull.

If a sample is a vertex of convex hull, there must have a direction vector w on which the vertex vector **x** has the largest length among all the samples. For example in Figure 2, on direction vector **w**, the projection of boundary sample **x** is the longest sample.

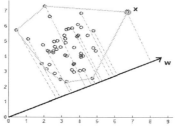

Fig. 2. Subclass convex hull sampling process

That is, for a sample x, if $\exists \mathbf{w}$ satisfy Eq.(3) , then x is the boundary sample of convex hull of all the samples.

$$\mathbf{x} = \arg\max_{x_i}(\mathbf{x}_i^T \mathbf{w}) \tag{3}$$

Based on the definition of boundary sample of convex hull, we can prove that the two sampling measures, convex distance sampling and subspace sampling, just select the boundary sample in each iteration process. As the chosen sample by two sampling measures is the furthest one to the chosen body(convex hull or subspace),see Figure 3 (a) and (b), along the direction of selected sample **x** and its projection point **x'** on chosen body, the chosen sample has the longest projection scale. That is, there exists a direction vector $\mathbf{w} = \mathbf{x} - \mathbf{x'}$, to make selected sample x have longest length, so we can proven that the selected samples by the two sampling measures are both the boundary samples of convex hull.

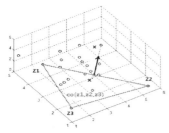
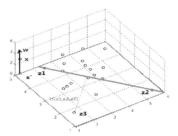

(a)   Convex distance Sampling (CS)   (b)Subspace distance Sampling(SS)

Fig. 3. Subclass convex hull sampling process

3   Motivation

In above two sampling methods, the estimated convex hulls by chosen sample directly decide the accuracy of SVM classifiers. For SVM's speed, the smaller chosen set the faster its runs, but for recognition accuracy the larger convex hull usually means better performance. What size of the convex hull is suitable for SVM is a key problem. Current methods in [16][17] control sampling by given expected number of chosen samples, which is difficult to decide how many samples suitable for SVM. In this paper, in order to solve the adaptive sampling control problem, we consider the following two aspects.

(1) We can utilize the approximate errors to adaptively control the scale of convex hull for enhanced sampling. From the view of sample distribution, the effect of approximating convex hull by chosen samples is important for data expression and classification. In a class, the nearer distance from samples to the chosen convex hull, the better chosen set. The furthest sample to chosen set just reflects the maximal approximation error, so absorbing such sample means to quickly lower the maximal error. When the maximal approximate error reaches an expected value, all samples are not far away from the estimated convex hull, sampling will be stopped. So we hope to find a suitable error bound to control the scale of sampling, and make chosen samples express the class within the error. In this paper, some statistical distances of sample distribution will be considered to roughly estimate the approximation error bounds.

(2) We can use a classification feedback mechanism to make the choice for the approximation error. An important purpose of sampling is also to keep high accuracy for classifier. For an expected approximation error of sampling, we can validate its impact on classifier by testing classification accuracies on known training set. Classification accuracies on training set not only verify the representation capacity of chosen set, but reflect the validity of classifier trained by chosen set. In this paper, we adopt a classification feedback mechanism to adjust the approximation error for sampling.

Considering above two aspects, this paper firstly give a rough expected error bounds by statistical distances of sample distribution, and then within the error scope adaptively choose the best value which leads the highest accuracy on training data. Based on the frame, we enhance the two distance sampling methods for SVM classifiers. Experimental results on two face databases show that our sampling strategy can decide suitable scale of convex structure sampling for effective classification.

4   Algorithm

4.1   Approximation Error Bound

In actual classification problem, each data set has respective distribution characteristics. Thus to find a suitable approximate error for a class, we should utilize some statistical distances of data distribution. In order to choose fewer samples to represent class convex hull, meanwhile limit the approximate error of chosen set for all samples within an expected value, we consider two kinds of statistical distance to estimate a rough limit of error. If taking the distances of any two samples as distribution features, we use the average of any two samples' distance as the upper error limit, and use the minimal distance among them as the lower error limit. In the following, we present the two error bound.

Given a multi-classification dataset $S=\{S_1, S_2, ..., S_c\}$, we firstly estimate the upper error bound.

$$d_i^U = \frac{2}{n_i(n_i - 1)} \sum_{\forall p,q, p \neq q \in Si} \|\mathbf{x}_p - \mathbf{x}_q\|, \quad i = 1,...,c. \tag{4}$$

where c is number of class, n_i is number of samples in the ith class. In each class, the average distance of any two samples describes the density of sample distribution, which is a rough scale tolerated for estimate error of the class. For a sample, the region around within d_i^U has higher probability to be shared with the same class samples.

Based on such distribution features of each class, we take an overall average from all class as our sampling upper error bound. The upper error bound ε_U is an average distribution of all class:

$$\varepsilon_U = \frac{1}{c} \sum_{i=1}^{c} d_i^U \tag{5}$$

Using ε_U as the upper error bound of sampling means that chosen samples should represent other samples within error no more than ε_U, that is, the maximal distance of all samples to the approximated convex hull spanned by chosen samples should be smaller than the upper bound ε_U.

Secondly, we compute the estimation of the lower error bound for sampling. For each class, the minimal distance of two samples is written as:

$$d_i^L = \min_{p \neq q, [p,q] \in Si} \|\mathbf{x}_p - \mathbf{x}_q\|, \quad i = 1,...,c. \tag{6}$$

We use the minimal distance of any two samples in a class as the lower error bound, because the minimal distance d_i^L can be taken as the minimum area of same class. We think that a sample at least owns the class region within the minimum distance d_i^L, in where homogeneous samples usually emerge with high probability. In a class, we think the area around the convex hull of chosen samples within distance d_i^L must belong to this class. So if the approximate error reaches the lower error bound, the data can be represented by chosen samples perfectly. In our paper, we take the average value of d_i^L from all the class as the lower error bound. The average lower error bound ε_L is

$$\varepsilon_L = \frac{1}{c} \sum_{i=1}^{c} d_i^L \tag{7}$$

Overall, our expected approximate error scope of sampling is between ε_L and ε_U, and we will find the best ε within the range $[\varepsilon_L, \varepsilon_U]$.

4.2   Adaptive Sampling Control

From above obtained error range $[\varepsilon_L, \varepsilon_U]$, we extract a sequence of values $\varepsilon_L = \varepsilon_1 < \varepsilon_2 < \cdots < \varepsilon_m = \varepsilon_U$ with equal interval to verify their impacts on classification accuracies. We begin with the largest error ε_m to select samples for training, and then run classification accuracy on training set. If the accuracy can not reach an expectation, the sampling procedure will be continued with the next smaller error ε_{m-1}, until finding ε_i to touch the expected classification accuracy. If there no ε_i satisfy the accuracy expectation, we choose the best ε_i with the highest accuracy on training set. We hope to find an ε_i as bigger as possible, because the bigger it is, the smaller the chosen set is. That is also the reason of our search from the upper error to the lower error. Compared with old sampling measures, we add error control and classification feedback mechanism on convex structure sampling. Figure 4separately gives the comparison of older and newer frames.

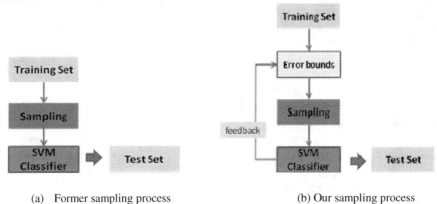

(a)   Former sampling process   (b) Our sampling process

Fig. 4. The comparison of sampling frames

Based on the frame of Figure 4 (b), we develop two adaptive sampling measures. In the following we give our enhanced sampling algorithm frame. With different distance measure of sampling, we can conduct two enhanced sampling algorithms, enhance convex distance sampling (ECS) and enhanced subspace distance sampling (ESS).In our paper, we extend the distance measures, convex distance in Eq.(1), subspace distance in Eq.(2), ε_U and ε_L in Eq.(4) and (6), all with kernel trick.

Table 1. Enhanced Convex Hull Sampling Algorithm

Algorithm: Enhanced Convex Hull Sampling

1: Input multi-class training set $S = \{S_1, S_2, ..., S_c\}$.

2: Estimate error scope $[\varepsilon_L, \varepsilon_U]$.

3: Extract merrors with equal interval $\Delta = (\varepsilon_U - \varepsilon_L)/m$.

4: Initialization

    chosen set: $S_i' = \{\}$, i=1,...,c ,

    number of chosen sample in S_i' : $l_i = 0$, i=1,...,c,

    expectional accuracy: acc,

    Initialize $\varepsilon = \varepsilon_U$; $R = 0$;

5: for each class

6: Select two furthest samples as initial chosen set:

$$S_i' = \{z_1, z_2 | [x_a, x_b] = \underset{x_p, x_q \in S_i}{\arg\max} \|\phi(x_p) - \phi(x_q)\|_2^2 \} \; ; \; l = 2 ;$$

7: while($maxd > \varepsilon$ & $l < n$)

    (n is sample's number in set S_i)

8: for $\forall x \in S_i \setminus S_i'$

    $dx = d^2(x, co(S_i'))$ in   ECS algorithm

    $dx = d^2(x, H(S_i'))$ in ESS algorithm

    end

9: $maxd = \underset{\forall x \in S_i \setminus S_i'}{\max} dx$; $z_{l+1} = \underset{\forall x \in S_i \setminus S_i'}{\arg\max} d$, $l_i = l$;

10: $S_i' = S_i' \cup z_{l+1}$; $l = l+1$;

    end

    end

11: Train SVM classifiersby chosen $S' = \{S_1', S_2', ..., S_c'\}$.

12: Test classifierson original training set $S = \{S_1, S_2, ..., S_c\}$,

    andget classification accuracy R' ;

13: If $R' > R$ then $R = R'$, $\varepsilon = \varepsilon'$, $A = S'$;

14: If $R' > acc$ thengo to 16;

    else $\varepsilon' = \varepsilon' - \Delta$;

15: If $\varepsilon' \geq \varepsilon_L$ then goto 5;

16: Output A, , ε , l_i , R .

5 Experiments

In this section, our enhanced sampling methods are experimentally evaluated for linear kernel SVM and RBF kernel SVM on MIT-CBCL face database [20] and Head pose face dataset [21]. We mainly conduct three experiments: (i) test two enhanced sampling methods with some given errors ε without adaptive process; (ii) test our presented enhanced sampling; (iii) compare different SVMs with various sampling methods. Our experiments are all carried out on Matlab 7.0. We use Matlab optimization toolbox to solve quadratic programming tasks involved in our experiments.

5.1 Experiments on MIT-CBCL Face Database

The experimental MIT-CBCL face database contains 3240 face images from 10 different persons, and each person has 324 synthetic faces generated at a resolution of 200×200 under varying pose and illumination. We divide the dataset into two sets, 2430 training samples (243 training images per class) and 810 test samples. The images are all normalized by bicubic interpolation to 16×16 pixels.

In order to clearly analyze the relation between the approximated errors ε and sampling, we firstly test some sampling on given error ε without accuracy feedback process. For a given error ε, we conduct convex distance sampling and subspace distance sampling until the maximal distance between candidate sample and chosen body (chosen convex hull or subspace) is smaller than the given error ε. With some given errors, the number of total chosen samples from all the class (Size), the ratio of chosen sample on (Ratio), the accuracy of classifier on test set (Accu) are given in Table 2.

Table 2. Experiments with Given Errors on MIT-CBCL

	Convex Distance Sampling				Subspace Distance Sampling			
	ε	Size	Ratio(%)	Accu(%)	ε	Size	Ratio(%)	Accu(%)
	2	28	1.15	92.84	1	38	1.56	92.22
Linear	1	55	2.26	99.75	0.7	45	1.85	96.79
Kernel	0.9	58	2.39	99.88	0.6	50	2.06	98.52
	0.8	**67**	**2.76**	**100**	**0.5**	**58**	**2.39**	**100**
RBF	0.04	45	1.85	99.75	0.06	29	1.19	90.37
kernel	0.03	54	2.22	99.75	0.05	34	1.40	98.40
($\sigma=6$)	**0.025**	**58**	**2.39**	**100**	0.04	35	1.44	99.14
	0.02	76	3.13	100	**0.03**	**48**	**1.98**	**100**

From above results, we can draw a conclusion, the smaller error ε, the more chosen samples and the better trained classifiers. However, actually we don't know what error is suitable for sampling. Our enhanced sampling methods can adaptively discovery a suitable error bound for sampling and make a balance between the size of training and classification accuracies. According to above results, we can verify our methods by the experience values in Table 2, for example, 0.8 and 0.025 are separately the best for linear and RBF convex hull distance sampling, 0.5 and 0.03 are the best for subspace distance sampling methods. These best errors can help us to analysis the effects of our enhanced sampling methods in the following experiments.

We conduct our enhanced sampling methods, and the experimental results of enhanced convex distance sampling and enhanced subspace distance sampling are separately given in Table 3 and 4, where ε is the best error by our method, 'Size' is the total number of chosen samples by all classifiers, 'Accu' is recognition accuracy on test set, 'Ts' is the sampling time, and 'T_{SVM}' is the execution time for SVM, and 'T' is the total time. In our experiment, with the purpose of getting higher accuracy, the expected accuracy is set to 100%.

Table 3. Enhanced Convex distance Sampling onMIT-CBCL

	ε	Size	Ratio (%)	Accu(%)	Ts (s)	Tsvm (s)	T(s)
Linear	0.564	81	3.46	100	56.48	1.16	57.64
RBF	0.022	67	2.76	100	92.03	3.89	95.92

Table 4. Enhanced Subspace Sampling on MIT-CBCL

	ε	Size	Ratio (%)	Accu(%)	Ts (s)	Tsvm (s)	T(s)
Linear	0.299	82	3.37	100	19.14	1.08	20.22
RBF	0.030	48	1.98	100	48.44	2.97	51.41

The results about the lower and upper error $[\varepsilon_L, \varepsilon_U]$ by two kernel functions are [0.034, 2.419] and [0.00002, 0.067] separately. For RBF kernel, ε_L is 0.00002 and we get approximation as 0. The extracted ten error values with equal intervals are given in Table 5.

Table 5. Ten Extracted Errors for Samplingon MIT-CBCL

Linear	2.419,	2.154,	1.889,	1.624,	1.359,	1.094,	0.829,	0.564,	0.299,
RBF	0.067,	0.060,	0.052,	0.045,	0.037,	0.030,	0.022,	0.015,	0.007,

We also compare our enhanced sampling method with un-sampled SVMs and other two inter-class sampling SVMs (Hausdorff distance method and k-means based method). Un-sampling SVM is the standard SVM with any sampling process. Hausdorff distance measure [8] and Clustering-based algorithm [9] are both inter-class sampling methods, which need to sampling for each binary SVM. For k-class problem, the two inter-class algorithms need to be conducted k(k-1)/2 times on SVM binary tree, while our enhanced sampling algorithm ECS and HS only need k times. Under linear kernel and RBF kernel, the comparisons are shown in Table 6.

Table 6. Comparisons of different methodson MIT-CBCL

Method	Linear kernel			RBF kernel		
	Size	Accu (%)	T_{SVM} (s)	Size	Accu (%)	T_{SVM} (s)
SVM	2430	100	3810.2	2430	100	4039.3
k-means	494	100	18.34	229	100	7.75
Huadroff	730	100	5.36	730	100	17.66
Enhanced CS	81	100	1.16	67	100	3.89
Enhanced HS	82	100	1.08	48	100	2.97

5.2   Experiments on Head Pose Dataset

Head Pose Face Database [22] includes 15 persons of color images, and each person contains 2 series of 93 images with different poses. Our experiments are conducted on the first series of images. The first series includes 15 persons (93 images per person), wearing glasses or not and having various skin color. The pose, or head orientation is determined by 2 angles (vertical and horizontal), which varies from -90 to +90 degrees. Images are 384×288 size in JPEG format. We divide the images into training and test set. The training set includes t 1215 images (81 images per class), and the test set contain 12 images of each person (total 180 images). The images are all normalized by bicubic interpolation to 16×16 pixels.

Firstly, with some given errors ε we also perform two sampling methods, the results is in Table 7.

Table 7. Experiments with Given Errors on Head Pose Dataset

	Convex Distance Sampling				Subspace Distance Sampling			
	ε	Size	Ratio (%)	Accu (%)	ε	Size	Ratio (%)	Accu (%)
Linear Kernel	5	68	5.60	91.67	3	118	9.71	96.11
	3	132	10.86	97.22	2	183	15.06	99.44
	2	**218**	**17.94**	**100**	**1.9**	197	**16.21**	**100**
RBF kernel ($\sigma=6$)	0.3	69	5.68	90.00	1.2	297	24.44	100
	0.2	118	9.71	96.11	0.2	114	9.38	97.22
	0.13	**202**	**16.63**	**100**	**0.12**	**208**	**17.12**	**100**

View from above results in Table 7, we can obtain similar conclusion with MIT-CBCL dataset. With the decrease of given errors, the accuracies of SVM become higher. We also can refer to the best error values to analyze our extracted errors of following enhanced sampling algorithms.

On Head Pose training set, the experimental results of our two enhancedsampling methods are separately given in Table 8 and Table 9.

Table 8. Enhanced Convex distance Sampling onHead Pose

	ε	Size	Ratio(%)	Accu(%)	Ts (s)	Tsvm(s)	T (s)
Linear	1.827	241	19.84	100	86.906	3.078	89.984
RBF	0.099	281	23.13	100	180.251	6.953	187.204

Table 9. Enhanced Subspace Sampling on Head Pose Dataset

	ε	Size	Ratio(%)	Accu(%)	Ts (s)	Tsvm(s)	T (s)
Linear	1.023	335	27.57	100	24.406	5.578	29.984
RBF	0.099	260	21.40	100	81.625	5.859	87.484

The estimated error bound $[\varepsilon_L, \varepsilon_U]$ with two different kernels are separately [0.219, 7.456] and [0.014, 0.397], and we extract ten errors with equal intervals during the bound are given in Table10.

Table 10. Ten Errors for Sampling on Head Pose Dataset

Linear	7.456,	6.652,	5.848,	5.044,	4.239,	3.435,	2.631,	1.827,
RBF	0.397,	0.355,	0.312,	0.270,	0.227,	0.184,	0.142,	0.099,

The comparisons of our enhanced sampling SVMs with un-sampled SVM, Hausdorff distance-based sampling SVM[8] and k-means based sampling SVM[9], under linear kernel and RBF kernel, are separately shown in Table 11.

Table 11. Comparisons of different methodson Head Pose

Method	Linear kernel			RBF kernel		
	Size	Accu (%)	T_{SVM} (s)	Size	Accu (%)	T_{SVM} (s)
SVM	1215	100	243.59	1215	100	240.31
k-means	1210	99.44	46.06	1215	100	43.77
Huadroff	1203	98.89	60.03	1189	98.89	48.48
Enhanced CS	241	100	3.08	281	100	6.95
Enhanced HS	335	100	5.58	260	100	5.86

5.3   Experimental Analysis

Approximate Error Analysis

From the results of MIT-CBCL dataset in Table 2 and Head Pose dataset in Table 7, we can see that, with ε decrease, the number of chosen samples is increasing, meanwhile the accuracy of SVM becomes better and better. When error ε reaches a certain value, the classification accuracy achieves the highest. So it is a feasible idea for data reduction by controlling the sampling error.

Our Adaptive Sampling Strategy Analysis

In our experiments, we preliminarily estimate the rough error bound by sample distribution distance features, and then search a better error in it. The error must be as large as possible and can support the highest classification accuracy.

Table 5 and Table 10 show that the estimated error bounds of each method can effectively reduce the scope of error searching, and also cover the best given errors (see Table 2 and 7). With an equal interval searching step, we can basically find the equal level error with the best given error in Table 2 and 7. For example, in Table 3, the obtained best error ε by our adaptive convex hull sampling is 0.564 and 0.022, and the given best values in Table 2 are 0.8 and 0.025.

From the results of sampling shown in Table 2and 7, our two algorithms can choose fewer samples to represent original data well and preserve 100% accuracy for prediction. As run time of subspace distance measure is smaller than that of convex hull distance, our adaptive subspace sampling method is total much faster than adaptive convex hull distance sampling method. But it is worthwhile to note that the accuracy of error ε and sampling time is directly relative to search step. The more searching steps and the smaller interval, the better error ε found, but running time of searching also becomes longer. In our experiments, we adopt 10 steps to find the best error ε, by which the ratio of chosen samples is around 3% on MIT-CBCL dataset (Table3,4) and no more than 28% on Head pose dataset (Table 8, 9).

Experimental Comparison Analysis

Compared with other methods (see Table 6 and 11), our adaptive sampling algorithms use much fewer samples to complete a multi-class SVM training problem. That not only save the memory occupation of training set, but also greatly speed the running of SVMs. For example, on MIT-CBCL dataset (see Table 6), for 100% accuracy of SVM, k-means based sampling methods involve 494 and 229 samples, and Huadroff distance methods choose 730 samples, in contrast, our two enhanced algorithms only select no more than 82 samples to represent original training data. From the results of running time, all the compared sampling SVM is much faster than un-sampling SVMs. In Table 6, run time of un-sampling SVMs are all more than 3810s, whereas the time of any sampling SVM is less than 19s, our adaptive sampling SVMs are only 1.16s, 1.08s, 3.89s and 2.97s. Similar results are observed on Head Pose dataset (see Table 11). Overall the experimental results on above two datasets show that our proposal strategy can adaptively find a suitable error by learning known data to control the sampling for SVM on face recognition.

6   Conclusion

In this paper, we present enhanced sample selection methods by introducing sampling estimation errors and classification feedback control in convex structure sampling of SVM. By adaptively learning a suitable estimate error on training set, our two proposed enhanced sampling algorithm scan automatically control the stop of sampling and effectively approximate original convex hull for SVM classifier. Experiments on MIT-CBCL face database and Head Pose face dataset show that our enhanced sampling method scan automatically control the sampling and support a suitable scale of training set for SVMs, which ensure the quality of samples for SVMs, meanwhile greatly improve the speeds of SVM.

Acknowledgment. This work was supported by Strategic Priority Research Program of Chinese Academy of Sciences (No.XDA06030200), National Nature Science Foundation of China (No. 61202226).

References

1. Vapnik, V.N.: The Nature of Statistical Learning Theory. Springer, New York (1995)
2. Klinkenberg, R., Joachims, T.: Detecting Concept Drift with Support Vector Machines. In: ICML, pp. 487–494 (2000)
3. Lin, K.P., Chen, M.S.: On the Design and analysis of the Privacy-Preserving SVM Classifier. IEEE Transactions on Knowledge and Data Engineering 23(11), 1704–1717 (2011)
4. Ferras, M., Leuing, C.C., Barras, C., Gauvain, J.L.: Comparison of speaker adaptation methods as feature extraction for SVM-based speaker recognition. IEEE Transactions on Audio, Speech, and Language Processing 18(6), 1366–1378 (2010)
5. Moustakidis, S., Mallinis, G., Koutsias, N., Theocharis, J.B., Petridis, V.: SVM-based fuzzy decision trees for classification of high spatial resolution remote sensing images. IEEE Transaction on Geoscience and Remote Sensing 50(1), 149–169 (2012)
6. Lee, Y.J., Huang, S.Y.: Reduced Support Vector Machines: A Statistical Theory. IEEE Transactions on Neural Networks 18(1), 1–13 (2007)
7. Shin, H., Cho, S.: Invariance of neighborhood relation under input space to feature space mapping. Pattern Recognition Letters 26, 707–718 (2005)

8. Wang, J., Neskovic, P., Cooper, L.N.: Training data selection for support vector machines. In: Wang, L., Chen, K., S. Ong, Y. (eds.) ICNC 2005. LNCS, vol. 3610, pp. 554–564. Springer, Heidelberg (2005)

9. Almeida, M.B., Braga, A.P., Braga, J.P.: Svm-km: speeding svms learning with a priori cluster selection and k-means. In: Proc. Sixth Brazilian Symposium on Neural Networks, pp. 162–167 (2000)

10. Koggalage, R., Halgamuge, S.: Reducing the number of traning samples for fast support vector machine classification. Neural Information Processing-Letters and Reviews 2(3), 57–65 (2004)

11. Kim, H., Park, H.: Data reduction in support vector machines by a kernelized ionic interaction model. In: Proc. Fourth SIAM International Conference on Data Mining (SDM 2004), pp. 507–511 (2004)

12. Xia, J.T., He, M.Y., Wang, Y.Y., Yan, F.: A fast training algorithm for support vector machine via boundary sample selection. In: IEEE Int. Conf. Neural Networks & Signal Processing, Nanjing, China, December 14-17, pp. 20–22 (2003)

13. Tong, S., Koller, D.: Support vector machine active learning with applications to text classification. In: Proc. of the 17th Int. Conf. on Machine Learning (June 2000)

14. Tong, S., Chang, E.: Support vector machine active learning for image retrieval. In: Proceedings of the Ninth ACM International Conference on Multimedia, pp. 107–118 (2001)

15. Cheng, J., Wang, K.: Active learning for image retrieval with Co-SVM. Pattern Recognition, 330–334 (2006)

16. Zhou, X.F., Jiang, W.H., Tian, Y.J., Shi, Y.: Kernel Subclass Convex Hull Sample Selection Method for SVM on Face Recognition. Neurocomputing 73, 2234–2246 (2010)

17. Zhou, X.F., Shi, Y.: Subspace distance-based sampling method for SVM. In: IEEE Proceeding of International Conference on Data Mining, ICDM 2010w, pp. 1289–1296 (2010)

18. Bennett, K., Bredensteiner, E.: Duality and Geometry in SVM Classifiers. In: Proc. Seventeenth International Conference on Machine Learning, pp. 57–64 (2000)

19. Keerthi, S.S., Shevade, S.K., Bhattacharyya, C., Murthy, K.R.K.: A fast iterative nearest point algorithm for support vector machine classifier design. IEEE Transactions on Neural Networks 11(1), 124–136 (2000)

20. Weyrauch, B., Huang, J., Heisele, B., Blanz, V.: Component-based Face Recognition with 3D Morphable Models. In: First IEEE Workshop on Face Processing in Video, Washington, D.C. (2004)

21. Pose, H., Gourier, N., Hall, D., Crowley, J.L.: Estimating Face Orientation from Robust Detection of Salient Facial Features. In: Proceedings of Pointing 2004, ICPR, International Workshop on Visual Observation of Deictic Gestures, Cambridge, UK (2004)

22. Crisp, D.J., Burges, C.J.C.: A geometric interpretation of v-SVM classifiers. In: Solla, S.A., Leen, T.K., Muller, K.R. (eds.) Advances in Neural Information Processing System, vol. 12, pp. 244–251. MIT Press, Cambridge (2000)

Author Index